Simulation Modeling

Edited by Constantin Volosencu
and Cheon Seoung Ryoo

Published in London, United Kingdom

IntechOpen

Supporting open minds since 2005

Simulation Modeling
http://dx.doi.org/10.5772/intechopen.95666
Edited by Constantin Volosencu and Cheon Seoung Ryoo

Contributors
Konstantin Suslov, Nikolai Voropai, Dmitriy Gerasimov, Ekaterina Serdyukova, Naser Zaeri, Yuriy V. Vasylenko, Atsushi Niida, Watal M. Iwasaki, Abdellah Lamnii, Ahmed Zidna, Mohamed Yassir Nour, Mohamed Jeyar, Fatima Oumellal, Mirko Djelosevic, Goran Tepic, Tien M. Manh Nguyen, Noureddine Bouteraa, Habib Djourdem, Kushal Bhattacharyya, Satyanarayan Patel, Cheon Seoung Ryoo

First published in London, United Kingdom, 2022 by IntechOpen
IntechOpen is the global imprint of INTECHOPEN LIMITED, registered in England and Wales, registration number: 11086078, 5 Princes Gate Court, London, SW7 2QJ, United Kingdom
Printed in Croatia

British Library Cataloguing-in-Publication Data
A catalogue record for this book is available from the British Library

Additional hard and PDF copies can be obtained from orders@intechopen.com

Simulation Modeling
Edited by Constantin Volosencu and Cheon Seoung Ryoo
p. cm.
Print ISBN 978-1-83969-683-1
Online ISBN 978-1-83969-684-8
eBook (PDF) ISBN 978-1-83969-685-5

For EU product safety concerns:
IN TECH d.o.o., Prolaz Marije Krucifikse Kozulić 3, 51000 Rijeka, Croatia, info@intechopen.com or visit our website at intechopen.com.

Meet the editors

Prof. Dr. Constantin Voloşencu graduated as an engineer from the Traian Vuia Polytechnic Institute of Timisoara and obtained a doctorate from the Politehnica University of Timisoara. He is currently a full professor in the Department of Automation and Applied Informatics, Politehnica University of Timisoara. He is an author of ten books, editor of 12 books, and author of seven book chapters and has more than 160 papers published in journals and conference proceedings and 27 patents. He is also manager of research grants, an editor in chief and a member of international journal editorial boards, a former plenary speaker, a member of scientific committees, and a chair at international conferences. His research is in the fields of control systems, control of electric drives, fuzzy control systems, neural network applications, fault detection and diagnosis, sensor network applications, monitoring of distributed parameter systems, and power ultrasound applications. He has developed automation equipment for machine tools, spooling machines, high-power ultrasound processes, and more.

Dr. Cheon Seoung Ryoo is a professor in mathematics at Hannam University. He received his Ph.D. in mathematics from Kyushu University. Dr. Ryoo is the author of several research articles in numerical computations with guaranteed accuracy. Also, he has contributed to the field of scientific computing, p-adic functional analysis, and analytic number theory. More recently, he has been working with quantum calculus, special functions, differential equations, and dynamical systems.

Contents

Preface

This preface serves to introduce the book Simulation Modeling to the general readership, providing a brief summary of the topics presented in this edited volume. The book promotes new research results in the field of simulation modeling – the process of creating and analyzing digital prototypes of physical models to predict their performances in the real world. With the development of computers and numerical approximation methods, simulation modeling has become a means of verifying various scientific and engineering theories, based on methods, such as digital prototypes for design, analysis calculations, algorithms, or finite element analysis. Simulation modeling helps scientists and designers predict circumstances, such as, under what conditions, in which ways a part could fail, and at what loads can it withstand, based on analyses on the approximate working conditions by applying the simulation software. Simulation modeling allows scientists and designers to avoid the repeated building of multiple physical prototypes to analyze new models of real study objects. Using these techniques, scientists can optimize their solutions to analyzed problems, can select materials and methods that meet practical requirements, simulate parts failure and identify a condition that causes them, can assess extreme environmental conditions, which is not so easy for practical tests on real life, can verify hand calculations, and can validate the safety and survival of physical model. Simulation modeling has also a large utilization in nontechnical domains, such as biomedicine, economics, society, and others.

The book aims to present some recent examples of scientific works and applications of simulation modeling, in a wide range of domains, which reduce human intervention in real life, using predetermining mathematic models, based on computers and algorithms. These examples are using various theories and intelligent methods with applications in biomedicine, engineering, or economics, for monitoring and diagnosis, to take optimal decisions in various applications based on practical measurement data. The book is a way to disseminate researchers' contributions in the field. This project presents applications that bring to analysis the advantages of simulation modeling by a high scientific level of theoretical and practical attainments. Researchers in different domains developed new theories and methods that enhance human understanding and improve the specialist ability to implement high-performance solutions. Studies that emphasize the implementation of theoretical methodologies in practice are presented. The authors have published work examples and case studies that resulted from their researches in the field. The readers get new solutions and answers to questions related to the emerging simulation modeling in various applications.

The book is structured in six sections, which refer to the following thematic areas: Mathematics is considered an important field in the theory of knowledge, mathematical activity consisting of discovering and proving properties of abstract issues with axioms, which is being widely used in science to model phenomena and enable the extraction of variable predictions from experimental laws. Biomedicine is a branch of medical science that applies biological and physiological principles to clinical practice, developing treatments by validating biological research, with biomedical disciplines, such as, medical biology, bioinformatics, virology, pathology, and many others. Systems of systems is a collection of task-oriented systems that pool their resources and capabilities together to create a new and more complex system that offers more functionality and performance than simply the sum of the constituent systems. Materials science and engineering (MSE) is an interdisciplinary field that covers the discovery of new solid materials and phenomenological observations of a material's properties and performance in metallurgy and mineralogy, incorporating elements of physics and chemistry respectively. Electric power systems are networks of electrical components deployed to supply, transfer, and use electric power, an example being the electric grid that provides power to homes and industries within an extended area. And the economy, the domain for production, distribution, and consumption of goods and services analyze the behavior and interactions of economic agents and how they work.

This book contains 11 chapters, mentioned as follows in the exact order in which they appear in the edited volume. The first section contains the first three chapters: The first chapter is presenting a study on the numerical verification method of solutions for elliptic variational inequalities. The second chapter developed an algebraic hyperbolic spline quasi-interpolation scheme for solving Burgers–Fisher equations. And the third chapter presents a study of the nonlinear boundary value problem. A method often used in simulation and modeling is the finite element method. It is used in numerical solving by approximating the differential equations. Typical domains of interest are the traditional fields of structural analysis, heat transfer, fluid flows, mass transport, and electromagnetic potential. The second section contains the fourth and fifth chapter that are respectively interesting studies on agent-based modeling and analysis of cancer evolution and on using artificial intelligence to process data for diagnosis of coronavirus disease (COVID-19. Simulation modeling has become important in biomedicine as a strategy to understand and predict the trajectory and the spread of diseases, as a support for clinical decisions. Cancer is a somatic evolutionary process characterized by the accumulation of mutations, which contribute to tumor growth, clinical progression, immune escape, and drug-resistance development. Simulation modeling can be used to analyze the dynamics of tumors and to make inferences about the evolutionary history of tumor data. The new COVID-19 pandemic is a global health emergency. In this regard, recent advances in artificial intelligence may be used as tools to process-related data into knowledge through computational-based models for diagnosis and providing treatment. The third section contains the sixth chapter, a theoretical study on advanced modeling, simulation, and analysis approaches to support complex space systems, gaming, and decision support system using a systems-of-systems perspective. The fourth section contains three chapters: Chapter seven analyzes thermomechanical properties of ceramic composites using object-oriented finite element analysis. The eighth chapter is an investigating study of strain effect on cleavage fracture for reactor pressure vessel material. And the ninth chapter is presenting a simulation model for fragmentation risk assessment to a cylindrical tank explosion. The fifth section contains the tenth chapter, a simulation modeling of integrated multi-carrier energy systems in the concept of an energy hub. The sixth section contains the 11th chapter, a simulation modeling for finding the limits of economic development lending without a financial crisis.

The chapters were edited and published following a rigorous selection process, with only a small number of the proposed chapters being selected for publication.

The editors thank the authors for their excellent contributions in the field and understanding during the process of editing.

The editors wish to thank the entire staff of the publishing house that contributed to the editorial process in any manner or capacity.

The publishing provided a set of editorial standards that ensured the quality of the scientific level of relevance of accepted chapters.

Constantin Voloşencu
Department of Automation and Applied Informatics,
Politehnica University Timisoara,
Timisoara, Romania

Cheon Seoung Ryoo
Department of Mathematics,
Hannam University,
Daejeon, South Korea

Section 1

Mathematics

Chapter 1

Numerical Verification Method of Solutions for Elliptic Variational Inequalities

Cheon Seoung Ryoo

Abstract

In this chapter, we propose numerical techniques which enable us to verify the existence of solutions for the free boundary problems governed by two kinds of elliptic variational inequalities. Based upon the finite element approximations and explicit a priori error estimates for some elliptic variational inequalities, we present effective verification procedures that, through numerical computation, generat a set which includes exact solutions. We describe a survey of the previous works as well as show newly obtained results up to now.

Keywords: numerical verification method, variational inequalities, error estimates, fixed point formulation, newton-like method, finite element method

1. Introduction

Numerical verification methods of solutions for differential equations have been the subject of extensive study in recent years and much progress have been made both mathematically and computationally [1–23]. However, for some problems governed by the elliptic variational inequalities, there are very few approaches. As far as we know, it is hard to find any applicable methods except for those of Nakao and Ryoo [13, 24–46].

The authors have studied for several years the numerical verification method of solutions for elliptic variational inequalities using finite element method and the constructive error estimates combining with Schauder's and Banach's fixed point theorem. Several results in our research are already published in [13, 24–46]. In this chapter, we briefly overview our resent research results including works not yet published.

The outline of this chapter is as follows. In Section 2, the two types of elliptic variational inequalities are considered. In Subsection 2.1, we describe the elliptic variational inequalities and give a fixed point formulation to prove the existence of solutions. In Subsections 2.2 and 2.3, the main tool of the verification method is explained at an abstract level. In Subsection 2.2, we present a simple iteration method for numerical verification of solutions for the elliptic variational inequalities. We construct the concepts of rounding and rounding error for functions and present a computer algorithm to construct the set satisfying the verification conditions. However, it is difficult to apply the method in Subsection 2.2 to a problem in which an associated operator is not retractive in a neighborhood of the solution, because it is based upon a simple iteration method. In Subsection 2.3, we propose another approach to overcome such a difficulty. This method can be applied to general elliptic variational inequalities without any retraction property of the associated operator. We introduce a Newton-like operator and reformulate the problem using it. Particularly, special emphasis is placed on the way to devise the Newton-like operator for a kind of non-differentiable map which defines the original problem. We introduce a computational verification condition. In order to show a concrete usage of the tool, in Section 3, we present an application to some problems governed by the elliptic variational inequalities. Many difficulties remain to be overcome in the construction of general techniques applicable to a broader range of problems. However, the authors have no doubt that investigation along this line will lead to a new approach employing numerical methods in the field of existence theory of solutions for various variational inequalities that appear in mathematical analysis.

2. Elliptic variational inequalities

The theory of elliptic variational inequalities has become a rich source of inspiration in both mathematical and engineering sciences. Elliptic variational inequalities are an effective tool for studying the existence of solutions of constrained problems arising in mechanics, optimization and control, operation research, engineering science, etc. [47–52]. It is the aim of this chapter to introduce a numerical technique to verify the solutions for elliptic variational inequalities. The basic approach of this technique consists of the fixed point formulation of elliptic variational inequalities and construction of the function set, on computer, satisfying the validation condition of a certain infinite dimensional fixed point theorem. For fixed point formulation, we consider a candidate set which possibly contains a solution. In order to get such a candidate set, we divide the verification procedure into two phases: one is the computation of a projection into a closed convex subset of some finite dimensional subspace (rounding); the other is the estimation of the error for the projection (rounding error). Combining these methods with some iterative technique, the exact solution can be enclosed by sum of rounding parts, which is a subset of finite dimensional space, and the rounding error, which is indicated by a nonnegative real number. These two procedures enable us to treat infinite dimensional problems as finite procedures, thta is, by computer.

Notations

- V: real Hilbert space with scalar product $(\cdot,\cdot)$ and associated norm $\|\cdot\|$,
- V^*: the dual space of V,
- $a(\cdot,\cdot) : V \times V \to \mathbf{R}$ is a bilinear, continuous and V-*elliptic* from on $V \times V$.

A bilinear form $a(\cdot,\cdot)$ is said to be V-*elliptic* if there exists a positive constant α such that $a(v,v) \geq \alpha\|v\|^2, \forall v \in V$..

In general we do not assume $a(\cdot,\cdot)$ to be symmetric, since in some applications nonsymmetric bilinear forms may occur naturally.

- $L : V \to \mathbf{R}$ continuous, linear functional,
- $K$is a closed convex nonempty subset of V,

- $j(\cdot) : V \to \mathbf{R} \cup \{\infty\}$ is a convex lower semicontinuous (l.s.c) and proper functional ($j(\cdot)$ is proper if $j(v) > -\infty, \forall v \in V$ and $j \neq +\infty$).

The two types of elliptic variational inequalities.
We consider two classes of elliptic variational inequalities.

- Elliptic variational inequalities of the first kind: *Find $u \in V$ such that u is a solution of the problem*

$$a(u, v - u) \geq L(v - u), \forall v \in K, u \in K.$$

- Elliptic variational inequalities of the second kind: *Find $u \in V$such that u is a solution of the problem*

$$a(u, -u) + j(v) - j(u) \geq L(v - u), \forall v \in V, u \in V.$$

2.1 The problem and the fixed point formulation

Let us first set a few notations [1, 47, 49, 50, 53–61]. In what follows we shall make use of the Sobolev spaces $W^{k,p}(\Omega)$ of functions which possess generalized derivatives integrable with the pth power up to and including the kth order. For $p = 2$, we shall write $W^{k,p}(\Omega) = H^k(\Omega), H^0(\Omega) = L^2(\Omega)$. Further, we introduce the scalar product in $L^2(\Omega)$ by

$$(f,g) = \int_\Omega f(x)g(x)dx.$$

The norm in $H^k(\Omega)$ will be denoted by $\|\cdot\|_{H^k(\Omega)}$. The symbol $|\cdot|_{H^k(\Omega)}$ will stand for the seminorm,

$$|u|_{H^k(\Omega)} = \left(\sum_{|\alpha|=k} \|D^\alpha u\|^2_{L^2(\Omega)}\right)^{\frac{1}{2}}, \quad \|u\|_{H^k(\Omega)} = \left(\sum_{j=0}^{k} |u|^2_{H^j(\Omega)}\right)^{\frac{1}{2}}.$$

Let V be a real Hilbert space with a scalar product $(\cdot,\cdot)_V$ and an associated norm $\|\cdot\|_V$, V^* its dual space. K denotes a nonempty closed convex subset of $V, a(\cdot,\cdot) : V \times V \to \mathbf{R}$ is a bilinear, symmetric, continuous and elliptic form of V, $a(\cdot,\cdot) : V \times V \to \mathbf{R}$ is a bilinear, symmetric, continuous and elliptic form of $V \times V$; that is, there exist constants $\alpha > 0$, and $\beta > 0$ such that$a(u,v) \leq \alpha\|u\|_V\|v\|_V, \forall u, v \in V$ and $a(v,v) \geq \beta\|v\|^2_V, \forall v \in V$. The pairing between V and V^* is denoted by $< \cdot, \cdot >$. Let Λ be a canonical isomorphism from V^* onto V defined, for $g \subset V^*$, by $<g, v> = (\Lambda g, v)_V, \forall v \subset V$. We can easily see that $\|\Lambda\|_{V^*} = \|\Lambda^{-1}\|_V = 1$. Now, let us consider the following variational inequality:

$$\text{Find } u \in K \text{ such that } a(u, v-u) \geq <f(u), v-u>, \forall v \in K, \tag{1}$$

where f is a nonlinear operator such that $f(u) \in V^*$..
In order to obtain a fixed point formulation of variational inequality (1) we need the following standard result.

Lemma 1. *Let K be a closed convex subset of V. Then $u = P_K\omega$, the projection of ω on K, if and only if*

$$u \in K : (u - \omega, v - u)_V \geq 0, \forall v \in K. \tag{2}$$

For some constant $\rho > 0$, let us define a mapping $G : V \rightarrow V$ by

$$G(u) = P_K \Lambda \Phi(u), \tag{3}$$

where $u \in V$, $\Phi(u) \in V^*$ is defined by

$$< \Phi(u), v > = (u, v)_V - \rho a(u, v) + \rho < f(u), v >, \forall v \in V. \tag{4}$$

For some constant $\rho > 0$, problem (1) can be written as

$$(u, v - u)_V - \{(u, v - u)_V - \rho a(u, v - u) + \rho < f(u), v - u >\} \geq 0, \forall v \in K.$$

Using (4) in the above inequality, problem (1) is equivalent to that of finding $u \in K$ such that

$$(u - \Lambda\Phi(u), v - u)_V \geq 0, \forall v \in K. \tag{5}$$

By (2) and (5), we now have the following fixed point problem for the operator G:

$$u = P_K \Lambda \Phi(u) = G(u). \tag{6}$$

Under appropriate conditions on the space V and the operator $G : V \rightarrow V$(e.g., continuity, compactness), which usually have to be verified by theoretical means, fixed point theorem yields the existence of a solution u of the problem (1) in some suitable set $U \subset V$, provided that

$$G(U) \subset U. \tag{7}$$

In order to compute an explicit inclusion, we must therefore construct U explicitly. For the numerical verification of condition (7), we have to use interval analysis on many levels between basic interval arithmetic and functional analysis. For the appropriate and suitable choice of the operator f, the form $a(\cdot, \cdot)$, and the convex set K; one encounters problems governed by the elliptic variational inequality as special cases from the problem (1) [48–52]. Inbrief, it is clear that the problem (1) is the most common. Up to now, devising a verification technique for the problem (1) is still an open problem. It is an important and interesting area of future research to find the numerical inclusion methods for the problem (1) by using (6). In this paper, we suppose that $V \subset L^2(\Omega)$ and the nonlinear map $f(\cdot) : V \rightarrow L^2(\Omega)$ satisfies the following assumptions.

A1. f is a continuous map from V to $L^2(\Omega)$.

A2. For each bounded subset $W \subset V$, $f(W)$ is also bounded in $L^2(\Omega)$.

If we restrict the nonlinear map f as above, then it can be shown that the problem (1) can be characterized by a class of variational inequality of the type,

$$\text{find } u \in K \text{ such that } a(u, v - u) \geq (f(u), v - u), \forall v \in K. \tag{8}$$

The problem (8) has the restricted condition; even so (8) is an important and very useful class of nonlinear problems arising in mathematical physics, mechanics, engineering sciences, etc. In Section 3, we briefly consider a particular example of interest in applications. Another example is given in [13, 24–46]. In the special case in which $K \equiv V$, (8) yields the variational theory of the boundary value problems for partial differential equations. We will discuss existence and inclusion methods for problem (8). These are methods providing the existence of a solution of the problem (8) within explicitly computable bounds. As we have seen before, the transformation of problem (8) into some fixed point formulation (6) can be carried out in the same way. In a conclusion problem (8) is equivalent to the fixed point problem of finding $u \in K$ such that

$$u = S(u), \tag{9}$$

where S denotes a specific operator, not necessarily the same as in (6). In particular for a given problem, we reduced the problem (8) to the fixed point formulation (9) and the continuity and compactness of S is discussed. For this reason, we shall say nothing about this problem for which we refer to [13, 24–46]. In order to simplify argument we assume that S is a continuous and compact operator. Since S is continuous and compact, as a result of Schauder's fixed point theorem, if there exists a nonempty, bounded, convex, and closed subset U such that $S(U) \subset U$, then there exists a solution of $u = S(u)$ in U. In Sections 2.2 and 2.3, we describe how to construct U explicitly.

2.2 Verification by a simple iteration method

In this subsection, we describe a simple iteration method for numerical verification of solutions for elliptic variational inequalities. In order to treat functions and variational inequalities in the infinite dimensional space V by computer, we introduce two concepts, rounding and rounding error. Now, let V_h be a finite dimensional subspace of V dependent on $h (0 < h < 1)$ and let K_h be a nonempty closed convex subset of V_h. Usually, V_h is taken to be a finite element subspace with mesh size h. For the sake of simplicity, we shall define K_h, an approximate subset of K, by $K_h = V_h \cap K$. K_h is a closed convex subset of V_h. In practical applications, the construction of K_h is one of the difficulties presented by variational inequalities. For a given problem, several approximations are available. For a general study of the approximation of convex sets, we refer the reader to the work of Mosco [51]. We define the projection P_{K_h} from V into K_h [49, 50]. That is, $v_h = P_{K_h}(u)$, the projection of u into K_h, is defined as follows:

$$u = S(u), v_h \in K_h : \quad (v_h, \zeta - v_h)_V \geq (u, \zeta - v_h)_V, \quad \forall \zeta \in K_h. \tag{10}$$

To verify the existence of a solution of (9), we determine a set W for a bounded, convex, and closed subset $U \subset V$ as

$$W = \{v \in V : v = S(u), u \in U\}.$$

From Schauder's fixed point theorem, if $W \subset U$ holds, then there exists a solution of (8) in the set U. Our goal is to find a set U which includes W. For any subset $W \subset V$, we define $R(W) \subset K_h$ by the projection of V to K_h, which is called the rounding of W. Additionally, we define RE(W), the rounding error of W, as a subset of V so that $W \subset R(W) + RE(W)$ holds. Using $R(W) + RE(W)$ instead of W, the verification condition becomes

$$R(W) + RE(W) \subset U. \tag{11}$$

Let us describe the procedure more concretely. First, we consider the auxiliary problem: given $g \in L^2(\Omega)$,

$$\text{find } u \in K \text{ such that } a(u, v-u) \geq (g, v-u), \forall v \in K. \tag{12}$$

We note that, by well known result [49], there is a unique element u which satisfies (12).

Secondly, we define the approximate problem corresponding to (12) as

$$a(u_h, v_h - u_h) \geq (g, v_h - u_h), \forall v_h \in K_h, u_h \in K_h \tag{13}$$

and (13) admit one and only one solution [49]. Error estimates for the variational inequalities can be found in [48, 49, 52], etc. Now, using (10), (12), (13) and error estimates, we make the following assumption.

A3. For each $u \in V$, there exists a positive constant C, independent of u and h, such that

$$\|u - P_{K_h} u\|_V \leq Ch\|g\|_{L^2(\Omega)}. \tag{14}$$

In order to verify the solutions numerically, it is necessary to determine the constant C that appears in a priori error estimations; this constant will be discussed later.

In order to construct the set U satisfying the verification condition (11) in a computer, we use an iterative procedure, that is, the sequential iteration. We propose a computer algorithm to obtain the set U which satisfies the condition (11).

(1) First, we obtain an approximate solution $v_h^{(0)} \in K_h$ to (8) by an appropriate method. Set $U_h^{(0)} = \left\{v_h^{(0)}\right\}$ and $\alpha_0 = 0$..

(2) Next we will define $R\left(W^{(i)}\right)$ and $RE\left(W^{(i)}\right)$ for $i \geq 0$, where $W^{(i)}$ is the set defined as follows:

$$W^{(i)} = \left\{v^{(i)} \in V : v^{(i)} = S\left(u^{(i)}\right), \quad u^{(i)} \in U^{(i)}\right\}.$$

$R\left(W^{(i)}\right)$ is defined by the subset of K_h which consists of all the elements $v_h^{(i)} \in K_h$ such that

$$a\left(v_h^{(i)}, \psi - v_h^{(i)}\right) \geq \left(f\left(u^{(i)}\right), \psi - v_h^{(i)}\right), \quad \forall \psi \in K_h, \tag{15}$$

holds for some $u^{(i)} \in U^{(i)}$. Note that $R\left(W^{(i)}\right)$ can be enclosed by $R\left(W^{(i)}\right) \subset \sum_{j=1}^{M} A_j \phi_j$, where $A_j = \left[\underline{A_j}, \overline{A_j}\right]$ are intervals, $\left\{\phi_j\right\}_{j=1}^{M}$ is a basis of V_h, and $M = \dim V_h$. For details of the interval calculation, we refer the reader to Nakao [6, 7, 12]. Next $RE\left(W^{(i)}\right)$ is defined as

$$RE\left(W^{(i)}\right) = \left\{v \in V : \|v\|_V \leq Ch \sup_{u^{(i)} \in U^{(i)}} \|f\left(u^{(i)}\right)\|_{L^2(\Omega)}\right\}. \tag{16}$$

Here, C is the same constant as in (14). Hence, $W^{(i)} \subset R\left(W^{(i)}\right) + RE\left(W^{(i)}\right)$ holds.
(3) Check the verification condition:

$$R\left(W^{(i)}\right) + RE\left(W^{(i)}\right) \subset U^{(i)}. \tag{17}$$

If the condition is satisfied, then $U^{(i)}$ is the desired set, and a solution to (8) exists in $W^{(i)}$, and hence in $U^{(i)}$.

(4) If the condition is not satisfied, we continue the simple iteration by using $\delta-$ *inflation*; that is, let δ be a certain positive constant given beforehand, and take

$$\alpha_{i+1} = Ch\sup_{u^{(i)}\in U^{(i)}} \|f(u^{(i)})\|_{L^2(\Omega)} + \delta,$$
$$[\alpha_{i+1}] = \{v \in V : \|v\|_V \leq \alpha_{i+1}\},$$
$$U_h^{(i+1)} = \sum_{j=1}^{M} \left[\underline{A_j} - \delta,\ \overline{A_j} + \delta\right]\phi_j,$$
$$U^{(i+1)} = U_h^{(i+1)} + [\alpha_{i+1}],$$

and then go back to the second step. The reader may refer to [26–46] for the details. If the condition (17) is satisfied, in our inclusion method of solutions for (9), the solution u is enclosed in the set $U^{(i)}$, which we call 'a candidate set' of the form $U^{(i)} = U_h^{(i)} + [\alpha_i]$.

2.3 Verification by a Newton-like method

The significance of a Newton-like operator was already pointed out in [29, 43]. Hence we will not discuss it in detail here. In Subsection 2.1, numerical verification of solutions for elliptic variational inequalities using a finite element method have been discussed only for simple iteration method. The method proposed in Subsection 2.2 is such that $\left\{\left(U_h^{(i)}, \alpha_i\right)\right\}$ always converges to the limit value $\{(U_h, \alpha)\}$ from an arbitrary initial value $\left\{\left(U_h^{(0)}, \alpha_0\right)\right\}$ if S in (9) is retractive operator (we refer to Zeidler [59–61] for the definition of retraction), while no convergence can generally be expected if S is not retractive operator. Briefly, for not retractive operator in the neighborhood of the solution, it is difficult to use the previous scheme proposed in Subsection 2.2. To overcome such a difficulty, in this section, we newly formulate a verification method using the Newton-like method. This approach enables us to remove the restriction in Subsection 2.2 to the retraction property of the operator in the neighborhood of the solution. Namely, this technique can be applied to general variational inequalities without any retraction property of the associated operator S. We refer to [29, 43] for a detailed study of the properties of the Newton-like Method.

In this subsection, we use the notation of Section 2.2. We assume that $K_h = V_h \cap K$ is a closed convex cone with vertex at 0 and K_h^* its dual. We note that K_h^* is also a closed convex cone with vertex at 0, which is the only point common to K_h and K_h^*. From (10) it follows that K_h^* is the set of points whose projections into K_h is 0. We need some additional lemma.

Lemma 2. *Any $u \in V$ can be uniquely decomposed into the sum of two orthogonal elements. That is,*

$$u = P_{K_h}u \oplus (I - P_{K_h})u = P_{K_h}u \oplus P_{K_h^*}u.$$

Here, $\oplus$ denotes the sum of two orthogonal elements in the sense of V..

Note that (9) can be rewritten as the following decomposed form in K_h and K_h^*:

$$\begin{cases} P_{K_h}u = P_{K_h}S(u), \\ (I - P_{K_h})u = (I - P_{K_h})S(u). \end{cases} \tag{18}$$

In order to formulate a Newton-like verification condition for (18), we need a Fréchet derivative of the operator S. For most of the variational inequalities, the S in

(9) is not Fréchet differentiable at all. Therefore, in order to use a Newton-like type method, a major difficulty in numerically solving the fixed point formulation $u = S(u)$ is the treatment of the non-differentiable operator S. We need a suitable modification of the Fréchet derivative of S. Using some techniques, we can devise the approximate Fréchet derivative of S. Hence we shall assume that $\widetilde{D}S(u)$ is the approximate Fréchet derivative of the $S(u)$ at u as the linear operator. Let $\widetilde{D}S(u)$ be designated as the Fréchet-like derivative of S at u.

To consider the Newton-like operator for (18), we define the nonlinear operator $N_h : V \to V_h$ as

$$N_h(u) \equiv P_{K_h}u - \left[I - \widetilde{D}S(u_h)\right]_h^{-1} \left(P_{K_h} - P_{K_h}S\right)(u).$$

Here I is the identity operator and $\left[I - \widetilde{D}S(u_h)\right]_h^{-1}$ denotes the inverse on V_h of the restriction operator $\left[I - \widetilde{D}S(u_h)\right]\Big|_{V_h}$. Note that we will verify the existence of the inverse operator $\left[I - \widetilde{D}S(u_h)\right]_h^{-1}$ from the nonsingularity of the matrix corresponding to $\left[I - \widetilde{D}S(u_h)\right]\Big|_{V_h}$ in actual calculations.

Next we define the operator $T : V \to V$ as follows:

$$T(u) \equiv N_h(u) + \left(I - P_{K_h}\right)S(u). \tag{19}$$

Then T is considered as the Newton-like operator for the former part of (18), but as the simple iterative operator for the latter part. T becomes a compact and continuous map on V by properties of S. Using some techniques, for a given problem we can not only define the Newton-like operator, but also devise a Newton-like Method. Furthermore, we obtain the following proposition and theorem.

Proposition 3. *Given the assumption that $N_h(u) \in K_h$,*

$$u = S(u) \Leftrightarrow u = T(u). \tag{20}$$

Theorem 4. *If there exists a nonempty, bounded, convex, and closed subset $U \subset K$ such that $T(U) = \{T(u) | u \in U\} \subset U$, then by the Schauder fixed point theorem, there exists a solution $u \in U$ of $u = S(u)$.*

When we decompose the set U as $U = U_h \oplus U_\perp$ in Theorem 8.1, where $U_h \subset K_h$ and $U_\perp \subset K_h^*$, the verification condition can be written by

$$\begin{cases} N_h(U) \subset U_h, \\ \left(I - P_{K_h}\right)S(U) \subset U_\perp. \end{cases} \tag{21}$$

Here, U_h is represented as the linear combination of the base functions of V_h with interval coefficients, whereas $U_\perp$ is the intersection of K_h^* with a ball in V. That is,

$$U_h = \left\{ \varphi_h \in K_h : \varphi_h = \sum_{j=1}^{M} A_j \phi_j \ \text{ with } \ a_j \in \left[\underline{A_j}, \overline{A_j}\right] \right\},$$

$$U_\perp = \left\{ \varphi \in K_h^* \ : \ \|\varphi\|_V \le \alpha \right\},$$

respectively.

Note that $N_h(U)$ can be directly computed from U_h and $U_\perp$ with additional information on the a priori error estimates. On the other hand, $\left(I - P_{K_h}\right)S(U)$ is evaluated using (14), by the following constructive error estimates for the finite approximate solution of variational inequality (8):

$$\|(I - P_{K_h})S(U)\|_V \leq Ch \sup_{u \in U} \|f(u)\|_{L^2(\Omega)}.$$

Therefore, the former condition in (21) is validated as the inclusion relations of corresponding coefficient intervals; the latter part can be checked by comparing two nonnegative real numbers.

Next we show a computer algorithm to construct the set U which satisfies the verification condition (21). In order to realize it, we use the iteration method described in Subsection 2.2. Similarly to that in Subsection 2.2, we now generate the following iteration sequence $\left\{\left(U_h^{(n)}, \alpha_n\right)\right\}$ for $n = 0, 1, 2, \cdots$. For $n \geq 1$, the δ-inflation of $\left(U_h^{(n-1)}, \alpha_{n-1}\right)$ is denoted by $\left(\widetilde{U}_h^{(n-1)}, \widetilde{\alpha}_{n-1}\right)$. Next, for the set $\widetilde{U}^{(n-1)} = \widetilde{U}_h^{(n-1)} \oplus [\widetilde{\alpha}_{n-1}]$, define $\left(U_h^{(n)}, \alpha_n\right)$ by

$$\begin{cases} U_h^{(n)} \supset N_h\left(\widetilde{U}^{(n-1)}\right), \\ \alpha_n = Ch \sup\limits_{u \in \widetilde{U}^{(n-1)}} \|f(u)\|_{L^2(\Omega)}. \end{cases} \tag{22}$$

Finally, the verification condition in a computer is given by the following theorem. The proof of Theorem 4 will be given here for the sake of completeness; it is based on Proposition 3 and Schauder's fixed point theorem.

Theorem 5. *For an integer N, if two relationships*

$$U_h^{(N)} \subset \widetilde{U}_h^{(N-1)} \quad \textit{and} \quad \alpha_N < \widetilde{\alpha}_{N-1} \tag{23}$$

hold, then there exists a solution u of (8) in $U_h^{(N)} \oplus [\alpha_N]$. Here, the first term of (21) means the strict inclusion in the sense of each coefficient interval of $U_h^{(N)}$ and $\widetilde{U}_h^{(N-1)}$.

3. Applications

The study for the numerical verification method for elliptic variational inequalities has been still made less progress than for the differential equation case. The author's method in the present chapter can be also applied, in principal, to the verification of solutions of the practical problems. Namely, in Section 3.1, we first give, a slightly detailed descroption of the basic principle and formulation of our numerical verification method for the solution of obstacle problems with a homogeneous condition. This should be an appropriate introduction to another applications of our idea. The basic approach of the method consisits of the fixed point formulation of the problems and construction of the function set, in a computer, satisying the validation condition of a certain infinite dimensional fixed point theorem. We also mention that it is possible to extend the method to more general problems with non-homogenerous obstacles. Moreover, in order to apply our method to the problem whose associated operator is not retractive in a neighborhood of the solution, a Newton-like method is introduced. Next, in Section 3.2, we apply our method to another type of free boundary problem with appears in the elasto-plastic deformation theory. This problem causes some properties of non-smoothness in tha associated finite dimensional equations. But, we can also overcome such a difficulty by appling the solution method for non-smooth problems developed by [29, 32, 33]. In the Section 3.3, we briefly remark that our enclosure method can also be applied to the so-called simplified Signorini problem which is a simplified version of a problem accurring in the elasticity theory [43]. Finally, in Section 3.4, we show the way to apply our approach to elliptic variational inequalities of the second kind appearing in the flow problems of a viscos-plastic fulied in a pipe.

3.1 Obstacle problems

We introduce the verification method for solutions of the obstacle problem which is known as a free boundary problem to cahracterize the contacted zone by an obstacle ψ in an elastic membrane region.

3.1.1 Homogeneous case

Here, 'homogeneous' stands for the case that obstacle $\psi \equiv 0$ in the whole domain.

3.1.1.1 Basic formulation of verification

Though the basic idea of verification is given in other places [26–28], in order to keep the paper as self-contained as possible, we describe rather detailed formulation and verification procedure for the present case.

Let Ω be a bounded convex domain in $\mathbb{R}^n$, $1 \le n \le 2$, with piecewise smooth boundary $\partial\Omega$. We set $V \equiv H_0^1(\Omega) = \{v \in H^1(\Omega) : v|_{\partial\Omega} = 0\}$ and

$$a(u,v) = (\nabla u, \nabla v)$$

which is adopted as the inner product on V, where $(\cdot,\cdot)$ stands for the inner product on $L^2(\Omega)$. We define $K := \{v \in V : v \ge 0 \ \ a.e. \ on \ \ \Omega\}$..

First, we note that, by well-known result [49], for any $g \in L^2(\Omega)$, the problem:

$$a(u, v-u) \ge (g, v-u), \quad \forall v \in K, \quad u \in K, \tag{24}$$

has a unique solution $u \in V \cap H^2(\Omega)$, and the estimate

$$|u|_{H^2(\Omega)} \le \ \|g\|_{L^2(\Omega)} \tag{25}$$

holds [49], where $|w|_{H^2}$ implies the semi-norm of w in $H^2(\Omega)$ defined by

$$|w|^2_{H^2(\Omega)} \equiv \sum_{i,j=1}^{n} \| \frac{\partial^2 w}{\partial x_i \partial x_j} \|^2_{L^2(\Omega)}.$$

Now consider the following elliptic variational inequalities with nonlinear right-hand side;

$$\begin{cases} \text{Find } \ w \in K \quad \text{such that} \\ a(w, v-w) \ge (f(w), v-w), \quad \forall v \in K. \end{cases} \tag{26}$$

We take an appropriate finite dimensional subspace V_h of V for $0 < h < 1$. Usually, V_h is taken to be a finite element subspace with mesh size h. We then define K_h, an approximation of K, by

$$K_h = V_h \cap K = \{ v_h | v_h \in V_h, \quad v_h \geq 0 \ \text{ on } \ \overline{\Omega} \}.$$

We also define the projection P_K from V onto K. That is, $v = P_K(w)$, the projection of $w \in V$ into K, is defined as the unique solution of the following problem:

$$v \in K : \quad a(v, \zeta - v) \geq a(w, \zeta - v), \quad \forall \zeta \in K. \tag{27}$$

And define the projection P_{K_h} from V onto K_h. That is, $v_h = P_{K_h}(w)$, the projection of w into K_h, is defined as follows:

$$v_h \in K_h : \quad a(v_h, \zeta - v_h) \geq a(w, \zeta - v_h), \quad \forall \zeta \in K_h. \tag{28}$$

Now, as one of the approximation properties of K_h, assume that.

For each $w \in K \cap H^2(\Omega)$, there exists a positive constant C_1, independent of h, such that

$$\|w - P_{K_h} w\|_V \leq C_1 h |w|_{H^2(\Omega)}. \tag{29}$$

Here, C_1 has to be numerically determined. For example, it is known that we may take $C_1 = \frac{\sqrt{5}}{\pi}$ for the linear element in one dimensional case [27]. Furthermore, it will be readily seen that the same constant can be taken for the two dimensional bilinear element from the consideration on the proof of Theorem 5.1 in [27]. To verify the existence of a solution of (26) in a computer, we use the fixed point formulation.

First, note that, for each $w \in V$, there exists a unique $F(w) \in V$ such that

$$(\nabla F(w), \nabla v) = (f(w), v), \quad \forall v \in V, \tag{30}$$

which also implies that

$$\begin{cases} -\Delta F(w) = f(w) & \text{in } \Omega, \\ F(w) = 0 & \text{on } \partial\Omega. \end{cases} \tag{31}$$

Then the map $F : V \to V$ is compact. By (30), the problem (26) is equivalent to finding $w \in V$ such that

$$a(w, v - w) \geq a(F(w), v - w), \qquad \forall v \in K. \tag{32}$$

Using the definition (27) and (32), we now have the following fixed point problem for the compact operator $P_K F$.

$$\text{Find } \exists w \in V \ \text{ such that } \ w = P_K F(w). \tag{33}$$

3.1.1.2 Verification condition

We introduce two concepts, rounding and rounding error, which enable us to deal with the infinite dimensional problem by finite procedures, that is, in a computer.

Now we define the dual cone of K_h by

$$K_h^* = \{w \in V : a(w,v) \leq 0, \quad \forall v \in K_h\},$$

and note that K_h^* is also closed convex cone in V with vertex at 0 which is the only point common to K_h and K_h^*. From (28) it follows that K_h^* is the set of points whose projections into K_h is 0.

Lemma 6. *Any $w \in V$ can be uniquely decomposed into the sum of two orthogonal elements. That is,*

$$w = P_{K_h} w \oplus (I - P_{K_h}) w = P_{K_h} w \oplus P_{K_h^*} w.$$

Here, $\oplus$ denotes the sum of two orthogonal elements in the sense of V.

For any $w \in V$, we now define the rounding $R(P_K F(w)) \in K_h$ by the solution of the following problem:

$$a(R(P_K F(w)), v_h - R(P_K F(w))) \geq (f(w), v_h - R(P_K F(w))), \quad \forall v_h \in K_h.$$

Next, for any subset $W \subset V$, we define the rounding $R(P_K FW) \subset K_h$ by

$$R(P_K FW) = \{w_h \in K_h : w_h = R(P_K F(w)), \quad w \in W\}.$$

Usually, $R(P_K FW)$ is enclosed and represented as a linear conbination of the base functions in V_h with interval coefficients.

Moreover, for $W \subset V$, we define $RE(P_K FW)$, the rounding error of $P_K FW$, as a subset of K_h^*, that is,

$$RE(P_K FW) = \{v \in K_h^* : \|v\|_V \leq C_0 h \|f(W)\|_{L^2}\}, \tag{34}$$

where

$$\|f(W)\|_{L^2} \equiv \sup_{w \in W} \|f(w)\|_{L^2}.$$

Here, $C_0 \equiv C_1 C_2$, where C_1 is the same positive constant as in (29), and C_2 is determined by the following regularity estimate for the solution to (24) of the form

$$|u|_{H^2} \leq C_2 \|g\|_{L^2}. \tag{35}$$

Thus we may take as $C_2 = 1$ for the present case from (25). Then, we have

$$P_K F(w) - R(P_K F(w)) \in RE(P_K F(w)), \quad \forall w \in W.$$

Therefore, the following verification condition is obtained by Schauder's fixed point theorem.

Lemma 7. *If there exists a nonempty, bounded, convex, and closed subset $W \subset K$ such that*

$$R(P_K FW) \oplus RE(P_K FW) \subset W, \tag{36}$$

then there exists a solution of $w = P_K F(w)$ in W.

We sometimes refer the above set W as *a candidate set*, which we generate in computer so that it satisfies the condition (36).

3.1.1.3 Verification procedures

We describe the method to find a set W satisfying (36) in th ebelow.

Consider the following approximate solution $w_h \in K_h$ of (24):

$$a(w_h, v_h - w_h) \geq (g, v_h - w_h), \quad \forall v_h \in K_h, \quad w_h \in K_h. \tag{37}$$

Since the bilinear form $a(\cdot, \ \cdot)$ is symmetric, (37) is reduced to the quadratic programming problem:

$$\min_{v \in K_h} \left[\frac{1}{2} a(v, v) - (g, v) \right]. \tag{38}$$

Let $\left\{\phi_j\right\}_{j=1\cdots M}$ be a basis of V_h with usual linear functions such that $\phi_j(x) \geq 0, \quad \forall x \in \Omega$ and satisfying

$$\phi_j(x_i) = \begin{cases} 1, & i = j, \\ 0, & i \neq j, \end{cases}$$

where x_i is an interior node of the finite element mesh. Then (38) reduces to the following vector form:

$$\min_{w \geq 0} \left[\frac{1}{2} w' D w - P' w \right], \tag{39}$$

where $w \geq 0$ means the componentwise relation. Here, $D := (d_{ij})_{1 \leq i,j \leq M}$ with $d_{ij} = \left(\nabla \phi_i, \nabla \phi_j\right)$, and w is the coefficient vector with $\left\{\phi_j\right\}$ of the function v in (38). Also, $P := \left(\left(g, \phi_j\right)\right)_{1 \leq j \leq M}$.

Furthermore, we define for any $\alpha \in R^+$, nonnegative real number, we set

$$[\alpha] \equiv \{\phi \in K_h^*; \quad \|\phi\|_V \leq \alpha\}.$$

Then, for a given candidate set $W = W_h \oplus [\alpha]$ with $W_h \subset K_h$, the computation of the rounding $R(P_K F W)$ reduces to enclose an interval vector $Z = (Z_j)$ and $Y = (Y_j)$ satisfying the following nonlinear system of equations [27]:

$$\begin{cases} Y - DZ = -\left(f(W), \phi_j\right), & 1 \leq j \leq M, \\ Y_j Z_j = 0, \quad 1 \leq j \leq M. \end{cases} \tag{40}$$

Here, $\left(f(W), \phi_j\right)$ is evaluated as an interval B_j such that $\left\{\left(f(w), \phi_j\right) | w \in W\right\} \subset B_j$. In order to solve (40) with guaranteed accuracy, we use some interval approaches for nonlinear system of equations [19, 20]. Thus, using the solution of (40), we can enclose the set $R(P_K F W)$ in (36). Combining this with (34), we can successfully compute the left-hand side of (36) for any candidate set $W = W_h \oplus [\alpha]$.

Thus we can present a computational verification condition. In the actual computation, we use an iterative procedure with *δ-inflation* technique to find the set W satisfying (36). Several numerical examples for verification are presented in [27] for one dimensional problem using linear finite element.

3.1.2 Non-homogeneous case

In this subsection, we consider the two-dimensional case. In order to verify solutions numerically, it is necessary to determine some constants that appear in the a priori error estimates. For the non-homogeneous case, we define $K := \{v \in V : v \geq \psi \text{ a.e. on } \Omega\}$, where ψ is a given $H^2(\Omega)$ function such that $\psi \leq 0$ on $\partial\Omega$ and is not identically equal to 0. Let Ω be a square with side 1 and let $\mathcal{T}_h$ be the uniform triangulation of Ω. We introduce $\Sigma_h = \{p; \quad p \in \overline{\Omega}, p \text{ is a vertex of } T \in \mathcal{T}_h\}$ and define the approximate V_h of $H_0^1(\Omega)$ by $V_h = \{v_h; \quad v_h \in H_0^1(\Omega) \cap C^0(\overline{\Omega}), v_h|_T \in P_1, \forall T \in \mathcal{T}_h\}$. Here, $v_h|_T$ denotes the restriction of v_h to T and P_1 representing the space of polynomials in two variables of degree ≤ 1. It is then quite natural to approximate K by

$$K_h = \{v_h \in V_h; v_h(p) \geq \psi(p), \forall p \in \Sigma_h\}.$$

Note that, in general, $K_h \neq V_h \cap K$. Then, P_K and P_{K_h} are similarly defined as before, and we also have the constructive error estimates of the form, $\forall v_h \in K_h$ and $\forall v \in K$,

$$\|u_h - u\|_{H_0^1(\Omega)} \leq C(g, \psi, h), \tag{41}$$

where,

$$C(g,\psi,h) \leq \sup_{g \in L^2(\Omega)} \sqrt{(0.494)^2 h^2 |u|_{H^2}^2 + 2\left(\|g\|_{L^2} + \|Au\|_{L^2}\right)\left((0.494)^2 h^2 |u|_{H^2} + 6h^2 |\psi|_{H^2}\right)}.$$

We provide a numerical example of verification in the two-dimensional case according to the procedures described in the previous section. Let $\Omega = (0,1) \times (0,1)$. We consider the case $f(u) = Ku + \sin \pi x \sin 2\pi y$ and $\psi = \sin \pi x \sin \pi y$. For simplicity, we only consider the uniform mesh here. First, we divide the domain into small triangles with a uniform mesh size h and choose the basis of V_h as the pyramid functions.

The execution conditions are as follows (**Figures 1–3**):

$K = 0.1, \quad \dim V_h = 10$

Obstacle function $\psi = \sin \pi x \sin \pi y$

the outline of ψ is shown in Figure 1.

Initial value : $u_h^{(0)} =$ Galerkin approximation, $\alpha_0 = 0$

the outline of $u_h^{(0)}$ is shown in Figure 2.

Illustration of contact zone between obstacle

and approximate solution is shown in Figure 3.

Extension parameters : $\delta = 10^{-5}$.

Results are as follows:

Iteration numbers for verification : 2

$H_0^1(\Omega)$ – error bound : 0.15437

Maximum width of coefficient intervals in $\left\{A_j^{(N)}\right\} = 0.00001$.

Detailed arguments and with numerical examples are presented in [42].

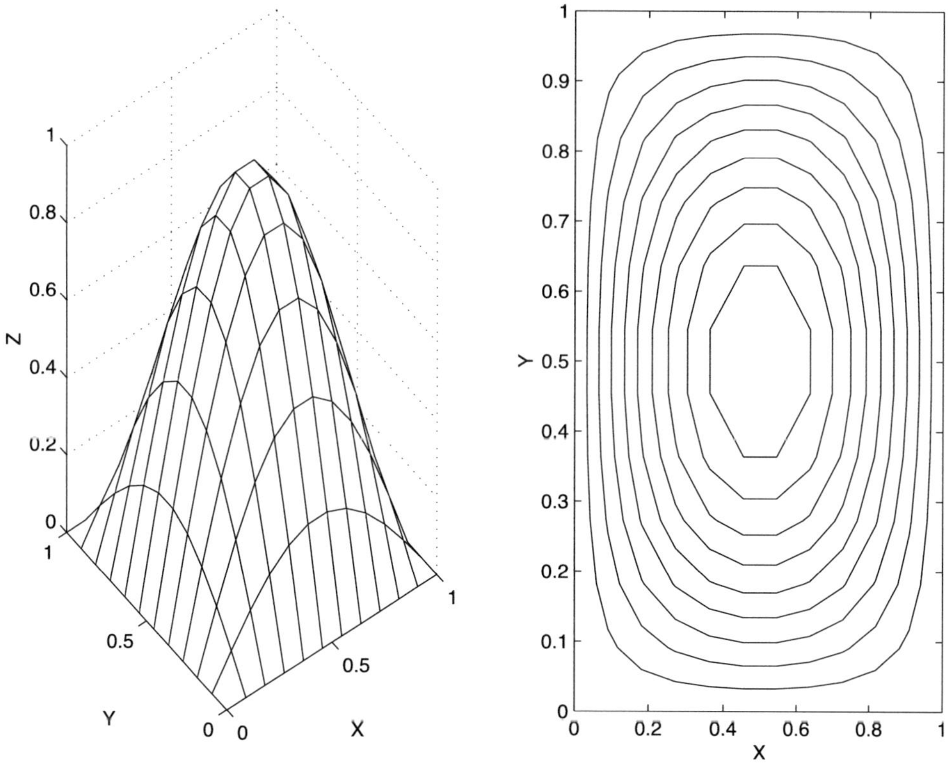

Figure 1.
Obstacle function ψ.

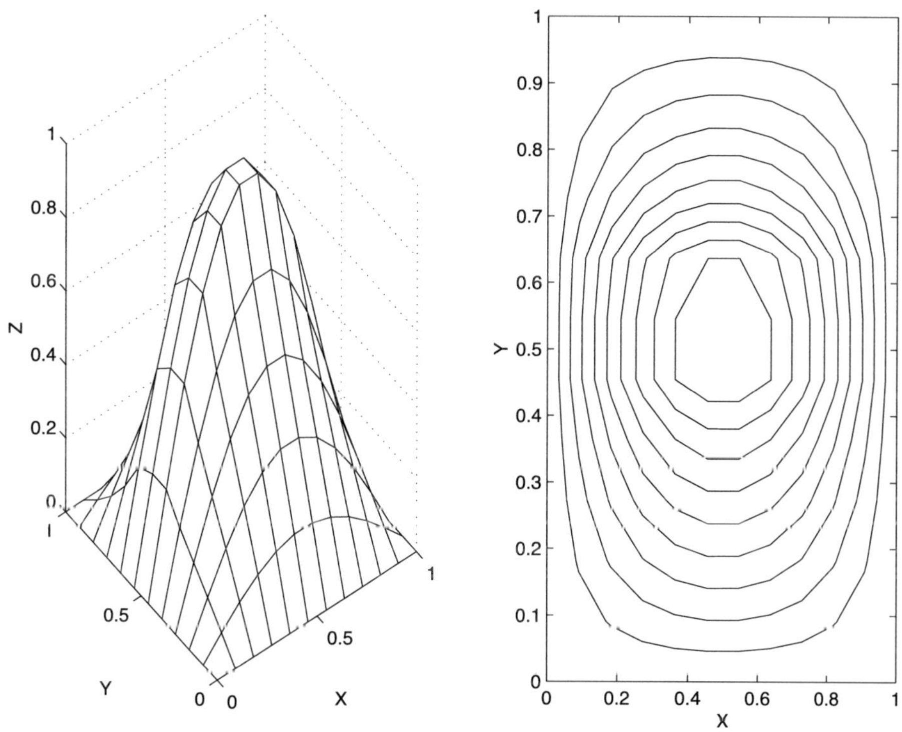

Figure 2.
Approximate solution $u_h^{(0)}$.

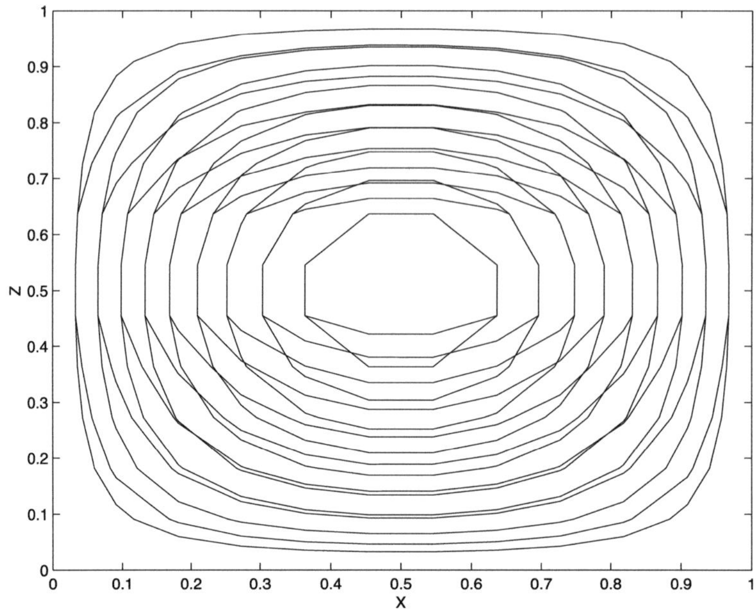

Figure 3.
Illustration of the contact zone.

3.1.3 A Newton-type verification method

The idea of the enclosure method for solutions of obstacle problems is based upon simply sequential iterations for the original fixed point operator $P_K F$. Therefore, it is difficult to apply the method to the problem of which associated operator is not retractive in a neighborhood of the solution. In order to overcome such a difficulty, we introduce an another formulation using a Newton-like operator. The essential point is the way to devise the Newton-like operator for a kind of non-differentiable map which defines the original problem.

To formulate a Newton-type verification condition, we need a Fréchet derivative of the operator $P_K F$. However, $P_K F$ is not Fréchet differentiable at all. Therefore, we define the approximate Fréchet-like derivative $\widetilde{D}_K F(u_h)$ on V_h for some $u_h \in K_h$ instead of the Fréchet derivative. Assume that $\left\{\phi_j\right\}_{j=1\cdots M}$ is a basis of V_h, where $M = \dim V_h$, such that $\phi_j(x) \geq 0$ on Ω and satisfying

$$\phi_j(x_i) = \begin{cases} 1, & i = j, \\ 0, & i \neq j, \end{cases}$$

where x_i is an interior node of the finite element mesh.

And, for $v_h \in V_h$, we represent it such as

$$v_h = \sum_{j=1}^{M} v_{hj} \phi_j.$$

Here, $\left(v_{hj}\right)_{j=1,\cdots,M}$ is called as the coefficient vector of v_h. Now we take a fixed subset $N_0 \subset \{1, 2, \cdots, M\}$, define V_{h,N_0}, the closed subspace of V_h, by

$$V_{h,N_0} = \left\{v_h | v_h \in V_h, v_{hj} = 0 \ \ for \ \ j \notin N_0\right\}.$$

And let P_{h,N_0} be a H_0^1-projection from V onto V_{h,N_0} defined by

$$a(u - P_{h,N_0}u, v) = 0, \quad \forall v \in V_{h,N_0}, P_{h,N_0}u \in V_{h,N_0}.$$

In order to define $\widetilde{D}_K F(u_h) : V_h \to V_{h,N_0}$, we differentiate the first equation of (40) in W at $W = u_h$ to get, for arbitrary $\delta \in V_h$,

$$\partial Y^* - D\partial Z^* = -\left\{\left(f'(u_h)\delta, \phi_j\right)\right\}_{1\le j\le M}. \tag{42}$$

Here, $\partial Y^* = \left(\widetilde{Y}_j^*\right)_{1\le j\le M}$ and $\partial Z^* = \left(\widetilde{Z}_j^*\right)_{1\le j\le M}$, where $\widetilde{Y}_j^* = 0$ for $j \in N_0$ and $\widetilde{Z}_j^* = 0$ for $j \notin N_0$, respectively.

Then we define the approximate Fréchet-like derivative of $P_K F(u)$ at $u = u_h$, as the linear map $\widetilde{D}_K F(u_h) : V_h \to V_{h,N_0}$ such that, for each $\delta \in V_h$,

$$\widetilde{D}_K F(u_h)(\delta) := \sum_{j=1}^{M} \widetilde{Z}_j^* \phi_j.$$

We now assume that.

A4. The restriction to V_{h,N_0} of the operator $P_{h,N_0}\left[I - \widetilde{D}_K F(u_h)\right] : V_h \to V_{h,N_0}$ has the inverse operator

$$\left[P_{h,N_0} - \widetilde{D}_K F(u_h)\right]_h^{-1} : V_{h,N_0} \to V_{h,N_0}.$$

Here, I means the identity map on V_h.

By using the above approximate Fréchet-like derivative, we define the Newton-like operator $N_h : V \to V_h$ by

$$N_h(w) \equiv P_{K_h}w - \left[P_{h,N_0} - \widetilde{D}_K F(u_h)\right]_h^{-1} P_{h,N_0}(P_{K_h} - P_{K_h}P_K F)(w)).$$

Next we define the operator $T : V \to V$ as follows:

$$T(w) \equiv N_h(w) + (I - P_{K_h})P_K F(w).$$

Then T becomes a compact map on V and it follows the fixed point problem $w = P_K F w$ is equivalent to $w = T(w)$. Detailed arguments and with numerical examples are presented in [35].

3.2 Elasto-plastic torsion problems

In this subsection, we consider an enclosure metnod of solutions for elasto-plastic torsion problems governed by an elliptic variational inequalities [25, 32, 33]. The nonlinear elasto-plastic torsion problem is defined as the same type elliptic variational inequalities as (26) with

$$K := \left\{v \in H_0^1(\Omega); |\nabla v| \le 1 \quad \text{a.e. on } \Omega\right\}. \tag{43}$$

As is well known [56, 58], two sub-domains Ω_p and Ω_e defined by

$$\Omega_p = \{x; x \in \Omega, |\nabla u| = 1\},$$

and

$$\Omega_e = \Omega \backslash \Omega_p = \{x; x \in \Omega, |\nabla u| < 1\}$$

correspond to the plastic and elastic regions, respectively. The elastic region Ω_e and the plastic region Ω_p are not known beforehand and should be determined, therefore $\partial\Omega_e \cap \partial\Omega_p$ is actually the free boundary of the problem (26). The problem (26) has been formulated as the problem of finding u satisfying

$$\begin{cases} -\Delta u = f(u) & \text{in} \quad \Omega_e, \\ |\nabla u| = 1 & \text{in} \quad \Omega_p, \\ u = 0 & \text{on} \quad \partial\Omega. \end{cases} \tag{44}$$

The finite dimensional convex subset K_h is also defined similarly as before:

$$K_h := V_h \cap K = \{v_h | \ v_h \in V_h, \ |\nabla v_h| \le 1 \ \text{a.e. on} \ \Omega\}. \tag{45}$$

In order to formulate the verification procedure, we need a verified computational method for solving the finite dimensional part (rounding) and a constructive estimates for infinite dimensional part (rounding error) as in the previous subsection.

Following [49, 56], we define the Lagrangian functional $\mathcal{L}$ associated with (1) by

$$\mathcal{L}(v,\mu) = \frac{1}{2}\int_\Omega |\nabla v|^2 dx - (g,v) + \frac{1}{2}\int_\Omega \mu\left(|\nabla v|^2 - 1\right)dx.$$

It follows, from [49, 56], that if L has a saddle point $\{u,\lambda\} \in H_0^1(\Omega) \times L_+^\infty(\Omega)$, then u is a solution of (1), where $L_+^\infty(\Omega) = \{q \in L^\infty(\Omega); q \ge 0 \ a.e. \ in \ \Omega\}$. We use the Uzawa algorithm to solve (1). Thus we can claculate the rounding $R(P_K F(W))$, for a candidate set W, by solving the following problem with guaranteed error bounds:

$$\begin{cases} \text{Find} \ \{u_h, \lambda_h\} \in K_h \times \Lambda_h \quad \text{such that} \\ \lambda_h = \max\left[\lambda_h + \rho\left(|\nabla u_h|^2 - 1\right), 0\right] \quad \text{with} \ \rho > 0. \\ \int_\Omega (1+\lambda_h)\nabla u_h \cdot \nabla v_h dx = (f(W), v_h), \forall v_h \in V_h, \quad u_h \in V_h, \end{cases} \tag{46}$$

The problem (46) can be formulated as a system of nonlinear and nonsmooth (nondifferentiable) equations. A verification method for nonsmooth equations by a generalized Krawczyk operator is studied in [1, 55]. We briefly describe the method presented by [55] in the below.

We consider the following equivalent system of nonlinear(and nondifferentiable) equation to (46) for a fixed $w \in W$

$$H(x) = 0. \tag{47}$$

Here, we assume that $H : R^n \to R^n$ is locally Lipschitz continuous. The equivalence means that x^* solves (46) if and only if x^* solves (47). The method is based on the mean value theorem for local Lipschitz functions of the form

$$H(x) - H(y) \in co\partial H([x])(x-y), \quad \text{for all} \ x, y \in [x],$$

where [x] stands for an interval vector, "co" denotes the convex hull, and ∂H the generalized Jacobian in Clarke's sense [57], which is also considered as a slope function, and

$$co\partial H([x]) := co\{V \in \partial H(x);\quad x \in [x]\}.$$

Let $[L_{[x]}]$ be an interval matrix such that $co\partial H([x]) \subseteq [L_{[x]}]$. Then for any $x, y \in [x] \subseteq R^n$ it holds that $H(x) - H(y) \in [L_{[x]}](x - y)$.

Then an interval operator for nonsmooth equations is defined by

$$G(x, A, [x]) := x - A^{-1}H(x) + (I - A^{-1}[L_{[x]}])([x] - x). \tag{48}$$

The mapping $G(x, A, [x])$ is called a generalized Krawczyk operator. Therefore, the verification condition of solutions for (46) in $[x]$ is given by

$$G(x, A, [x]) \subseteq [x] \subset D.$$

Thus, we can compute the solution of (46) with guaranteed accuracy. That is, we can enclose the rounding $R(P_K F(U))$. On the other hand, in order for the calculation of the rounding error $RE(P_K F(U))$, the similar arguements can also be applied for one dimensional problem. Actually, we can prove that the same constant $C_0 = \frac{\sqrt{5}}{\pi}$ is also valid for the present problem in one dimensinal case, which implies that we can give a verification procedure besed on the same principle as before [25, 32, 33]. In [33], we extended the approach to the numerical proof of existence of solutions for elasto-plastic torsion problems as well as gave a numerical example for one dimensional case. The verification method in [33] is based on the generalized Krawczyk operator for solving a system of nonsmooth (nondifferentiable) equations. In order to use the generalized Krawczyk operator, we need to calculate the Jacobian. In that case, we need some complicated techniques. However, in many cases, calculating the generalized Jacobian is very difficult. To overcome such difficulties, we proposed a numerical verification method without using the generalized Krawczyk operator. This method is attractive, since calculating the generalized Jacobian is not required in the computational performance. Furthermore, up to know, our verification methods are mainly based on the enclosure of solutions in the sense of L^2 or H^1 norms. We considered a numerical verification method with guaranteed L^∞ error bounds for the solution of elasto-plastic torsion problem.

3.3 Simplified Signorini problems

A simplified Signorini problem is also given by the elliptic variational inequalities of the form (26) with

$$K := \{v \in H_0^1(\Omega);\quad v \geq 0 \quad \text{on} \quad \partial\Omega|\} \tag{49}$$

and

$$a(u, v) = \int_\Omega \nabla u \cdot \nabla v dx + \int_\Omega uv dx. \tag{50}$$

where

$$\nabla u \cdot \nabla v = \frac{\partial u}{\partial x_1}\frac{\partial v}{\partial x_1} + \frac{\partial u}{\partial x_2}\frac{\partial v}{\partial x_2}.$$

As well known, the solution u of this elliptic variational inequalities can be characterized as a solution of the following free boundary problem finding u and two subsets Γ_0 and Γ_+ such that $\Gamma_0 \cup \Gamma_+ = \partial\Omega$ and $\Gamma_0 \cap \Gamma_+ = \emptyset$

$$\begin{cases} -\Delta u + u = f(u) \quad \text{in} \quad \Omega, \\ u = 0 \ \text{ on } \ \Gamma_0, \dfrac{\partial u}{\partial n} \geq 0 \ \text{ on } \ \Gamma_0, \\ u > 0 \ \text{ on } \ \Gamma_+, \dfrac{\partial u}{\partial n} = 0 \ \text{ on } \ \Gamma_+, \end{cases} \tag{51}$$

where $\frac{\partial}{\partial n}$ the outer normal derivative on $\partial\Omega$. In the present case, the approximation subspace K_h is taken as

$$K_h := V_h \cap K = \{v_h | \quad v_h \in V_h, \quad v_h \geq 0 \quad \text{on } \partial\Omega\}. \tag{52}$$

For a candidate set W, the computation of rounding $R(P_K F(W))$ is also reduced to the quadratic programming problem as in the Section 3.1 [56].

Since the constant C_2 in (25) is easily estimated as $C_2 = 1$, the standard approximation property of the interpolation by K_h gives a constructive error estimates to compute the rounding error $RE(P_K F(W))$. For a simplified Signorini problem [43], we constructed a computing algorithm which automatically encloses the solution within guaranteed error bounds. In particular, the method proposed in [43] enables us to verify the free boundary of a simplified Signorini problem, which has been impossible so far. Concerning the numerical verification of solutions for elliptic variational inequalities, we would like to mention that the inclusion method described in this article can be applied to the solution of the elliptic variational inequalities on large space domains.

3.4 Some other problems

In this subsection, we show that our idea of verification method can also be applied to the elliptic variational inequalities of the second kind.

Now, we define the functional $j(v) = \int_\Omega |\nabla v| dx$. We consider the following problem of the flow of a viscous plastic fluid in a pipe:

$$\begin{cases} \text{Find } \ u \in H_0^1(\Omega) \qquad \text{such that} \\ a(u, v - u) + j(v) - j(u) \geq (f(u), v - u), \quad \forall v \in H_0^1(\Omega). \end{cases} \tag{53}$$

As in the previous section, we consider the following auxiliary problem associated with (53) for a given $g \in L^2(\Omega)$:

$$a(u, v - u) + j(v) - j(u) \geq (g, v - u), \forall v \in H_0^1(\Omega), u \in H_0^1(\Omega). \tag{54}$$

By the well known result, we have the following lemma.

Lemma 8. There exists a unique solution $u \in H_0^1(\Omega) \cap H^2(\Omega)$ of (54) for any $g \in L^2$, such that

$$\|u\|_{H^2(\Omega)} \leq \hat{C} \|g\|_{L^2(\Omega)}.$$

When we denote the solution u of (54) by $u = Ag$ and define the composite map F on $H_0^1(\Omega)$ by $F(u) \equiv Af(u)$, which is a little bit of different from the previously appeared symbol F in Section 2, we have.

Theorem 9. F is compact on $H_0^1(\Omega)$ and the problem (53) is equivalent to the fixed point problem

$$u = F(u).$$

Proof. First, for a bounded subset $U \subset L^2(\Omega)$, we show that $AU \subset H_0^1(\Omega)$ is relatively compact. Secondly, prove that $A : L^2(\Omega) \to H_0^1(\Omega)$ is continuous. By Lemma 3, $AU \subset H^2(\Omega) \cap H_0^1(\Omega)$ and AU is bounded in $H^2(\Omega)$. Since U is bounded in $L^2(\Omega)$, by the Sobolev imbedding theorem, we have AU is relatively compact in $H_0^1(\Omega)$. Next, for arbitrary $f_1, f_2 \in L^2(\Omega)$, setting $u_1 = Af_1$ and $u_2 = Af_2$, by using (54), we obtain

$$a(u_1, u_2 - u_1) + j(u_2) - j(u_1) \geq (f_1, u_2 - u_1),$$
$$a(u_2, u_1 - u_2) + j(u_1) - j(u_2) \geq (f_2, u_1 - u_2).$$

With the above inequalities, we obtain $a(u_2 - u_1, u_2 - u_1) = -a(u_1, u_2 - u_1) + a(u_2, u_2 - u_1) \leq j(u_2) - \mathcal{T}_h$ Hence, by the Poincaré inequality, we have

$$\|u_2 - u_1\|^2_{H_0^1(\Omega)} \leq \|f_2 - f_1\|_{L^2(\Omega)} \|u_2 - u_1\|_{L^2(\Omega)} \leq \overline{C} \|f_2 - f_1\|_{L^2} \|u_2 - u_1\|_{H_0^1(\Omega)}.$$

Therefore, we obtain

$$\|u_2 - u_1\|_{H_0^1(\Omega)} \leq \overline{C} \|f_2 - f_1\|_{L^2(\Omega)}.$$

That is, A is Lipschitz continuous as a map $L^2(\Omega) \to H_0^1(\Omega)$. Hence A is compact. The latter half in the theorem is straightforward from the definition of F.

We now define the approximate problem corresponding to (54) as

$$a(u_h, v_h - u_h) + j(v_h) - j(u_h) \geq (g, v_h - u_h), \forall v_h \in V_h, u_h \in V_h. \tag{55}$$

In order to apply our verification method to enclose the solutions of (53), we need a guaranteed computation of the exact solution of the problem (55), *a rounding procedure*, as well as the constructive error estimates between the solution of (54) and (55), *rounding error estimates*.

A major difficulty in solving the problem (55) numerically is the processing of the nondifferentiable term $j(u) = \int_\Omega |\nabla u| dx$. One approach is the method of Lagrange multiplier on that term, whose continuous version is as follows [56].

Let us define $\Lambda = \{q \mid q \in L^2(\Omega) \times L^2(\Omega), |q(x)| \leq 1 \quad a.e. \ x \in \Omega\}$ with $|q(x)| = \sqrt{q_1(x)^2 + q_2(x)^2}$. Then the solution u of (54) is equivalent to the existence of q satisfying

$$\begin{cases} a(u, v) + \int_\Omega q \cdot \nabla v = (g, v), \forall v \in H_0^1(\Omega), u \in H_0^1(\Omega), \\ q \cdot \nabla u = |\nabla u| \ \ a.e. \ \ , q \in \Lambda. \end{cases} \tag{56}$$

Moreover, it is known that (56) is equivalent to the following problem:

$$\begin{cases} a(u, v) + \int_\Omega q \cdot \nabla v = (g, v), \forall v \in H_0^1(\Omega), u \in H_0^1(\Omega), \\ q = \dfrac{q + \rho \nabla u}{\sup(1, |q + \rho \nabla u|)}. \end{cases} \tag{57}$$

Here ρ is a positive constant. Let $\mathcal{T}_h$ be a triangulation of Ω, and let define L_h and Λ_h (approximation of $L^\infty(\Omega) \times L^\infty(\Omega)$ and Λ, respectively) by

$$L_h = \left\{ q_h | q_h = \sum_{\tau \in \mathcal{T}_h} q_\tau \chi_\tau, q_\tau \in R^2 \right\} \text{ and } \Lambda_h = \Lambda \cap L_h, \text{ respectively,}$$

where χ_τ is the characteristic function of τ.

Then our first purpose, computing the rounding $RF(U)$, is to enclose the solution of the following approximation problem of (57):

$$\begin{cases} a(u_h, v_h) + \int_\Omega q_h \cdot \nabla v_h = (g, v_h), \forall v_h \in V_h, u_h \in V_h, \\ q_h = \dfrac{q_h + \rho \nabla u_h}{\sup(1, |q_h + \rho \nabla u_h|)}. \end{cases} \tag{58}$$

The Eq. (58) leads to a kind of finite dimensional, nonlinear but nondifferentiable problem. We use a slope function method proposed by Rump [18–20] to enclose the solutions of (58) with $g = f(W)$ for a candidate set W. On the other hand, the rounding error $RE(F(U))$ can be computed by using the following constructive error estimates:

Theorem 10. Let u and u_h be solutions of (54) and (55), respectively. If $g \in L^2(\Omega)$, then there exists a constant $C(h)$ such that

$$\|u_h - u\|_{H_0^1(\Omega)} \le C(h) \|g\|_{L^2(\Omega)}.$$

Here, we may take $C(h) = \frac{\sqrt{5}}{\pi} h$ for the linear element in one dimensional case, and C is also numerically estimated such that $C(h) \approx O\left(h^{\frac{1}{2}}\right)$ for the two dimensional linear element. A proof of this theorem is described in Ryoo and Nakao [34]. Thus we can also implement the verification algorithm for the solution of (53) as in the previous section. For details on this subsection, please refer to Ref. [47].

4. Conclusions

We have surveyed numerical verification methods for differential equations, especially around partial differential equations, variational inequalities and the author's works. But the period of this research is shorter than the history of the numerical methods for differential equations by computer and we can say it is still in the stage of case studies. Indeed, recently, this kind of studies have been referred little by little for practical applications in PDEs and variational inequalities but there are many open problems to be resolve. Therefore, we can make no safe prediction that these approaches will grow into really useful methods for various kinds of equations and variational inequalities in mathematical analysis. Also, since the program description of the verification algorithm is very complicated in general, there is another problem like software technology associated with assurance for the correctness of the verification program itself. Actually, some of the mathematician would not give credit the computer assisted proof in analysis as correct as they believe the theoretical proof, which might cause a kind of seriously emotional problem in the methodology of mathematical sciences. And there is another difficulty from the huge scale of numerical computations which often exceed the capacity of the concurrent computing facilities.

However, in the twenty-first century, the computing environment would make more and more rapid progress, which should be beyond conception in the present state. In any case, a realistic study for partial differential equations and variational inequalities should be the future subject of the numerical computations with guaranteed accuracy. The authors believe that numerical methods with guaranteed accuracy for differential equations and variational inequalities would highly improve the reliability in the numerical simulation of the complicated phenomena in both mathematical and engineering sciences.

Acknowledgements

This work was supported by the National Research Foundation of Korea (NRF) grant funded by the Korea government (MEST) (No. 2017R1A2B4006092).

Mathematics Subject Classification (2000)

65N15; 65G20

Author details

Cheon Seoung Ryoo
Department of Mathematics, Hannam University, Daejeon, Korea

*Address all correspondence to: ryoocs@hnu.kr

References

[1] Chen X, Yamamoto T. On the convergence of some quasi-Newton methods for nonlinear equations with nondifferentiable operators. Computing. 1992;**48**:87-94

[2] Minamoto T, Nakao MT. Numerical verifications of solutions for nonlinear parabolic equations in one-space dimensional case. Reliable Computing. 1997;**3**:137-147

[3] Minamoto T. Numerical verifications of solutions for nonlinear hyperbolic equations in one-space dimensional case. Applied Mathematics Letters. 1997;**10**:91-96

[4] Minamoto T, Yamamoto N, Nakao MT. Numerical verification method for solutions of the perturbed Gelfand equation. Methods and Applications of Analysis. 1998;7(200): 251-262

[5] Nagatou K, Yamamoto N, Nakao. An approach to the numerical verification of solutions for nonlinear elliptic problems with local uniqueness. Numerical Functional Analysis and Optimization. 1999;**20**:543-565

[6] Nakao MT. A numerical approach to the proof of existence of solutions for elliptic problems. Japan Journal of Industrial and Applied Mathematics. 1988;**5**:313-332

[7] Nakao MT. A numerical verification method for the existence of weak solutions for nonlinear boundary value problems. Journal of Mathematical Analysis and Applications. 1992;**164**: 489-507

[8] Nakao MT, Yamamoto N. Numerical6verifications of solutions for elliptic equations with strong nonlinearity. Numerical Functional Analysis and Optimization. 1991;**12**: 535-543

[9] Nakao MT, Watanabe Y. On computational proofs of the existence of solutions to nonlinear parabolic problems. Journal of Computational and Applied Mathematics. 1994;**50**:401-410

[10] Nakao MT, Lee SH, Ryoo CS. Numerical verification of solutions for elasto-plastic torsion problems. Computers & Mathematics with Applications. 2000;**39**(200):195-204

[11] Nakao MT, Yamamoto N, Kimura S. On the best constant in the error bound for the H_0^1-projection into piecewise polynomial space. Journal of Approximation Theory. 1998;**93**: 491-500

[12] Nakao MT, Yamamoto N, Watanabe Y. A posteriori and constructive a priori error bounds for finite element solutions of Stokes equations. Journal of Computational and Applied Mathematics. 1998;**91**:137-158

[13] Nakao MT, Ryoo CS. Numerical verifications of solutions for variational inequalities using Newton-like method. Information. 1999;**2**:27-35

[14] Oishi S. Numerical verification of existence and inclusion of solutions for nonlinear operator equations. Journal of Computational and Applied Mathematics. 1995;**60**:171-185

[15] Plum M. Computer-assisted existence proofs for two-point boundary value problems. Computing. 1991;**46**: 19-34

[16] Plum M. Numerical existence proofs and explicit bounds for solutions of nonlinear elliptic boundary value problems. Computing. 1992;**49**:25-44

[17] Plum M. Computer assisted enclosure methods for elliptic differential equations. Journal of Linear

Algebra and Its Applications. 2001;**324**: 147-187

[18] Rump SM. INTLAB-INTerval LABoratory. In: Csendes T, editor. Developments in Reliable Computing. Amsterdam: Kluwer Academic Publishers; 1999. pp. 77-104

[19] Rump SM. In: Kulish UW, Miranker WL, editors. Solving Algebraic Problems with High Accuracy, A New Approach to Scientific Computation. New York: Academic Press; 1983. pp. 51-120

[20] Rump SM. In: Herzberger J, editor. Verification Methods for Dense and Sparse Systems of Equations, Topics in Validated Computations. Amsterdam: North-Holland; 1994. pp. 63-136

[21] Tsuchiya T, Nakao MT. Numerical verification of solutions of para me trized nonlinear boundary value problems with turning points. Japan Journal of Industrial and Applied Mathematics. 1997;**14**:357-372

[22] Watanabe Y, Nakao MT, Yamamoto N. Verified computation of solutions for nondifferentiable elliptic equations related to MHD equilibra. Nonlinear Analysis. 1997;**28**:577-587

[23] Yamamoto N. A numerical verification method for solutions of boundary value problems with local uniqueness by Banach's fixed point theorem. SIAM Journal on Numerical Analysis. 1998;**35**:2004-2013

[24] Agarwal RP, Ryoo CS. Numerical verifications of solutions for obstacle problems. Computing Supplementum. 2001;**15**:9-19

[25] Lee SH, Ryoo CS. A numerical verification of the approximate solutions for elasto-plastic torsion problem. Kyungpook Mathematical Journal. 1999;**39**:149-157

[26] Ryoo CS. A computational verification method of solution with uniqueness for obstacle problems. Bulletin of Informatics and Cybernetics. 1998;**30**:133-144

[27] Ryoo CS, Nakao MT. Numerical verification of solutions for variational inequalities. Numerische Mathematik. 1998;**81**:305-320

[28] Ryoo CS. A priori estimates for the finite element approximation of an obstacle problem. Journal of Applied Mathematics and Computing. 2000;**7**: 175-181

[29] Ryoo CS. Numerical verification of solutions for obstacle problems using Newton-like method. Computers and Mathematics with Applications. 2000; **39**:185-194

[30] Ryoo CS. Numerical verification of solutions for some unilateral problems. Applied Mathematics Letters. 2000;**13**: 13-17

[31] Ryoo CS. A computational verification method of exiatence of solution for elastoplastic torsion problem with uniqueness. Applied Mathematics Letters. 2000;**13**:25-29

[32] Ryoo CS. Computational proofs of the existence of solutions to elasto-plastic torsion problems. Mathematical and Computer Modelling. 2000;**32**: 289-298

[33] Ryoo CS. Verified computation of solutions for obstacle problems with guaranteed L^∞ error bound. Applied Mathematics Letters. 2001;**14**:769-773

[34] Ryoo CS, Nakao MT. Numerical verification of solutions for variational inequalities of the second kind. Computers and Mathematics with Applications. 2002;**43**:1371-1380

[35] Ryoo CS, Agarwal RP. Numerical inclusion methods of solutions for

variational inequalities. International Journal for Numerical Methods in Engineering. 2002;**54**:1535-1556

[36] Ryoo CS, Song H, Kim SD. Numerical verification of solutions for some unilateral boundary value problems. Computers and Mathematics with Applications. 2002;**44**:787-797

[37] Ryoo CS. Solving obstacle problems with guaranteed accuracy. Computers and Mathematics with Applications. 2003;**45**:823-834

[38] Ryoo CS, Nakao MT. Numerical verification of solutions for obstacle problems. Journal of Computational and Applied Mathematics. 2003;**161**:405-416

[39] Ryoo CS. A numerical verification of solutions of free boundary problems. Computers and Mathematics with Applications. 2004;**48**:429-435

[40] Ryoo CS. Verification for existence of solutions for some generalized obstacle problems. Applied Mathematics Letters. 2005;**18**:635-640

[41] Ryoo CS. Numerical enclosure methods to verify the existence of solutions to variational inequalities. Proceeding of the Jangjeon Mathematical Society. 2005;**8**:219-228

[42] Ryoo CS. An approach to the numerical verification of solutions for obstacle problems. Computers and Mathematics with Applications. 2007;**53**:842-850

[43] Ryoo CS. Numerical verification of solutions for Signorini problems using Newton-like method. International Journal for Numerical Methods in Engineering. 2008;**73**:1181-1196

[44] Ryoo CS. A numerical approach to the proof of existence of solutions for some generalized obstacle problems. Applied Mathematics and Computation. 2010;**216**:3365-3369

[45] Ryoo CS. An approach to the numerical verification of solutions for variational inequalities using Schauder fixed point theory. Boundary Value Problems. 2014;**2014**:235

[46] Ryoo CS. Verified computations of solutions for some unilateral boundary value problems for second order equations, Journal of. Applied Mathematics and Informatics. 2021;**39**:295-302

[47] Bazaraa MS, Shetty CM. Nonlinear Programming. New York: John Wiley; 1979

[48] Falk RS. Error estimates for the approximation of a class of variational inequalities. Mathematics of Computation. 1974;**28**:963-971

[49] Glowinski R. Numerical Methods for Nonlinear Variational Problems. New York: Springer; 1984

[50] Kinderlehrer D, Stampacchia G. An Introduction to Variational Inequalities and Their Applications. New York: Academic Press; 1980

[51] Mosco U. Convergence of convex sets and of solutions of variational inequalities. Advances in Mathematics. 1969;**3**:510-585

[52] Mosco U, Strang G. One sided approximation and variational inequalities. Bulletin of the American Mathematical Society. 1974;**80**:308-312

[53] Adams RA. Sobolev Spaces. New York: Academic Press; 1975

[54] Adams E, Kulisch U. Scientific Computing with Automatic Result Verification. San Diego: Academic Press; 1993

[55] Chen X. A verification method for solutions of nonsmooth equations. Computing. 1997;**58**:281-294

[56] Ciarlet PG. The Finite Element Method for Elliptic Problems. Amsterdam: Noth–Holland; 1978

[57] Clarke FH. Optimization and Nonsmooth Analysis. New York: John Wiley; 1983

[58] Collatz L. The Numerical Treatment of Differential Equations. Berlin-Heidelberg: Springer; 1960

[59] Zeidler E. Applied Functional Analysis: Applications to Mathematical Physics. Vol. **108**. Berlin, Germany: Springer-Verlag; 1995

[60] Zeidler E. Applied Functional Analysis: Main Principles and Their Applications. Vol. **109**. Berlin, Germany: Springer-Verlag; 1995

[61] Zeidler E. Nonlinear Functional Analysis and its Applications I. Berlin, Germany: Springer-Verlag; 1986

Chapter 2

An Algebraic Hyperbolic Spline Quasi-Interpolation Scheme for Solving Burgers-Fisher Equations

Mohamed Jeyar, Abdellah Lamnii, Mohamed Yassir Nour, Fatima Oumellal and Ahmed Zidna

Abstract

In this work, the results on hyperbolic spline quasi-interpolation are recalled to establish the numerical scheme to obtain approximate solutions of the generalized Burgers-Fisher equation. After introducing the generalized Burgers-Fisher equation and the algebraic hyperbolic spline quasi-interpolation, the numerical scheme is presented. The stability of our scheme is well established and discussed. To verify the accuracy and reliability of the method presented in this work, we select two examples to conduct numerical experiments and compare them with the calculated results in the literature.

Keywords: Burgers-Fisher equation, Algebraic Hyperbolic Spline, Quasi-interpolation

1. Introduction

The utilization of quasi-interpolation methods has been advanced in several fields of numerical analysis. This method can yield directly to solutions and does not require the solution of any linear system. In general, quasi-interpolation methods have attracted much attention because of their potential for solving partial differential equations [1–3], curve and surface fitting, integration, differentiation, and so on. In [2], Foucher and Sablonnière developed some collocation methods based on quadratic spline quasi-interpolants for solving the elliptic boundary value problems. In [4], Bouhiri et al. have used the cubic spline collocation method to solve a two-dimensional convection-diffusion equation. Generally, the problems involving Burger's equation arise in several important applications throughout science and engineering, including fluid motion, gas dynamics, [5] transfer and number theory [6].

In literature, recent developments in the resolution of the nonlinear Burger's-Fisher equation have been achieved. In a recent study [7], Mohammadi developed a stable and accurate numerical method, based on the exponential spline and finite difference approximations, to solve the generalized Burgers'-Fisher equation. The main advantage of the last method is its simplicity. Kaya et al. [8] presented numerical simulation and explicit solutions of the generalized Burgers-Fisher. Ismail et al. [9] used the Adomian decomposition method for the solutions of Burger-Huxley and Burgers-Fisher equations. In [10] Mickens proposed a non-standard finite difference scheme for the Burgers-Fisher equation. A compact finite difference method for the generalized Burgers-Fisher equation was proposed by Sari et al. [11]. Khattak [12] presented a computational radial basis function method for the Burgers-Fisher equation and some various powerful mathematical methods such as factorization method [13], tanh function methods [6, 14], spectral collocation method [15, 16] and variational iteration method [17]. In [18], the fractional-order Burgers-Fisher and generalized Fisher's equations have been solved by using the Haar wavelet method. Recently, in Ref. [19] discontinuous Legendre wavelet Galerkin method is presented for the numerical solution of the Burgers-Fisher and generalized Burgers-Fisher equations. It consists to combines both the discontinuous Galerkin and the Legendre wavelet Galerkin methods. In [20], Zhu and Kang presented a numerical scheme to solve the hyperbolic conservation laws equation based on cubic B-spline quasi-interpolation. Nonlinear partial differential equations are encountered in a variety of domains of science. Burgers- Fisher equation is a well nonlinear equation because it combines the reaction, convection and diffusion mechanisms. The sticky tag of this equation is called Burgers-Fisher because it gathers the properties of the convective phenomenon from the Burgers equation and the diffusion transport as well as the reaction mechanism from the Fisher equation. This equation shows an exemplary model to express the interaction between the reaction mechanisms, convection effect and diffusion transport. For current applications, Burgers-Fisher equation is much known in financial mathematics, physics, applied mathematics.

In this work, we consider the generalized Burger's-Fisher equation ([9]) of the form:

$$\frac{\partial u}{\partial t}+\alpha u^{\delta}\frac{\partial u}{\partial x}=\frac{\partial^2 u}{\partial^2 x}+\beta u\left(1-u^{\delta}\right),\qquad x\in\Omega=[0,1],\qquad t\geq 0 \tag{1}$$

with the initial condition

$$u(x,0)=\left\{\frac{1}{2}+\frac{1}{2}\tanh\left(\frac{-\alpha\delta}{2(\delta+1)}\right)x\right\}^{\frac{1}{\delta}},\qquad x\in\Omega=[0,1] \tag{2}$$

and the boundary conditions

$$u(0,t)=\left\{\frac{1}{2}+\frac{1}{2}\tanh\left[\frac{\alpha\delta}{2(\delta+1)}\left(\frac{\alpha}{\delta+1}+\frac{\beta(\delta+1)}{\alpha}\right)t\right]\right\}^{\frac{1}{\delta}},\qquad t\geq 0 \tag{3}$$

and

$$u(1,t)=\left\{\frac{1}{2}+\frac{1}{2}\tanh\left[\frac{-\alpha\delta}{2(\delta+1)}\left(1-\left(\frac{\alpha}{\delta+1}+\frac{\beta(\delta+1)}{\alpha}\right)t\right)\right]\right\}^{\frac{1}{\delta}},\qquad t\geq 0 \tag{4}$$

The exact solution of Eq. (1) (presented in [9]) is given by:

$$u(x,t)=\left\{\frac{1}{2}+\frac{1}{2}\tanh\left[\frac{-\alpha\delta}{2(\delta+1)}\left(x-\left(\frac{\alpha}{\delta+1}+\frac{\beta(\delta+1)}{\alpha}\right)t\right)\right]\right\}^{\frac{1}{\delta}}. \tag{5}$$

Our main purpose in this chapter is to use the univariate quasi-interpolant associated with the algebraic hyperbolic B-spline of order 4 for solving the Burgers-Fisher Eqs. (10). Firstly, we approximate first and second-order partial derivatives by those of the algebraic hyperbolic spline $\mathcal{Q}_4 u(x_i,t_n)$ quasi-interpolant. Then, we use this derivatives to approximate $\left(\frac{\partial u}{\partial x}\right)_i^n$ and $\left(\frac{\partial^2 u}{\partial^2 x}\right)_i^n$. The resulting system can be solved using

MATLAB's ode solver. More precisely, we provide a powerful numerical scheme applying a hyperbolic quasi interplant used in [21] to solve Burger's Fisher equation. This method produces better results compared to the results obtained by all the schemes in the literature, for example, those studied in [22, 23].

The chapter is organized as follows. Section 2 is dedicated to the description of the quasi-interpolation of the algebraic hyperbolic splines. Afterward, Section 3 is devoted to the presentation of numerical techniques to solve the Burger's-Fisher equation. The stability of the scheme has been studied in Section 4. In Section 5, two examples of the Burger's-Fisher equation are illustrated and compared to those obtained with some previous results. Finally, our conclusion is presented in Section 6.

2. Algebraic hyperbolic spline quasi-interpolation of order 4

In this section, we recall the results on hyperbolic spline quasi-interpolation that we will use to establish the numerical method (see [21]). Let $T = \{x_i = ih\}_{i=-\infty}^{+\infty}$ $(0 < h < \pi)$ be a set of knots which partition the parameter axis x uniformly.

For $k \geq 3$, the B-spline family that generates the space $\Gamma_k = \{\sinh(x), \cosh(x), 1, x, \cdots, x^{k-3}\}$ is called algebraic hyperbolic B-spline (for more details see [24]), which can be defined as for $k = 2$:

$$N_{0,2}(x) = \begin{cases} \dfrac{h \sinh(x)}{2(\cosh(h) - 1)}, & 0 \leq x < h, \\ \dfrac{h \sinh(2h - x)}{2(\cosh(h) - 1)}, & h \leq x < 2h, \\ 0, & \text{otherwise}, \end{cases} \tag{6}$$

$$N_{i,2}(x) = N_{0,2}(h, x - ih), \quad (i = 0, \pm 1, \pm 2, \cdots) \tag{7}$$

and for $k \geq 3$,

$$N_{i,k}(x) = \frac{1}{h} \int_{x-h}^{x} N_{i,k-1}(h, s)\, ds, \quad (i = 0, \pm 1, \pm 2, \cdots). \tag{8}$$

We apply the recursion formula (8) to get the algebraic hyperbolic B-spline of order 4, which is defined in Γ_4 as follows:

$$N_{i,4}(x) = \begin{cases} \dfrac{x - x_i + \sinh(x - x_i)}{2h(1 - \cosh(h))}, & x_i \leq x < x_{i+1}, \\ \dfrac{x - x_{i+2} - 2h\cosh(h) + 2(x - x_i)\cosh(h) + 2\sinh(x_{i+1} - x) + \sinh(x_{i+2} - x)}{2h(\cosh(h) - 1)}, & x_{i+1} \leq x < x_{i+2}, \\ \dfrac{x_{i+2} - x + 6h\cosh(h) - 2(x - x_i)\cosh(h) - \sinh(x_{i+2} - x) - 2\sinh(x_{i+3} - x)}{2h(\cosh(h) - 1)}, & x_{i+2} \leq x < x_{i+3}, \\ \dfrac{x - x_{i+4} + \sinh(x_{i+4} - x)}{2h(\cosh(h) - 1)}, & x_{i+3} \leq x < x_{i+4}, \\ 0, & \text{otherwise}. \end{cases} \tag{9}$$

According to [21], the univariate Quasi-Interpolant associated to the algebraic hyperbolic B-spline of order 4, can be expressed as operators of the form

$$Q_4^1 f(x) = \sum_{i=-3}^{n-1} \left(\overline{\nu}_h^1 f_{i+2} + \nu_h^1 (f_{i+1} + f_{i+3}) \right) N_{i,4}(x) \tag{10}$$

where $\nu_h^1 = \frac{1}{4} \operatorname{csch}\left(\frac{h}{2}\right)^2 (h \operatorname{csch}(h) - 1)$, $\overline{\nu}_h^1 = 1 - 2\nu_h^1$ and $f_i = f(x_i)$.

The error associated with the quadrature formula based on $Q_4^1 f$ is of order 5 as the following theorem describes.

Theorem 1 There exists a constant C_2 such that for all $f \in L_1^4([a,b])$ and for all partitions τ_h of $[a,b]$,

$$\|f - Q_4^1 f\|_\infty \leq C_2 h^5 \|\mathbf{L}_4 f\|_\infty, \tag{11}$$

with L_4 is an operator defined by: $\mathbf{L}_4 := \mathbf{D}^2(\mathbf{D}^2 - 1)$ and $\mathbf{L}_4 f = 0$ for all $f \in \Gamma_4$.

Proof The proof is almost the same as that of Theorem 15 in [21].

3. Numerical scheme using hyperbolic spline quasi-interpolation

For approximate derivatives of f by derivatives of $Q_4^1 f$ up to the order h^4, we can evaluate the value of f at x_i by

$$(Q_4^1 f)' = \sum_{j=-3}^{n-1} \left(\overline{\nu}_h^1 f_{j+2} + \nu_h^1\left(f_{j+1} + f_{j+3}\right)\right) N'_{j,4} \tag{12}$$

and

$$(Q_4^1 f)'' = \sum_{j=-3}^{n-1} \left(\overline{\nu}_h^1 f_{j+2} + \nu_h^1\left(f_{j+1} + f_{j+3}\right)\right) N''_{j,4}. \tag{13}$$

The values of $N'_{j,4}$ and $N''_{j,4}$ using the formula (9) are

$$N'_{i,4}(x) = \begin{cases} \dfrac{1 - \cosh(x - x_i)}{2h(1 - \cosh(h))}, & x_i \leq x < x_{i+1}, \\ \dfrac{1 + 2\cosh(h) - 2\cosh(x_{i+1} - x) - \cosh(x_{i+2} - x)}{2h(\cosh(h) - 1)}, & x_{i+1} \leq x < x_{i+2}, \\ \dfrac{2\cosh(x_{i+3} - x) + \cosh(x_{i+2} - x) - 2\cosh(h) - 1}{2h(\cosh(h) - 1)}, & x_{i+2} \leq x < x_{i+3}, \\ \dfrac{1 - \cosh(x_{i+4} - x)}{2h(\cosh(h) - 1)}, & x_{i+3} \leq x < x_{i+4}, \\ 0, & \text{otherwise}, \end{cases} \tag{14}$$

and

$$N''_{i,4}(x) = \begin{cases} \dfrac{\sinh(x - x_i)}{2h(1 - \cosh(h))}, & x_i \leq x < x_{i+1}, \\ \dfrac{2\sinh(x_{i+1} - x) + \sinh(x_{i+2} - x)}{2h(\cosh(h) - 1)}, & x_{i+1} \leq x < x_{i+2}, \\ \dfrac{-\sinh(x_{i+2} - x) - 2\sinh(x_{i+3} - x)}{2h(\cosh(h) - 1)}, & x_{i+2} \leq x < x_{i+3}, \\ \dfrac{\sinh(x_{i+4} - x)}{2h(\cosh(h) - 1)}, & x_{i+3} \leq x < x_{i+4}, \\ 0, & \text{otherwise}, \end{cases} \tag{15}$$

By using Eq. (10), the first derivative of algebraic hyperbolic spline quasi-interpolation at x_i for all $i\in\{2,\cdots,n-2\}$ is

$$\begin{aligned} Q_4^1 f'(x_i) &= \sum_{j=-3}^{n-1}\left(\overline{\upsilon}_h^1 f_{j+2} + \upsilon_h^1\left(f_{j+1}+f_{j+3}\right)\right)N'_{j,4}(x_i) \\ &= \left(\overline{\upsilon}_h^1 f_i + \upsilon_h^1\left(f_{i-1}+f_{i+1}\right)\right)N_{i-2,4}'(x_i) + \left(\overline{\upsilon}_h^1 f_{i+1} + \upsilon_h^1\left(f_i+f_{i+2}\right)\right)N'_{i-1,4}(x_i) \\ &= -\upsilon_h^1 f_{j-2} - \overline{\upsilon}_h^1 f_{i-1} + \overline{\upsilon}_h^1 f_{i+1} + \upsilon_h^1 f_{i+2} \end{aligned} \quad (16)$$

That is to say

$$Q_4^1 f'(x_i) = -\upsilon_h^1 f_{j-2} - \overline{\upsilon}_h^1 f_{i-1} + \overline{\upsilon}_h^1 f_{i+1} + \upsilon_h^1 f_{i+2} \quad (17)$$

and the second derivative of algebraic hyperbolic spline quasi-interpolation at x_i for all $i\in\{2,\cdots,n-2\}$ is

$$\begin{aligned} Q_4^1 f''(x_i) &= \sum_{j=-3}^{n-1}\left(\overline{\upsilon}_h^1 f_{j+2} + \upsilon_h^1\left(f_{j+1}+f_{j+3}\right)\right)N''_{j,4}(x_i) \\ &= \left(\overline{\upsilon}_h^1 f_i + \upsilon_h^1\left(f_{i-1}+f_{i+1}\right)\right)N''_{i-2,4}(x_i) + \left(\overline{\upsilon}_h^1 f_{i+1} + \upsilon_h^1\left(f_i+f_{i+2}\right)\right)N''_{i-1,4}(x_i) \end{aligned} \quad (18)$$

That is to say

$$Q_4^1 f''(x_i) = a_h\left(\upsilon_h^1 f_{i-2} + \left(-2\upsilon_h^1+\overline{\upsilon}_h^1\right)f_{i-1} + 2\left(\upsilon_h^1-\overline{\upsilon}_h^1\right)f_i + \left(-2\upsilon_h^1+\overline{\upsilon}_h^1\right)f_{i+1} + \upsilon_h^1 f_{i+2}\right) \quad (19)$$

with $a_h = \frac{sinh\,(h)}{2h(\,cosh\,(h)-1)}$.

Discretizing (1) in time we get

$$u_i^{n+1} - u_i^n + \alpha\left(u^\delta\right)_i^n \tau\left(\frac{\partial u}{\partial x}\right)_i^n = \tau\left(\frac{\partial^2 u}{\partial^2 x}\right)_i^n + \beta\tau u_i^n\left(1-\left(u^\delta\right)_i^n\right) \quad (20)$$

where u_i^n is the approximation of the value $u(x,t)$ at (x_i,t_n), $t_n = n\tau$ and τ is the time step with $0\le i\le M$ and $0\le n\le N$. Then, we use the derivatives of the algebraic hyperbolic spline $Q_4u(x_i,t_n)$ quasi-interpolant to approximate $\left(\frac{\partial u}{\partial x}\right)_i^n$ and $\left(\frac{\partial^2 u}{\partial^2 x}\right)_i^n$.

Assume that $U^n = \left(u_0^n, u_1^n, \cdots, u_M^n\right)$ is known for the non-negative integer n. We set unknown vectors as

$$\begin{cases} \left(\frac{\partial U}{\partial x}\right)^n = \left(\left(\frac{\partial u}{\partial x}\right)_0^n, \left(\frac{\partial u}{\partial x}\right)_1^n, \cdots, \left(\frac{\partial u}{\partial x}\right)_M^n\right) \\ \left(\frac{\partial^2 U}{\partial^2 x}\right)^n = \left(\left(\frac{\partial^2 u}{\partial^2 x}\right)_0^n, \left(\frac{\partial^2 u}{\partial^2 x}\right)_1^n, \cdots, \left(\frac{\partial^2 u}{\partial^2 x}\right)_M^n\right) \end{cases} \quad (21)$$

From the initial conditions ((3), (4)) and boundary conditions (2), we can compute the numerical solution of (1) step by step using the scheme (20) and formulas ((17), (19)).

According to (17), (19) and (21). the scheme (20) can be rewritten as

$$u_i^{n+1} = u_i^n + \alpha(u^\delta)_i^n \frac{\tau}{2h}\left(-\nu_h^1 u_{i-2}^n - \overline{\nu}_h^1 u_{i-1}^n + \overline{\nu}_h^1 u_{i+1}^n + \nu_h^1 u_{i+2}^n\right) + \frac{\tau}{2h} a_h \left(\nu_h^1 u_{i-2}^n + \left(-2\nu_h^1 + \overline{\nu}_h^1\right) u_{i-1}^n \right.$$
$$\left. +2\left(\nu_h^1 - \overline{\nu}_h^1\right) u_i^n + \left(-2\nu_h^1 + \overline{\nu}_h^1\right) u_{i+1}^n + \nu_h^1 u_{i+2}^n\right) + \beta\tau u_i^n \left(1 - \left(u^\delta\right)_i^n\right), \text{for all } i \in \{2, \cdots, M-2\} \tag{22}$$

with $a_h = \frac{sinh\,(h)}{2h(\,cosh\,(h)-1)}$.

This scheme is called the algebraic hyperbolic quasi-interpolation (AHQI) scheme.

4. Stability analysis

Sharma and Singh provided a method to study the stability of the nonlinear partial equation in [25], which we used in this section to study the stability of our scheme.

If we set $r = \frac{\tau}{2h}$, $\mathcal{A}_i^n = \alpha(u^\delta)_i^n$, $\mathcal{B}_i^n = \beta(u^\delta)_i^n$, then the scheme (22) becomes

$$u_i^{n+1} = \left(-\mathcal{A}_i^n r\nu_h^1 + ra_h\nu_h^1\right)u_{i-2}^n + \left(-\mathcal{A}_i^n r\overline{\nu}_h^1 + ra_h\left(\overline{\nu}_h^1 - 2\nu_h^1\right)\right)u_{i-1}^n + \left(1 + \beta\tau - \tau\mathcal{B}_i^n\right.$$
$$\left. +2ra_h\left(\nu_h^1 - \overline{\nu}_h^1\right)\right)u_i^n + \left(\mathcal{A}_i^n r\overline{\nu}_h^1 + ra_h\left(\overline{\nu}_h^1 - 2\nu_h^1\right)\right)u_{i+1}^n + \left(\mathcal{A}_i^n r\nu_h^1 + ra_h\nu_h^1\right)u_{i+2}^n. \tag{23}$$

If we move to the L-infinity norm then we obtain

$$\left\|U^{n+1}\right\|_L^\infty \le \underset{i}{Sup}\left| -\mathcal{A}_i^n r\nu_h^1 + ra_h\nu_h^1\right| \left\|U^n\right\|_L^\infty + \underset{i}{Sup}\left| -\mathcal{A}_i^n r\overline{\nu}_h^1 + ra_h\left(\overline{\nu}_h^1 - 2\nu_h^1\right)\right| \left\|U^n\right\|_L^\infty$$
$$+\underset{i}{Sup}\left|1 + \beta\tau - \tau\mathcal{B}_i^n + 2ra_h\left(\nu_h^1 - \overline{\nu}_h^1\right)\right| \left\|U^n\right\|_L^\infty$$
$$+\underset{i}{Sup}\left|\mathcal{A}_i^n r\overline{\nu}_h^1 + ra_h\left(\overline{\nu}_h^1 - 2\nu_h^1\right)\right| \left\|U^n\right\|_L^\infty + \underset{i}{Sup}\left|\mathcal{A}_i^n r\nu_h^1 + ra_h\nu_h^1\right| \left\|U^n\right\|_L^\infty. \tag{24}$$

If we set $\mathcal{M}_1^n = Sup_i\,|a_h - \mathcal{A}_i^n|$, $\mathcal{M}_2^n = Sup_i\,|a_h + \mathcal{A}_i^n|$, $\mathcal{M}_3^n = Sup_i\,|1 + \beta\tau - \tau\mathcal{B}_i^n|$, then the Eq. (24) becomes

$$\left\|U^{n+1}\right\|_L^\infty \le \left(\mathcal{M}_3^n + |r|\left(|\overline{\nu}_h^1|\left(6|a_h| + \left(\mathcal{M}_1^n + \mathcal{M}_2^n\right)\right) + |\nu_h^1|\left(2|a_h| + \left(\mathcal{M}_1^n + \mathcal{M}_2^n\right)\right)\right)\right)\left\|U^n\right\|_L^\infty. \tag{25}$$

It implies that the scheme is stable if

$$\mathcal{M}_3^n + |r|\left(|\overline{\nu}_h^1|\left(6|a_h| + \left(\mathcal{M}_1^n + \mathcal{M}_2^n\right)\right) + |\nu_h^1|\left(2|a_h| + \left(\mathcal{M}_1^n + \mathcal{M}_2^n\right)\right)\right) \le C, \tag{26}$$

with C is a finite positive constant.

5. Numerical results

In this section, the proposed quasi-interpolation splines collocation methods are tested for their validity for solving the generalized Burgers-Fisher equation with the initial condition (2) and the boundary conditions (3). Two different examples for the Burgers-Fisher equation are solved and the obtained results are compared with those presented in [22, 25]. To verify the accuracy and reliability of the present method in this article, we select two examples to conduct numerical experiments and compare them with the calculated results in the existing literature. That's why we divided this section into two subsections, in each subsection we compared our scheme (AHQI scheme) to each example by comparing their maximum error which is defined by

$$e = \underset{1 \le i \le M}{Max} |u_i^{exact} - u_i^{approach}|. \tag{27}$$

5.1 First example: MCN scheme

In the first example, we compared the maximum error of AHQI scheme with MCN scheme proposed in [22]. In **Table 1** we showed the maximum error of each scheme with different values of N with $\alpha = \beta = \delta = 1$ and we remarked that our method is better than that presented in [22], also we illustrated the

x	N = 10		N = 100		N = 1000	
	AHQI	MCN	AHQI	MCN	AHQI	MCN
	1×10^{-5}	1×10^{-4}	1×10^{-6}	1×10^{-5}	1×10^{-9}	1×10^{-8}
0.1	0.0442	0.0987	0.0248	0.0865	0.0128	0.2880
0.2	0.0923	0.1269	0.0760	0.1153	0.0384	0.2834
0.3	0.1399	0.1352	0.1225	0.1232	0.0727	0.2060
0.4	0.1869	0.1376	0.1662	0.1250	0.1293	0.1419
0.5	0.2329	0.1383	0.2064	0.1253	0.2180	0.1158
0.6	0.2778	0.1379	0.2394	0.1251	0.3439	0.1315
0.7	0.3213	0.1359	0.2595	0.1235	0.4929	0.1836
0.8	0.3633	0.1287	0.2458	0.1162	0.5740	0.2489
0.9	0.4037	0.2489	0.1225	0.0882	0.3241	0.2516

Table 1.
Values of errors by AHQI and MCN.

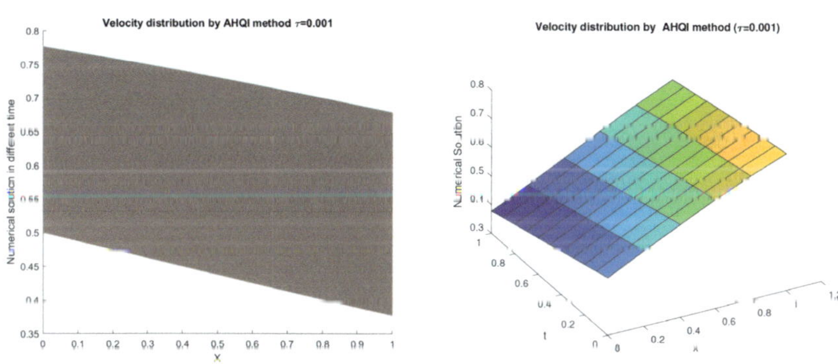

(a) Velocity distibution by AHQI method ($\tau = 0.001$) (b) Velocity distibution by AHQI method ($\tau = 0.001$)

Figure 1.
The behavior of numerical results of equation (1) by AHQI for $\tau = 0.001$.

numerical results of Eq. (1) by our method in **Figure 1(a)** and **(b)** for different values in space and time.

5.2 Second example: BSQI scheme

For the second example, we compared our scheme to BSQI scheme proposed in [25] for different values of α, β and τ: in the **Table 2** $\alpha = \beta = 0.001, \tau = 0.0001$ and in the **Table 3** $\alpha = \beta = 1, \tau = 0.0001$ with $M = 16$. In each table we calculate the maximum error for different values of δ and we remarked that for $\delta = 1$ and $\delta = 2$ our method is better than the other scheme but is close to it for $\delta = 3$. We also

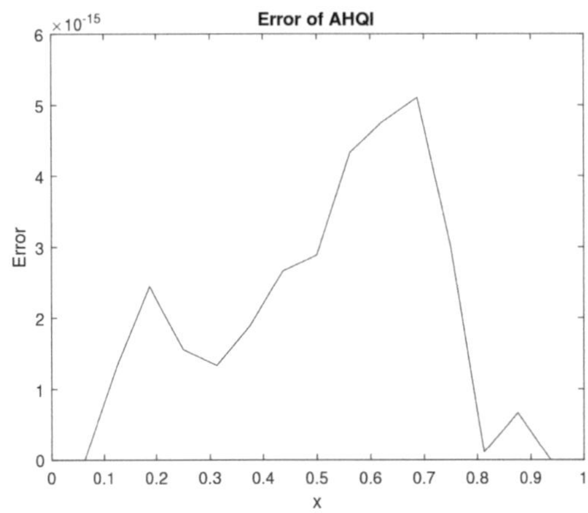

Figure 2.
The absolute errors for $\alpha = \beta = 0.001$, $\tau = 0.0001$, $\delta = 1$, $t = 1$.

t	$\delta = 1$		$\delta = 2$		$\delta = 4$	
	AHQI	BSQI	AHQI	BSQI	AHQI	BSQI
0.001	2.22044×10^{-16}	4.44089×10^{-16}	2.88657×10^{-15}	9.76996×10^{-15}	2.37587×10^{-14}	3.37508×10^{-14}
0.005	3.88578×10^{-16}	1.55431×10^{-15}	1.26565×10^{-14}	4.17888×10^{-13}	1.16573×10^{-13}	1.66422×10^{-13}
0.010	6.66133×10^{-16}	1.83187×10^{-15}	2.45359×10^{-14}	9.18154×10^{-14}	2.32702×10^{-13}	3.25628×10^{-13}
0.500	1.19904×10^{-14}	3.81917×10^{-14}	2.12041×10^{-13}	1.07303×10^{-12}	2.12041×10^{-12}	3.88234×10^{-12}
1.000	5.10702×10^{-15}	3.04201×10^{-14}	2.17048×10^{-13}	1.08047×10^{-12}	2.12352×10^{-12}	3.91087×10^{-12}

Table 2.
Error for various values of δ *and* x *with* $\alpha = \beta = 0.001$, $\tau = 0.0001$.

t	$\delta = 1$		$\delta = 2$		$\delta = 4$	
	AHQI	BSQI	AHQI	BSQI	AHQI	BSQI
0.2	3.99146×10^{-8}	5.55746×10^{-7}	1.26606×10^{-7}	2.56108×10^{-6}	1.76174×10^{-7}	1.76161×10^{-6}
0.4	7.94950×10^{-8}	9.05507×10^{-7}	1.74941×10^{-7}	4.24308×10^{-6}	1.82135×10^{-6}	4.17351×10^{-7}
0.6	1.76244×10^{-7}	2.18808×10^{-6}	2.71508×10^{-7}	3.56848×10^{-6}	1.32793×10^{-6}	2.42401×10^{-6}
0.8	2.34913×10^{-7}	2.93314×10^{-6}	3.79941×10^{-7}	1.46518×10^{-6}	7.62594×10^{-7}	2.35757×10^{-6}
1	2.47511×10^{-7}	3.01455×10^{-6}	3.98493×10^{-7}	5.54230×10^{-6}	3.79941×10^{-7}	1.44350×10^{-6}

Table 3.
Error for various values of δ *and* x *with* $\alpha = \beta = 1$, $\tau = 0.0001$.

illustrated the maximum error for $\alpha = \beta = 0.001, \delta = 1, \tau = 0.0001$ and $\alpha = \beta = 1, \delta = 1, \tau = 0.00001$ in $t = 1$ and in different values of space as the **Figures 1** and **2** respectively show. The results of our method for three different space size steps $(\delta = 1, 2, 4)$ and five different time size steps t are shown in **Tables 2** and **3**. It is very clear that a good agreement between the analytical solution and the present numerical results with a minimum error is obtained, and the error becomes clear when using a large size step for time and space.

6. Conclusion

In this work, a numerical scheme to solve the nonlinear Burgers -Fisher equation has been proposed using algebraic hyperbolic spline quasi-interpolation. The numerical scheme stability was well established. The scheme efficiency, as well as its accuracy, are justified by treating well-known examples in the literature, for each case the error is reported. We conclude that the scheme with algebraic hyperbolic spline quasi-interpolation can solve Burgers-Fisher equations since it produces reasonably good results, with high convergence with very small errors.

Author details

Mohamed Jeyar[1#], Abdellah Lamnii[2*#], Mohamed Yassir Nour[2,3#], Fatima Oumellal[2#] and Ahmed Zidna[3]

1 Faculty of Sciences, First Mohammed University, Oujda, Morocco

2 Hassan First University of Settat, Faculté des Sciences et Technique, LaboratoryMISI, Morocco

3 LGIPM, Université Lorraine, Metz, France

*Address all correspondence to: a_lamnii@yahoo.fr

# These authors contributed equally.

References

[1] R. Chen, Quasi-interpolation with radial basis function and application to solve partial differential equations, Ph. D Thesis, Fudan University, (2005).

[2] F. Foucher, P. Sablonnière, Quadratic spline quasi-interpolants and collocation methods, Math. Comp. Simul, **79** (2009, 3455-3465.

[3] C.G. Zhu, R.H. Wang, Numerical solution of burgers' equation by cubic b-spline quasiinterpolation, Appl. Math. Comput, **208** (2009), 260-272.

[4] S. Bouhiri, A. Lamnii, M. Lamnii Cubic quasi-interpolation spline collocation method for solving convection-diffusion equations, Mathematics and Computers in Simulation **164** (2019), 33-45.

[5] E. J. Parkes, B.R. Duffy, An automated tanh-function method for finding solitary wave solutions to non-linear evolution equations, Comput. Phys. Comm, **98** (1996), 288-300.

[6] A. M. Wazwaz, The tanh method for generalized forms of nonlinear heat conduction and Burgers-Fisher equations, Appl. Math. Comput, **169** (2005), 321-338.

[7] R. Mohammadi, Spline solution of the generalized Burgers'-Fisher equation, Applicable Analysis, **91** (2012), 2189-2215.

[8] D. Kaya, S. M. El-Sayed, A numerical simulation and explicit solutions of the generalized Burger-Fisher equation, Appl Math Comput, **152** (2004), 403-413.

[9] H.N.A. Ismail, K. Raslan, A.A.A. Rabboh: Adomian decomposition method for Burgers' Huxley and Burgers-Fisher equations, Appl, Math. Comput. **159** (2004), 291-301.

[10] R.E. Mickens, A.B. Gumel, Construction and analysis of a non-standard finite difference scheme for the Burgers-Fisher equation, J. Sound Vib, **257** (2002),791-797.

[11] M. Sari, G. Gürarslan, I. Dağ, A compact finite difference method for the solution of the generalized Burgers-Fisher equation, Numer.Methods Partial Differential Equations, **26** (2010), 125-134.

[12] A.J. Khattak, A computational meshless method for the generalized Burger's-Huxley equation, Appl. Math. Modelling, **33**(2009), 3718-3729.

[13] H. Fahmy, Travelling wave solutions for some time-delayed equations through factorizations, Chaos Soliton Fract, **38** (2008), 1209-1216.

[14] S. A. El-Wakil, M. A. Abdou, Modified extended tanh-function method for solving nonlinear partial differential equation, Chaos Soliton Fract, **31** (2007), 1256-1264.

[15] A. Golbabai, M. Javidi, A spectral domain decomposition approach for the generalized Burger's-Fisher equation, Chaos Soliton Fract, **39** (2009), 385-392.

[16] M. Javidi, Spectral collocation method for the solution of the generalized Burger-Fisher equation, Appl Math Comput, **174**(2006), 345-352.

[17] M. Moghimi, F. S. A. Hejazi, Variational iteration method for solving generalized Burger-Fisher and Burger equations, Chaos Soliton Fract, **33** (2007), 1756-1761.

[18] A. K. Gupta and S. Saha Ray, On the Solutions of Fractional Burgers-Fisher and Generalized Fisher's Equations Using Two Reliable Methods, International Journal of Mathematics

and Mathematical Sciences
Volume 2014, Article ID 682910.

[19] S. Kumar, S. Saha Ray,Numerical treatment for Burgers-Fisher and generalized Burgers-Fisher equations, Mathematical Sciences, **15** (2021), 21-28.

[20] C.G. Zhu, W.S. Kang, Applying cubic b-spline quasi-interpolation to solve hyperbolic conservation laws, UPB Sci. Bull., Series D, **72** (2010) 49-58.

[21] S. Eddargani, A. Lamnii, M. Lamnii, D. Sbibih, A. Zidna, Algebraic hyperbolic spline quasi-interpolants and applications, JCAM, **347** (2019), 196-209.

[22] V. Chandraker, A. Awasthi, S. Jayaraj, Numerical Treatment of Burger-Fisher equation, Procedia Technology, **25** (2016), 1217-1225.

[23] C.G. Zhu, W.S. Kang, Numerical solution of Burgers-Fisher equation by cubic B-spline quasi-interpolation, Applied Mathematics and Computation, **216** (2010), 2679-2686.

[24] Y. Lü, G. Wang, X. Yang, Uniform hyperbolic polynomial B-spline curves, Comput. Aided Geom. Design. **19** (2002), 379-393.

[25] K. K. Sharma, P. Singh, Hyperbolic partial differential-difference equation in the mathematical modeling of neuronal firing and its numerical solution, Applied Mathematics and Computation, **201** (2008), 229-238.

Chapter 3

A Study of Nonlinear Boundary Value Problem

Noureddine Bouteraa and Habib Djourdem

Abstract

In this chapter, firstly we apply the iterative method to establish the existence of the positive solution for a type of nonlinear singular higher-order fractional differential equation with fractional multi-point boundary conditions. Explicit iterative sequences are given to approximate the solutions and the error estimations are also given. Secondly, we cover the multi-valued case of our problem. We investigate it for nonconvex compact valued multifunctions via a fixed point theorem for multivalued maps due to Covitz and Nadler. Two illustrative examples are presented at the end to illustrate the validity of our results.

Keywords: Positive solution, Uniqueness, Iterative sequence, Green's function, Fractional differential equation and inclusion, Existence, Nonlocal boundary value problem, Fixed point theorem

1. Introduction

In this chapter, we are interested in the existence of solutions for the nonlinear fractional boundary value problem (BVP)

$$\begin{cases} D_{0^+}^{\alpha} u(t) + f(t, u(t)) = 0, & t \in (0,1), \\ u^{(i)}(0) = 0, i \in \{0,1,2, \dots, n-2\}, \ D_{0^+}^{\beta} u(1) = \sum_{j=1}^{p} a_j D_{0^+}^{\beta} u(\eta_j). \end{cases} \tag{1}$$

We also cover the multi-valued case of problem

$$\begin{cases} -D_{0^+}^{\alpha} u(t) \in F(t, u(t)), & t \in (0,1), \\ u^{(i)}(0) = 0, i \in \{0,1,2, \dots, n-2\}, \ D_{0^+}^{\beta} u(1) = \sum_{j=1}^{p} a_j D_{0^+}^{\beta} u(\eta_j), \end{cases} \tag{2}$$

where $D_{0^+}^{\alpha}$, $D_{0^+}^{\beta}$ are the standard Riemann-Liouville fractional derivative of order

$$\alpha \in (n-1, n], \beta \in [1, n-2] \ for \ n \in \mathbb{N}^* \ and \ n \geq 3,$$

where D_{0+}^{α}, D_{0+}^{β} are the stantard Riemann-Liouville fractional derivative of order $\alpha \in (n-1, n], \beta \in [1, n-2] for \ n \geq 3$, the function $f \in C((0,1) \times \mathbb{R}, \mathbb{R})$, the multifunction $F : [0,1] \times \mathbb{R} \to 2^{\mathbb{R}}$ are allowed to be singular at $t = 0$ and/or $t = 1$ and $a_j \in \mathbb{R}^+$, $j = 1, 2, \ldots, p$, $0 < \eta_1 < \eta_2 < \ldots < \eta_p < 1$, $for p \in \mathbb{N}^*$.

The first definition of fractional derivative was introduced at the end of the nineteenth century by Liouville and Riemann, but the concept of non-integer derivative and integral, as a generalization of the traditional integer order differential and integral calculus, was mentioned already in 1695 by Leibniz [1] and L'Hospital [2]. In fact, fractional derivatives provide an excellent tool for the description of memory and hereditary properties of various materials and processes. The mathematical modeling of systems and processes in the fields of physics, chemistry, aerodynamics, electrodynamics of complex medium, polymer rheology, Bode's analysis of feedback amplifiers, capacitor theory, electrical circuits, electro-analytical chemistry, biology, control theory, fitting of experimental data, involves derivatives of fractional order. In consequence, the subject of fractional differential equations is gaining much importance and attention. For more details we refer the reader to [1–6] and the references cited therein.

Boundary value problems for nonlinear differential equations arise in a variety of areas of applied mathematics, physics and variational problems of control theory. A point of central importance in the study of nonlinear boundary value problems is to understand how the properties of nonlinearity in a problem influence the nature of the solutions to the boundary value problems. The multi-point boundary conditions are important in various physical problems of applied science when the controllers at the end points of the interval (under consideration) dissipate or add energy according to the sensors located, at intermediate points, see [7, 8] and the references therein. We quote also that realistic problems arising from economics, optimal control, stochastic analysis can be modeled as differential inclusion. The study of fractional differential inclusions was initiated by El-Sayed and Ibrahim [9]. Also, recently, several qualitative results for fractional differential inclusion were obtained in [10–13] and the references therein.

The techniques of nonlinear analysis, as the main method to deal with the problems of nonlinear differential equations (DEs), nonlinear fractional differential equations (FDEs), nonlinear partial differential equations (PDEs), nonlinear fractional partial differential equations (FPDEs), nonlinear stochastic fractional partial differential equations (SFPDEs), plays an essential role in the research of this field, such as establishing the existence, uniqueness and multiplicity of solutions (or positive solutions) and mild solutions for nonlinear of different kinds of FPDEs, FPDEs, SFPDEs, inclusion differential equations and inclusion fractional differential equations with various boundary conditions, by using different techniques (approaches). For more details, see [14–37] and the references therein. For example, iterative method is an important tool for solving linear and nonlinear Boundary Value Problems. It has been used in the research areas of mathematics and several branches of science and other fields. However, Many authors showed the existence of positive solutions for a class of boundary value problem at resonance case. Some recent devolopment for resonant case can be found in [38, 39]. Let us cited few papers. In [40], the authors studied the boundary value problems of the fractional order differential equation:

$$\begin{cases} D_{0+}^{\alpha}u(t) = f(t, u(t)) = 0, \quad t \in (0,1), \\ u(0) = 0, \;\; D_{0+}^{\beta}u(1) = aD_{0+}^{\beta}u(\eta), \end{cases}$$

where $1 < \alpha \leq 2$, $0 < \eta < 1$, $0 < a, \beta < 1$, $f \in C([0,1] \times \mathbb{R}^2, \mathbb{R})$ and D_{0+}^{α}, D_{0+}^{β} are the stantard Riemann-Liouville fractional derivative of order α. They obtained the multiple positive solutions by the Leray-Schauder nonlinear alternative and the fixed point theorem on cones.

In 2020 Li et al. [41] consider the existence of a positive solution for the following BVP of nonlinear fractional differential equation with integral boundary conditions:

$$\begin{cases} \left({}^C D_{0+}^{q} u\right)(t) + f(t, u(t)) = 0, \quad t \in [0,1], \\ u''(0) = 0, \\ \alpha u(0) - \beta u'(0) = \int_0^1 h_1(s) u(s) ds, \\ \gamma u(1) + \delta\left({}^C D_{0+}^{\sigma} u\right)(1) = \int_0^1 h_2(s) ds, \end{cases}$$

where $2 < q \le 3$, $0 < \sigma \le 1$, $\alpha, \gamma, \delta \ge 0$, and $\beta > 0$ satisfying $0 < \rho(\alpha + \beta)\gamma + \frac{\alpha\delta}{\Gamma(2-\sigma)} < \beta\left[\gamma + \frac{\delta\Gamma(q)}{\Gamma(q-\sigma)}\right]$, $f : [0,1] \times [0, +\infty) \to [0, +\infty)$ and $h_i (i = 1, 2) : [0,1] \to [0, +\infty)$ are continuous. To obtain the existence results, the authors used the well-known GuoKrasnoselskiis fixed point theorem.

In 2017, Rezapour et al. [42] investigated a Caputo fractional inclusion with integral boundary condition for the following problem

$$\begin{cases} {}^c D^{\alpha} u(t) \in F\left(t, u(t), {}^c D^{\beta} u(t), u'(t)\right), \\ u(0) + u'(0) + {}^c D^{\beta} u(0) = {}_0^{\eta} u(s) ds, \\ u(1) + u'(1) + {}^c D^{\beta} u(1) = {}_0^{\nu} u(s) ds, \end{cases}$$

where $1 < \alpha \le 2$, $\eta, \nu, \beta \in (0,1)$, $F : [0,1] \times \mathbb{R} \times \mathbb{R} \times \mathbb{R} \to 2^{\mathbb{R}}$ is a compact valued multifunction and ${}^c D^{\alpha}$ denotes the Caputo fractional derivative of order α.

In 2018, Bouteraa and Benaicha [10] studied the existence of solutions for the Caputo fractional differential inclusion

$${}^c D^{\alpha} u(t) \in F(t, u(t), u'(t)), \quad t \in J = [0,1],$$

subject to three-point boundary conditions

$$\begin{cases} \beta u(0) + \gamma u(1) = u(\eta), \\ \quad u(0) = {}_0^{\eta} u(s) ds, \\ \quad \beta^c D^p u(0) + \gamma^c D^p u(1) = {}^c D^p u(\eta), \end{cases}$$

where $2 < \alpha \le 3$, $1 < p \le 2$, $0 < \eta < 1$, $\beta, \gamma \in \mathbb{R}^+$, $F : [0,1] \times \mathbb{R} \times \mathbb{R} \to 2^{\mathbb{R}}$ is a compact valued multifunction and ${}^c D^{\alpha}$ denotes the Caputo fractional derivative of order α.

In 2019, Ahmad et al. [43] investigated the existence of solutions for the boundary value problem of coupled Caputo (Liouville-Caputo) type fractional differential inclusions:

$$\begin{cases} {}^C D^{\alpha} x(t) \in F(t, x(t), y(t)), \quad t \in [0, T], \ 1 < \alpha \le 2, \\ {}^C D^{\beta} y(t) \in F(t, x(t), y(t)), \quad t \in [0, T], \ 1 < \beta < 2, \end{cases}$$

subject to the coupled boundary conditions:

$$\begin{aligned} x(0) &= \nu_1 y(T), \ x'(0) = \nu_2 y'(T), \\ y(0) &= \mu_1 x(T), \ y'(0) = \mu_2 x'(T), \end{aligned}$$

where ${}^{C}D^{\alpha}$, ${}^{C}D^{\beta}$ denote the Caputo fractional derivatives of order α and β respectively, $F, G : [0, T] \times \mathbb{R} \times \mathbb{R}$ are given multivalued maps, $P(\mathbb{R})$ is the family of all nonempty subsets of $\mathbb{R}$, and $\nu_i, \mu_i, i = 1, 2$ are real constants with $\nu_i\mu_i \neq 1, i = 1, 2$.

Inspired and motivated by the works mentioned above, we focus on the uniqueness of positive solutions for the nonlocal boundary value problem (1) with the iterative method and properties of$f(t, u)$, explicit iterative sequences are given to approximate the solutions and the error estimations are also given. We also cover the multi-valued case of problem (2) when the right-hand side is nonconvex compact valued multifunctions via a fixed point theorem for multivalued maps due to Covitz and Nadler.

The chapter is organized as follows. In Section 2, we present some notations and lemmas that will be used to prove our main results of problem (1) and we discuss the uniqueness of problem (1). Finally, we give an example to illustrate our result. In Section 3, we introduce some definitions and preliminary results about essential properties of multifunction that will be used in the remainder of the chapter and we present existence results for the problem (2) when the right-hand side is a nonconvex compact multifunction. We shall use the fixed point theorem for contraction multivalued maps due to Covitz and Nadler [44] to prove the uniqueness of solution of problem (2). Finally, we give an example to ascertain the main result.

2. Existence and uniqueness results for problem (2)

2.1 Preliminaries

In this section, we recall some definitions and facts which will be used in the later analysis. These details can be found in the recent literature; see [2, 4, 6, 45–47] and the references therein.

Let $AC^{i}([0, 1], \mathbb{R})$ denote the space of $i - times$ differentiable functions $u : [0, 1] \to \mathbb{R}$ whose $i - th$ derivative $u^{(i)}$ is absolutely continuous and $[\alpha]$ donotes the integer part of number α.

Definition 2.1. Let $\alpha > 0,\ \ n - 1 < \alpha < n,\ \ n = [\alpha] + 1$ and $u \in AC^{n}([0, \infty), \mathbb{R})$.

The Caputo derivative of fractional order α for the function $u : [0, +\infty) \to \mathbb{R}$ is defined by

$$
{}^{c}D^{\alpha}u(t) = \frac{1}{\Gamma(n-\alpha)}\int_{0}^{t}(t-s)^{n-\alpha-1}u^{(n)}(s)ds.
$$

The Riemann-Liouville fractional derivative order α for the function $u : [0, +\infty) \to \mathbb{R}$ is defined by

$$
D^{\alpha}_{0^+}u(t) = \frac{1}{\Gamma(n-\alpha)}\frac{d^n}{dt^n}\int_{0}^{t}(t-s)^{n-\alpha-1}u(s)ds, \quad t > 0,
$$

provided that the right hand side is pointwise defined in $(0, \infty)$ and the function $\Gamma : (0, \infty) \to \mathbb{R}$, defined by

$$
\Gamma(u) = \int_{0}^{\infty} t^{u-1}e^{-t}dt,
$$

is called Euler's gamma function.

Definition 2.2. The Riemann-Liouville fractional integral of order $\alpha > 0$ of a function $u : (0, \infty) \to \mathbb{R}$ is given by

$$I^{\alpha}u(t) = \frac{1}{\Gamma(\alpha)}\int_0^t (t-s)^{\alpha-1}u(s)ds, \quad t > 0,$$

provided that the right hand side is pointwise defined in $(0, \infty)$.

We recall in the following lemma some properties involving Riemann-Liouville fractional integral and Riemann-Liouville fractional derivative or Caputo fractional derivative which are need in Lemma 2.4.

Lemma 2.1. (([45], Prop.4.3), [46]) *Let $\alpha, \beta \geq 0$ and $u \in L^1(0,1)$. Then the next formulas hold.*

i. $\left(D^{\beta}I^{\alpha}u\right)(t) = I^{\alpha-\beta}u(t)$,

ii. $(D^{\alpha}I^{\alpha}u)(t) = u(t)$,

iii. $I^{\alpha}_{0^+}I^{\beta}_{0^+}u(t) = I^{\alpha+\beta}_{0^+}u(t)$.

iv. *If $\beta > \alpha > 0$, then $D^{\alpha}t^{\beta-1} = \frac{\Gamma(\beta)t^{\beta-\alpha-1}}{\Gamma(\beta-\alpha)}$. where D^{α} and D^{β} represents Riemann-Liouville's or Caputo's fractional derivative of order α and β respectively.*

Lemma 2.2 [47]. *Let $\alpha > 0$ and $y \in L^1(0,1)$. Then, the general solution of the fractional differential equation $D^{\alpha}_{0^+}u(t) + y(t) = 0, \ \ 0 < t < 1$ is given by*

$$u(t) = -\frac{1}{\Gamma(\alpha)}\int_0^t (t-s)^{\alpha-1}y(s)ds + c_1t^{\alpha-1} + c_2t^{\alpha-2} + \cdots + c_nt^{\alpha-n}, \quad 0 < t < 1,$$

where $c_0, c_1, \ldots, c_{n-1}$ are real constants and $n = [\alpha] + 1$.

Based on the previous Lemma 2.2, we will define the integral solution of our problem (1).

Lemma 2.3. *Let $\sum_{j=1}^{p} a_j\eta_j^{\alpha-\beta-1} \in [0,1), \alpha \in (n-1, n], \beta \in [1, n-2], n \geq 3$ and $y(\cdot) \in C[0,1]$. Then the solution of the fractional boundary value problem*

$$\begin{cases} D^{\alpha}_{0+}u(t) + y(t) = 0, \\ u^{(i)}(0) = 0, \quad i \in \{0, 1, 2, \ldots, n-2\}, \\ D^{\beta}_{0+}u(1) = \sum_{j=1}^{p} a_j D^{\beta}_{0+}u\left(\eta_j\right), \end{cases} \tag{3}$$

is given by

$$u(t) = \int_0^1 G(t,s)y(s)ds, \tag{4}$$

where

$$G(t,s) = g(t,s) + \frac{t^{\alpha-1}}{d}\sum_{j=1}^{p} a_j h\left(\eta_j, s\right), \tag{5}$$

$$g(t,s) = \frac{1}{\Gamma(\alpha)}\begin{cases} t^{\alpha-1}(1-s)^{\alpha-\beta-1} - (t-s)^{\alpha-1}, & 0 \le s \le t \le 1, \\ t^{\alpha-1}(1-s)^{\alpha-\beta-1}, & 0 \le t \le s \le 1, \end{cases} \tag{6}$$

$$h(t,s) = \frac{1}{\Gamma(\alpha)}\begin{cases} t^{\alpha-\beta-1}(1-s)^{\alpha-\beta-1} - (t-s)^{\alpha-\beta-1}, & 0 \le s \le t \le 1, \\ t^{\alpha-\beta-1}(1-s)^{\alpha-\beta-1}, & 0 \le t \le s \le 1, \end{cases} \tag{7}$$

where $d = 1 - \sum_{j=1}^{p} a_j \eta_j^{\alpha-\beta-1}$.

Proof. By using Lemma 2.2, the solution of the equation $D_{0+}^{\alpha}u(t) + y(t) = 0$ is

$$u(t) = -\frac{1}{\Gamma(\alpha)}\int_0^t (t-s)^{\alpha-1}y(s)ds + c_1 t^{\alpha-1} + c_2 t^{\alpha-2} + \cdots + c_n t^{\alpha-n},$$

where $c_1, c_2 \ldots, c_n$ are arbitrary real constants.

From the boundary condition in (1), one can $c_2 = c_3 \ldots = c_{n-2} = c_{n-1} = c_n = 0$. Hence

$$u(t) = -\frac{1}{\Gamma(\alpha)}\int_0^t (t-s)^{\alpha-1}y(s)ds + c_1 t^{\alpha-1}.$$

By the last above equation and Lemma 2.1 (i), we get

$$D_{0+}^{\beta}u(t) = \frac{1}{\Gamma(\alpha-\beta)}\left[c_1\Gamma(\alpha)t^{\alpha-\beta-1} - \int_0^t (t-s)^{\alpha-\beta-1}y(s)ds\right],$$

this and by $D_{0+}^{\beta}u(1) = \sum_{j=1}^{p} a_j D_{0+}^{\beta}u\left(\eta_j\right)$, we have

$$c_1 = \frac{1}{d\Gamma(\alpha)}\left[\int_0^1 (1-s)^{\alpha-\beta-1}y(s)ds - \sum_{j=1}^{p} a_j \int_0^{\eta_j}\left(\eta_j - s\right)^{\alpha-\beta-1}y(s)ds\right].$$

Then, the unique solution of the problem (1) is given by

$$u(t)=\frac{t^{\alpha-1}}{d\Gamma(\alpha)}\left[\int_0^1(1-s)^{\alpha-\beta-1}y(s)ds-\sum_{j=1}^{p}a_j\int_0^{\eta_j}\left(\eta_j-s\right)^{\alpha-\beta-1}y(s)ds\right]-\frac{1}{\Gamma(\alpha)}\int_0^t(t-s)^{\alpha-1}y(s)ds,$$

$$=\frac{1}{\Gamma(\alpha)}\left[\int_0^t\left[t^{\alpha-1}(1-s)^{\alpha-\beta-1}-(t-s)^{\alpha-1}\right]y(s)ds+\int_t^1 t^{\alpha-1}(1-s)^{\alpha-\beta-1}y(s)ds\right.$$

$$\left.+\frac{1-d}{d}\int_0^1 t^{\alpha-1}(1-s)^{\alpha-\beta-1}y(s)ds-\frac{t^{\alpha-1}}{d}\sum_{j=1}^{p}a_j\int_0^{\eta_j}\left(\eta_j-s\right)^{\alpha-\beta-1}y(s)ds\right]$$

$$=\int_0^1 g(t,s)y(s)ds+\frac{t^{\alpha-1}}{d}\sum_{j-1}^{p}a_j\left[\int_{\eta_j}^1\eta_j^{\alpha-\beta-1}(1-s)^{\alpha-\beta-1}y(s)ds\right.$$

$$\left.+\int_0^{\eta_j}\left[\eta_j^{\alpha-\beta-1}(1-s)^{\alpha-\beta-1}-\left(\eta_j-s\right)^{\alpha-\beta-1}\right]y(s)ds\right]$$

$$=\int_0^1 g(t,s)y(s)ds+\frac{t^{\alpha-1}}{d}\sum_{j=1}^{p}a_j\int_0^1 h\left(\eta_j,s\right)y(s)ds$$

$$=\int_0^1 G(t,s)y(s)ds.$$

The proof is completed. □

Lemma 2.4. *Let* $\sum_{j=1}^{p}a_j\eta_j^{\alpha-\beta-1}\in[0,1),\alpha\in(n-1,n],\beta\in[1,n-2],n\geq 3$. *Then, the functions* $g(t,s)$ *and* $h(t,s)$ *defined by* (6) *and* (7) *have the following properties:*

i. *The functions* $g(t,s)$ *and* $h(t,s)$ *are continuous on* $[0,1]\times[0,1]$ *and for all* $t,s\in(0,1)$

$$g(t,s)>0,\quad h(t,s)>0.$$

ii. $g(t,s)\leq\frac{t^{\alpha-1}}{\Gamma(\alpha)}$ *for all* $t,s\in[0,1]$.

iii. $g(t,s)\geq t^{\alpha-1}g(1,s)$ *for all* $t,s\in[0,1]$, *where*

$$g(1,s)=\frac{1}{\Gamma(\alpha)}\left[(1-s)^{\alpha-\beta-1}-(1-s)^{\alpha-1}\right]$$

From the above properties, we deduce the following properties:

iv. *The function* $G(t,s)\geq 0$ *is continuous on* $[0,1]\times[0,1]$ *and* $G(t,s)>0$ *for all* $t,s\in(0,1)$.

v. $\max_{t\in[0,1]}G(t,s)=G(1,s)$, *for all* $s\in[0,1]$, *where*

$$G(1,s)=g(1,s)+\frac{1}{d}\sum_{j=1}^{p}a_jh\left(\eta_j,s\right)\leq\frac{(1-s)^{\alpha-\beta-1}}{d\Gamma(\alpha)}.$$

Proof. It is easy to chek that (i), (v), (vi) holds. So we prove that (ii) is true. Note that (6) and $0\leq(1-s)^{\alpha-\beta-1}\leq1$. It follows that $g(t,s)\leq\frac{t^{\alpha-1}}{\Gamma(\alpha)}$ for all $t,s\in[0,1]$. It remains to prove (iii). We divide the proof into two cases and by (1), we have.

Case1. When $0\leq s\leq t\leq1$, we have

$$g(t,s)=\frac{1}{\Gamma(\alpha)}t^{\alpha-1}\left[(1-s)^{\alpha-\beta-1}-\left(1-\frac{s}{t}\right)^{\alpha-1}\right]\geq t^{\alpha-1}g(1,s).$$

Case2. When $0\leq t\leq s\leq1$, we have

$$g(t,s)=\frac{1}{\Gamma(\alpha)}t^{\alpha-1}(1-s)^{\alpha-\beta-1}\geq t^{\alpha-1}g(1,s).$$

Hence $g(t,s)\geq t^{\alpha-1}g(1,s)$ for all $t,s\in[0,1]$. □

2.2 Existence results

First, for the uniqueness results of problem (1), we need the following assumptions.

$(A_1)\, f(t,u_1)\leq f(t,u_2)$ for any $0<t<1,\quad 0\leq u_1\leq u_2$.

(A_2) For any $r\in(0,1)$, there exists a constant $q\in(0,1)$ such that

$$f(t,ru)\geq r^q f(t,u),\quad (t,u)\in(0,1)\times[0,\infty). \tag{8}$$

$(A_3)\ 0<\int_0^1 f(s,s^{\alpha-1})ds<\infty$.

We shall consider the Banach space $E=C[0,1]$ equipped with the norm $\|u\|=\max_{0\leq t\leq1}|u(t)|$ and let

$$D=\left\{u\in C^+[0,1]:\exists M_u\geq m_u\geq0,\backslash m_ut^{\alpha-1}\leq u(t)\leq M_ut^{\alpha-1}, for\, t\in[0,1]\right\}, \tag{9}$$

where

$$C^+[0,1]=\{u\in E:u(t)\geq0,t\in[0,1]\}.$$

In view of Lemma 2.3, we define an operator T as

$$(Tu)(t)=\int_0^1 G(t,s)y(s)ds, \tag{10}$$

where $G(t,s)$ is given by (5).

By (A_1) it is easy to see that the operator $T:D\to C^+[0,1]$ is increasing. Observe that the BVP (1) has a solution if and only if the operator T has a fixed point.

Obviously, from (A_1) we obtain

$$f(t,ru)\leq r^q f(t,u),\quad \forall r>1, q\in(0,1),\quad (t,u)\in(0,1)\times[0,\infty).$$

In what follows, we first prove $T:D\to D$. In fact, for any $u\in D$, there exist a positive constants $0<m_u<1<M_u$ such that

$$m_u s^{\alpha-1} \le u(s) \le M_u s^{\alpha-1}, \quad s \in [0,1].$$

Then, from $(A_1), f(t,u)$ non-decreasing respect to u and (A_2), we can imply that for $s \in (0,1), \ \ q \in (0,1)$

$$(m_u)^q f\left(s, s^{\alpha-1}\right) \le f(s, u(s)) \le (M_u)^q f\left(s, s^{\alpha-1}\right), \quad s \in (0,1). \tag{11}$$

From (11) and Lemma 2.4, we obtain

$$\begin{aligned} Tu(t) &= \int_0^1 g(t,s) f(s,u(s))ds + \frac{t^{\alpha-1}}{d} \sum_{j=1}^p a_j \int_0^1 h\left(\eta_j, s\right) f(s,u(s))ds, \\ &\le t^{\alpha-1}\left[\frac{1}{\Gamma(\alpha)} \int_0^1 f(s,u(s))ds + \frac{1}{d} \sum_{j=1}^p a_j \int_0^1 h\left(\eta_j, s\right) f(s,u(s))ds\right], \\ &\le t^{\alpha-1}\left[\frac{(M_u)^q}{\Gamma(\alpha)} \int_0^1 f\left(s, s^{\alpha-1}\right)ds + \frac{(M_u)^q}{d} \sum_{j=1}^p a_j \int_0^1 h\left(\eta_j, s\right) f\left(s, s^{\alpha-1}\right)ds\right], \quad t \in [0,1], \end{aligned} \tag{12}$$

and

$$\begin{aligned} Tu(t) &= \int_0^1 g(t,s) f(s,u(s))ds + \frac{t^{\alpha-1}}{d} \sum_{j=1}^p a_j \int_0^1 h\left(\eta_j, s\right) f(s,u(s))ds, \\ &\ge t^{\alpha-1}\left[\frac{1}{\Gamma(\alpha)} \int_0^1 g(1,s) f(s,u(s))ds + \frac{1}{d} \sum_{j=1}^p a_j \int_0^1 h\left(\eta_j, s\right) f(s,u(s))ds\right], \\ &\ge t^{\alpha-1}\left[\frac{(m_u)^q}{\Gamma(\alpha)} \int_0^1 g(1,s) f\left(s, s^{\alpha-1}\right)ds + \frac{(m_u)^q}{d} \sum_{j=1}^p a_j \int_0^1 h\left(\eta_j, s\right) f\left(s, s^{\alpha-1}\right)ds\right], \quad t \in [0,1]. \end{aligned} \tag{13}$$

Eqs. (12) and (13) and assumption (A_3) imply that $T: D \to D$.

Now, we are in the position to give the first main result of this chapter.

Theorem 1.1 Suppose $(A_1) - (A_3)$ hold. Then problem (1) has a unique, nondecreasing solution $u^* \in D$, moreover, constructing successively the sequence of functions

$$h_n(t) = \int_0^1 G(t,s) f(s, h_{n-1}(s))ds, \quad t \in [0,1], \quad n = 1, 2, \ldots, \tag{14}$$

for any initial function $h_0(t) \in D$, then $\{h_n(t)\}$ must converge to $u^*(t)$ uniformly on $[0,1]$ and the rate of convergence is

$$\max_{t \in [0,1]} |h_n(t) - u^*(t)| = O\left(1 - \theta^{q^n}\right), \tag{15}$$

where $0 < \theta < 1$, which depends on the initial function $h_0(t)$.

Proof. For any $h_0 \in D$, we let

$$l_{h_0} = \sup\{l > 0 : lh_0(t) \le (Th_0)(t), t \in [0,1]\}, \tag{16}$$

$$L_{h_0} = \inf\{L > 0 : Lh_0(t) \ge (Th_0)(t), t \in [0,1]\}, \tag{17}$$

$$m = \min\left\{1, (l_{h_0})^{\frac{1}{1-q}}\right\}, \quad M = \max\left\{1, (L_{h_0})^{\frac{1}{1-q}}\right\}, \tag{18}$$

and

$$u_0(t) = mh_0(t), \quad v_0(t) = Mh_0(t), \tag{19}$$

$$u_n(t) = Tu_{n-1}(t), \quad v_n(t) = Tv_{n-1}(t), \quad n = 0, 1, \ldots, . \tag{20}$$

Since the operator T is increasing, (A_1), (A_2) and (16)–(20) imply that there exist iterative sequences $\{u_n\}, \{v_n\}$ satisfying

$$u_0(t) \le u_1(t) \le \ldots \le u_n(t) \le \ldots \le v_n(t) \le \ldots \le v_1(t) \le v_0(t), \; t \in [0,1]. \tag{21}$$

In fact, from (19) and (20), we have

$$u_0(t) \le v_0(t), \tag{22}$$

$$\begin{aligned} u_1(t) = Tu_0(t) &= \int_0^1 G_1(t,s)f(s, mh_0(s))ds + \frac{t^{\alpha-1}}{d}\sum_{i=1}^{n} a_j \int_0^1 G_2\left(\eta_j, s\right)f(s, mh_0(s))ds, \\ &\ge m^q\left[\int_0^1 G_1(t,s)f(s,h_0(s))ds + \frac{t^{\alpha-1}}{d}\sum_{i=1}^{n} a_j \int_0^1 G_2\left(\eta_j,s\right)f(s,h_0(s))ds\right], \\ &\ge m^q Th_0(t) \ge mh_0(t) = u_0(t), \end{aligned} \tag{23}$$

and

$$\begin{aligned} v_1(t) = Tv_0(t) &= \int_0^1 G_1(t,s)f(s, Mh_0(s))ds + \frac{t^{\alpha-1}}{d}\sum_{i=1}^{n} a_j \int_0^1 G_2\left(\eta_j, s\right)f(s, Mh_0(s))ds, \\ &\le M^q\left[\int_0^1 G_1(t,s)f(s,h_0(s))ds + \frac{t^{\alpha-1}}{d}\sum_{i=1}^{n} a_j \int_0^1 G_2\left(\eta_j,s\right)f(s,h_0(s))ds\right] \\ &\le M^q Th_0(t) \le Mh_0(t) = v_0(t). \end{aligned} \tag{24}$$

Then, by (22)–(24) and induction, the iterative sequences $\{u_n\}, \{v_n\}$ satisfy

$$u_0(t) \le u_1(t) \le \ldots \le u_n(t) \le \ldots \le v_n(t) \le \ldots \le v_1(t) \le v_0(t), \; \forall t \in [0,1].$$

Note that $u_0(t) = \frac{m}{M}v_0(t)$, from (A_1), (10), (19) and (20), it can obtained by induction that

$$u_n(t) \ge \theta^{q^n} v_n(t), \; t \in [0,1], \; n = 0, 1, 2, \ldots, \tag{25}$$

where $\theta = \frac{m}{M}$.

From (21) and (25) we know that

$$0 \leq u_{n+p}(t) - u_n(t) \leq v_n(t) - u_n(t) \leq \left(1 - \theta^{q^n}\right) M h_0(t), \quad \forall n, p \in \mathbb{N}, \tag{26}$$

and since $\left(1 - \theta^{q^n}\right) M h_0(t) \to 0, \quad as \ n \to \infty$,this yields that there exists $u^* \in D$ such that

$$u_n(t) \to u^*(t), \quad (uniformly\, on\, [0,1]).$$

Moreover, from (26) and

$$0 \leq v_n(t) - u^*(t) = v_n(t) - u_n(t) + u_n(t) - u^*(t),$$

$$\leq \left(1 - \theta^{q^n}\right) M h_0(t) \to 0, \quad as \ n \to \infty,$$

we have

$$v_n(t) \to u^*(t), \quad (uniformly\, on\, [0,1]),$$

so,

$$u_n(t) \to u^*(t), \ v_n(t) \to u^*(t), \quad (uniformly\, on\, [0,1]). \tag{27}$$

Therefore

$$u_n(t) \leq u^*(t) \leq v_n(t), \quad t \in [0,1], n = 0, 1, 2, \ldots,, \tag{28}$$

From (A_1), (19) and (20), we have

$$u_{n+1}(t) = Tu_n(t) \leq Tu^*(t) \leq Tv_n(t) = v_{n+1}(t), n = 0, 1, 2, \ldots,.$$

This together with (27) and uniqueness of limit imply that u^* satisfy $u^* = Tu^*$, that is $u^* \in D$ is a solution of BVP (1) and (2).

From (19)–(21) and (A_1), we obtain

$$u_n(t) \leq h_n(t) \leq v_n(t), n = 0, 1, 2, \ldots,. \tag{29}$$

It follows from (26)–(29) that

$$|h_n(t) - u^*(t)| \leq |h_n(t) - u_n(t)| + |u_n(t) - u^*(t)|,$$

$$\leq |h_n(t) - u_n(t)| + |u^*(t) - u_n(t)|,$$

$$\leq 2|v_n(t) - u_n(t)|,$$

$$\leq 2M\left(1 - \theta^{q^n}\right)|h_0(t)|.$$

Therefore

$$\max_{t \in [0,1]} |h_n(t) - u^*(t)| \leq 2M\left(1 - \theta^{q^n}\right) \max_{t \in [0,1]} |h_0(t)|.$$

Hence, (15) holds. Since $h_0(t)$ is arbitrary in D we know that $u^*(t)$ is the unique solution of the boundary value problem (1) in D. □

We construct an example to illustrate the applicability of the result presented.
Example 2.1. *Consider the following boundary value problem*

$$\begin{cases} D_{0^+}^{\frac{5}{2}} u(t) + \dfrac{(u)^{\frac{2}{3}-\frac{1}{6}\cos(t)}}{\sqrt{t}} = 0, \quad t \in (0,1), \\ u(0) = u'(0) = 0, \ u'(1) = \dfrac{\sqrt{2}}{2} u'\left(\dfrac{1}{2}\right), \end{cases} \tag{30}$$

where $\alpha = \frac{5}{2}, \beta = 1, a_1 = \frac{\sqrt{2}}{2}, \eta_1 = \frac{1}{2}$ *and* $f(t,u) = \frac{(u)^{\frac{2}{3}-\frac{1}{6}\cos(t)}}{\sqrt{t}}$ *is increasing function with respect to u for all* $t \in (0,1)$, *so, assumption* (A_1) *satisfied.*

By simple calculation we have $d = 1 - \frac{\sqrt{2}}{2}\left(\sqrt{\frac{1}{2}}\right) = \frac{1}{2}$.

For any $r \in (0,1)$, *there exists* $q = \frac{1}{2} \in (0,1)$ *such that*

$$f(t,ru) = \frac{(ru)^{\frac{2}{3}-\frac{1}{6}\cos(t)}}{\sqrt{t}} \geq r^{\frac{1}{2}} \frac{(u)^{\frac{2}{3}-\frac{1}{6}\cos(t)}}{\sqrt{t}} = r^{\frac{1}{2}} f(t,u),$$

thus, $f(t,u)$ *satisfies* (A_2) *and is singular at* $t = 0$.
On the other hand,

$$\int_0^1 f\left(t, t^{2,5-1}\right) dt \leq \int_0^1 t^{\frac{1}{4}} dt = \frac{4}{5} < \infty,$$

so, assumption (A_3) *is satisfied.*

Hence, all the assumptions of Theorem 1.1 *are satisfied. Which implies that the boundary value* (30) *has an unique, nondecreasing solution* $u^* \in D$.

3. Existence result for inclusion problem (2)

We provide another result about the existence of solutions for the problem (2) by using the assumption of nonconvex compact values for multifunction. Our strategy to deal with this problem is based on the Covitz-Nadler theorem for the contraction multivalued maps [44] for lower semi-continuous maps with decomposable values.

First, we will present notations, definitions and preliminary facts from multivalued analysis which are used throughout this chapter. For more details on the multivalued maps, see the book of Aubin and Cellina [48], Demling [49], Gorniewicz [50] and Hu and Papageorgiou [51], see also [44, 48, 49, 52–54].

Here $(C[0,1], \mathbb{R})$ denotes the Banach space of all continuous functions from $[0,1]$ into $\mathbb{R}$ with the norm $\|u\| = \sup\{|u(t)| : \mathit{for\,all}\, t \in [0,1]\}$, $L^1([0,1], \mathbb{R})$, the Banach space of measurable functions $u : [0,1] \to \mathbb{R}$ which are Lebesgue integrable, normed by $\|u\|_{L^1} = \int_0^1 |u(t)| dt$.

Let (X,d) be a metric space induced from the normed space $(X, \|\cdot\|)$. We denote

$$P_0(X) = \{A \in P(X) : A \neq \phi\},$$
$$P_b(X) = \{A \in P_0(X) : A \text{ is bounded}\},$$
$$P_{cl}(X) = \{A \in P_0(X) : A \text{ is closed}\},$$
$$P_{cp}(X) = \{A \in P_0(X) : A \text{ is compact}\},$$
$$P_{b,cl}(X) = \{A \in P_0(X) : A \text{ is closed and bounded}\},$$

where $P(X)$ is the family of all subsets of X.

Definition 3.1. *A multivalued map* $G : X \to P(X)$.

1. $G(u)$ is convex (closed) valued if $G(u)$ is convex (closed) for all $u \in X$,

2. is bounded on bounded sets if $G(B) = \bigcup_{u \in B} G(u)$ is bounded in X for all $B \in P_b(X)$ i.e., $\sup_{u \in B}\{\sup\{|v|, v \in G(u)\}\} < \infty$,

3. has a fixed point if there is $u \in X$ such that $u \in G(u)$. The fixed point set of the multivalued operator G will be denote by Fix G.

Definition 3.2. A multivalued map $G : [0,1] \to P_{cl}(\mathbb{R})$ is said to be measurable if for every $y \in \mathbb{R}$ the function

$$t \mapsto d(y, G(t)) = \inf\{\|y - z\| : z \in G(t)\},$$

is measurable.

Definition 3.3. Let Y be a nonempty closed subset of a Banach space E and $G : Y \to P_{cl}(E)$ be a multivalued operator with nonempty closed values.

i. G is said to be lower semi-continuous (l.s.c) if the set $\{x \in X : G(x) \cap U \neq \phi\}$ is open for any open set U in E.

ii. G has a fixed point if there is $x \in Y$ such that $x \in G(x)$.

For each $u \in (C[0,1], \mathbb{R})$, define the set of selection of F by

$$S_{F,u} = \{v \in AC([0,1], \mathbb{R}) : v \in F(t, u(t)), \ \textit{for almost all} \ t \in [0,1]\}.$$

For $P(X) = 2^X$, consider the Pompeiu-Hausdorff metric (see [55]).

$H_d : 2^X \times 2^X \to [0, \infty)$ given by

$$H_d(A, B) = \max\left\{\sup_{a \in A} d(a, B), \sup_{b \in B} d(b, A)\right\},$$

where $d(a, B) = \inf_{b \in B} d(a, b)$ and $d(b, A) = \inf_{a \in A} d(a, b)$. Then $(P_{b,cl}(X), H_d)$ is a metric space and $(P_{cl}(X), H_d)$ is a generalized metric space see [8].

Definition 3.4. Let A be a subset of $[0,1] \times \mathbb{R}$. A is $L \otimes B$ measurable if A belongs to the $\sigma-$algebra generated by all sets of the $J \times D$, where J is Lebesgue measurable in $[0,1]$ and D is Borel measurable in $\mathbb{R}$.

Definition 3.5. A subset A of $L^1([0,1], \mathbb{R})$ is decomposable if all $u, v \in A$ and measurable $J \subset [0,1] = j$, the function $u\chi_J + v\chi_{j \setminus J} \in A$, where χ_J stands for the caracteristic function of J.

Definition 3.6. Let Y be a separable metric space and $N : Y \to P(L^1([0,1], \mathbb{R}))$ be a multivalued operator. We say N has property (BC) if N is lower semi-continuous (l.s.c) and has nonempty closed and decomposable values.

Let $F : [0,1] \times \mathbb{R} \to P(\mathbb{R})$ be a multivalued map with nonempty compact values. Define a multivalued operator

$$\Phi : C([0,1], \mathbb{R}) \to P(L^1([0,1], \mathbb{R})),$$

by letting

$$\Phi(u) = \{w \in L^1([0,1],\mathbb{R}) : w(t) \in F(t,u(t)) for a.e. t \in [0,1]\}.$$

Definition 3.7. The operator Φ is called the Niemytzki operator associated with F

. We say F is of the lower semi-continuous type (l.s.c type) if its associated Niemytzki operator Φ has (BC) property.

Definition 3.8. A multivalued operator $N : X \to P_{cl}(X)$ is called.

i. $\rho-$Lipschitz if and only if there exists $\rho > 0$ such that $H_d(N(u), N(v)) \leq \rho d(u,v)$ for each $u, v \in X$,

ii. a contraction if and only if it is $\rho-$Lipschitz with $\rho < 1$.

Lemma 3.1. ([44] Covitz-Nadler). *Let (X,d) be a complete metric space. If $N : X \to P_{cl}(X)$ is a contraction, then $FixN \neq \phi$, where $FixN$ is the fixed point of the operator N.*

Definition 3.9. A measurable multivalued function $F : [0,1] \to P(X)$ is said to be integrably bounded if there exists a function $g \in L^1([0,1], X)$ such that, for all $v \in F(t)$, $\|v\| \leq g(t)$ for a.e. $t \in [0,1]$.

Let us introduce the following hypotheses.

(A_4) $F : [0,1] \times \mathbb{R} \to P_{cp}(\mathbb{R})$ be a multivalued map verifying.

i. $(t,u) \mapsto F(t,u)$ is $L \otimes B$ measurable.

ii. $u \mapsto F(t,u)$ is lower semi-continuous for $a.e.$ $t \in [0,1]$.

(A_5) F is integrably bounded, that is, there exists a function $m \in L^1([0,1], \mathbb{R}^+)$ such that $\|F(t,u)\| = \sup\{|v| : v \in F(t,u)\} \leq m(t)$ for almost all $t \in [0,1]$.

Lemma 3.2. [56] *Let $F : [0,1] \times \mathbb{R} \to P_{cp}(\mathbb{R})$ be a multivalued map. Assume (A_4) and (A_5) hold. Then F is of the $l.s.c.$ type.*

Definition 3.10. A function $u \in AC^2([0,1], \mathbb{R})$ is called a solution to the boundary value problem (2) if u satisfies the differential inclusion in (2) $a.e.$ on $[0,1]$ and the conditions in (2).

Finally, we state and prove the second main result of this Chapter. We prove the existence of solutions for the inclusion problem (2) with a nonconvex valued right hand side by applying a fixed point theorem for multivalued maps due to Covitz and Nadler. For investigation of the problem (2) we shall provide an application of the Lemma 3.4 and the following Lemma.

Lemma 3.3. ([13]) *A multifunction $F : X \to C(X)$ is called a contraction whenever there exists $\gamma \in (0,1)$ such that $H_d(N(u), N(v)) \leq \gamma d(u,v)$ for all $u, v \in X$.*

Now, we present second main result of this section.

Theorem 1.2 Assume that the following hypothyses hold.

(H_1) $F : J \times \mathbb{R} \to P_{cp}(\mathbb{R})$ is an integrable bounded multifunction such that the map $t \mapsto F(t,u)$ is measurable,

(H_2) $H_d(F(t,u_1), F(t,u_2)) \leq m(t)|u_1 - u_2|$ for almost all $t \in J$ and $u_1, u_2 \in \mathbb{R}$ with $m \in L^1(J, \mathbb{R})$ and $d(0, F(t,0)) \leq m(t)$ for almost all $t \in J$. Then the problem (2) has a solution provided that

$$l = \int_0^1 G(1,s)m(s)ds < 1.$$

Proof. We transform problem (2) into a fixed point problem. Consider the operator $N : C[0,1] \to P(C[0,1],\mathbb{R})$ defined by

$$N(u) = \left\{ h \in X, \exists y \in S_{F,u} \backslash h(t) = \int_0^1 G(t,s)y(s)ds, t \in J \right\}, \tag{31}$$

where $G(t,s)$ defined by (5). It is clear that fixed points of N are solution of (2).

We shall prove that N fulfills the assumptions of Covitz-Nadler contraction principle.

Note that, the multivalued map $t \mapsto F(t,u(t))$ is measurable and closed for all $u \in AC^1([0,\infty))$ (e.g., [52] Theorem III.6). Hence, it has a measurable selection and so the set $S_{F,u}$ is nonempty, so, $N(u)$ is nonempty for any $u \in C([0,\infty))$.

First, we show that $N(u)$ is a closed subset of X for all $u \in AC^1([0,\infty),\mathbb{R})$. Let $u \in X$ and $\{u_n\}_{n\geq 1}$ be a sequence in $N(u)$ with $u_n \to u, as\ n \to \infty$ in $u \in C([0,\infty))$. For each n, choose $y_n \in S_{F,u}$ such that

$$u_n(t) = \int_0^1 G(t,s)y_n(s)ds.$$

Since F has compact values, we may pass onto a subsequence (if necessary) to obtain that y_n converges to $y \in L^1([0,1],\mathbb{R})$ in $L^1([0,1],\mathbb{R})$. In particular, $y \in S_{F,u}$ and for any $t \in [0,1]$, we have

$$u_n(t) \to u(t) = \int_0^1 G(t,s)y(s)ds,$$

i.e., $u \in N(u)$ and $N(u)$ is closed.

Next, we show that N is a contractive multifunction with constant $l < 1$. Let $u, v \in C([0,1],\mathbb{R})$ and $h_1 \in N(u)$. Then there exist $y_1 \in S_{F,u}$ such that

$$h_1(t) = \int_0^1 G(t,s)y_1(s)ds, \quad t \in J.$$

By (H_2), we have

$$H_d(F(t,u(t)),F(t,v(t))) \leq m(t)(|u(t) - v(t)|),$$

for almost all $t \in J$.

So, there exists $w \in S_{F,v}$ such that

$$|y_1(t) - w| \leq m(t)(|u(t) - v(t)|),$$

for almost all $t \in J$.

Define the multifunction $U : J \to P(\mathbb{R})$ by

$$U(t) = \{ w \in \mathbb{R} : |y_1(t) - w| \leq m(t)(|u(t) - v(t)|) \ for\ almost\ all\ t \in J \}.$$

It is easy to chek that the multifunction $V(\cdot) = U(\cdot) \cap F(\cdot, v(\cdot))$ is measurable (e.g., [52] Theorem III.4).

Thus, there exists a function $y_2(t)$ which is measurable selection for V. So, $y_2 \in S_{F,v}$ and for each $t \in J$, we have

$$|y_1(t) - y_2(t)| \leq m(t)(|u(t) - v(t)|).$$

Now, consider $h_2 \in N(u)$ which is defined by

$$h_2(t) = \int_0^1 G(t,s)y_2(s)ds, \quad t \in J,$$

and one can obtain

$$|h_1(t) - h_2(t)| \leq \int_0^1 G(t,s)|y_1(s) - y_2(s)|ds$$
$$\leq \int_0^1 G(1,s)m(s)|u(s) - v(s)|ds.$$

Hence

$$\|h_1(t) - h_2(t)\| \leq \|p\|_\infty \|u - v\| \left[\int_0^1 G(1,s)m(s)ds\right].$$

Analogously, interchanging the roles of u and v, we obtain

$$H_d(N(u), N(v)) \leq \|u - v\| \left[\int_0^1 G(1,s)m(s)ds\right].$$

Since N is a contraction, it follows by Lemma 3.1 (by using the result of Covitz and Nadler) that N has a fixed point which is a solution to problem (2). □

We construct an example to illustrate the applicability of the result presented.

Example 3.1. *Consider the problem*

$$-D^\alpha u(t) \in F(t, u(t)), \quad t \in [0,1], \tag{32}$$

subject to the three-point boundary conditions

$$u^{(i)}(0) = 0, i \in \{0,1\}, \ D^\beta_{0^+}u(1) = \sum_{j=1}^{2} a_j D^\beta_{0^+}u\left(\eta_j\right), \tag{33}$$

where $\alpha = \frac{5}{2}, \beta = 1\, a_1 = \frac{1}{2}, a_2 = \frac{3}{2}, \eta_1 = \frac{1}{16}, \eta_2 = \frac{5}{16}$ *and* $F(t,u(t)) : [0,1] \times \mathbb{R} \to 2^{\mathbb{R}}$ *multivalued map given by*

$$u \mapsto F(t,u) = \left(0, \frac{t|u|}{2(1+|u|)}\right), u \in \mathbb{R},$$

verifying (H_1).

Obviously,

$$\sup\{|f| : f \in F(t,u)\} \leq \frac{t+1}{2},$$

we have

$$H_d(F(t,u),F(t,v)) \leq \left(\frac{t+1}{2}\right)|u-v|, u,v \in \mathbb{R}, \quad t \in [0,1],$$

which shows that (H_2) *holds*
So, if $m(t) = \frac{t+1}{2}$ *for all* $t \in [0,1]$, *then*

$$H_d(F(t,u),F(t,v)) \leq m(t)|u-v|.$$

It can be easily found that $d = 1 - \frac{1}{2}\left(\frac{1}{16}\right)^{\frac{5}{2}} - \frac{3}{2}\left(\frac{5}{16}\right)^{\frac{5}{2}} = 0,9176244637.$
Finally,

$$l = \int_0^1 G(1,s)m(s)ds = 0,4636273746 < 1.$$

Hence, all assumptions and conditions of Theorem 1.2 *are satisfied. So, Theorem* 1.2 *implies that the inclusion problem* (32) *and* (33) *has at least one solution.*

4. Conclusions

This chapter concerns the boundary value problem of a class of fractional differential equations involving the Riemann-Liouville fractional derivative with nonlocal boundary conditions. By using the properties of the Green's function and the monotone iteration technique, one shows the existence of positive solutions and constructs two successively iterative sequences to approximate the solutions. In the multi-valued case, an existence result is proved by using fixed point theorem for contraction multivalued maps due to Covitz and Nadler. The results of the present chapter are significantly contribute to the existing literature on the topic.

Acknowledgements

The authors want to thank the anonymous referee for the thorough reading of the manuscript and several suggestions that help us improve the presentation of the chapter.

Conflict of interest

The authors declare no conflict of interest.

Author details

Noureddine Bouteraa[1,2*†] and Habib Djourdem[1,3*†]

1 Laboratory of Fundamental and Applied Mathematics of Oran (LMFAO), University of Oran1, Ahmed Benbella, Algeria

2 Oran Graduate School of Economics. Algeria

3 University of Ahmed Zabbana, Relizane, Algeria

*Address all correspondence to: bouteraa-27@hotmail.fr; djourdem.habib7@gmail.com

† These authors contributed equally.

References

[1] R. P. Agarwal, D. Baleanu, V. Hedayati and S. H. Rezapour, *Two fractional derivative inclusion problems via integral boundary condition*, Appl. Math. Comput. **vol.257** (2015), 205-212.

[2] V. Kac and P. Cheung, *Quantum Calculus*, Springer, New-York, 2002.

[3] V. Lakshmikantham and A. S. Vatsala, General uniqueness and monotone iterative technique for fractional differential equations. Appl. Math. Lett. 21 (8)(2008), 828-834.

[4] S. Miller and B. Ross, An introduction to the fractional calculus and fractional differential equations, John Wiley and Sons, Inc. New-York, (1993).

[5] W. Rudin, Functional Analysis, 2nd edn. International Series in Pure and Applied Mathematics. Mc Graw-Hill, New York (1991).

[6] S. G. Samko, A. A. Kilbas and O. I. Marichev, Fractional Integrals and Derivatives: Theory and Applications. Gordon & Breach, Yverdon (1993).

[7] F. Jarad, T. Abdeljaw and D. Baleanu, On the generalized fractional derivatiives and their Caputo modification. J. Nonl. Sci. Appl. 10 (5) (2017), 2607-2619.

[8] Y. Tian, Positive solutions to m-point boundary value problem of fractional differential equation. Acta Math. Appl. Sinica (Engl. Ser.) 29 (2013), 661-672.

[9] A.M.A. El-Sayed and A.G. Ibrahim, Multivalued Fractional differential equations of arbitrary orders. Springer-Verlag, Appl. Math. Comput. 68 (1995), 15 25.

[10] N. Bouteraa and S. Benaicha, Existence of solutions for nonlocal boundary value problem for Caputo nonlinear fractional differential inclusion, Journal of Mathematical Sciences and Modelling, 1 (1) (2018), 45-55.

[11] N. Bouteraa and S. Benaicha, Existence results for fractional differential inclusion with nonlocal boundary conditions, Ri. Mat. Parma, Vol.11(2020),181-206.

[12] A. Cernia, *Existence of solutions for a certain boundary value problem associated to a fourth order differential inclusion*, Inter. Jour. Anal. Appl. **vol.14** (2017), 27-33.

[13] S. K. Ntouyas, S. Etemad, J. Tariboon and W. Sutsutad, Boundary value problems for Riemann-Liouville nonlinear fractional diffrential inclusions with nonlocal Hadamard fractional integral conditions, Medittter. J. Math. 2015 (2015), 16 pages.

[14] N. Bouteraa and S. Benaicha, Triple positive solutions of higher-order nonlinear boundary value problems, Journal of Computer Science and Computational Mathematics, Volume 7, Issue 2, June 2017, 25-31.

[15] N. Bouteraa and S. Benaicha, Existence of solutions for three-point boundary value problem for nonlinear fractional equations, Analele Universitatii Oradia Fasc. Mathematica, Tom XXIV (2017), Issue No. 2, 109-119.

[16] S. Benaicha and N. Bouteraa, Existence of solutions for three-point boundary value problem for nonlinear fractional differential equations, Bulltin of the Transilvania University of Brasov, Serie III: Mathematics, Informtics, Physics. Volume 10 (59), No. 2-2017.

[17] N. Bouteraa and S. Benaicha, Existence of solutions for third-order three-point boundary value problem, Mathematica. 60 (83), N 0 1, 2018, pp. 21-31.

[18] N. Bouteraa and S. Benaicha, The uniqueness of positive solution for higher-order nonlinear fractional differential equation with nonlocal boundary conditions, Advances in the Theory of Nonlinear and it Application, 2(2018) No 2, 74-84.

[19] N. Bouteraa, S. Benaicha and H. Djourdem, Positive solutions for nonlinear fractional differential equation with nonlocal boundary conditions, Universal Journal of Mathematics and Applications, 1 (1) (2018), 39-45.

[20] N. Bouteraa and S. Benaicha, The uniqueness of positive solution for nonlinear fractional differential equation with nonlocal boundary conditions, Analele universitatii Oradia. Fasc. Matematica. Tom XXV (2018), Issue No. 2, 53-65.

[21] N. Bouteraa, S. Benaicha, H. Djourdem and N. Benatia, Positive solutions of nonlinear fourth-order two-point boundary value problem with a parameter, Romanian Journal of Mathematics and Computer Science, 2018, Volume 8, Issue 1, p.17-30.

[22] N. Bouteraa and S. Benaicha, Positive periodic solutions for a class of fourth-order nonlinear differential equations, Siberian Journal of Numerical Mathematics, Volume 22, No. 1 (2019), 1-14.

[23] N. Bouteraa, Existence of solutions for some nonlinear boundary value problems, [thesis]. University of Oran1, Ahmed Benbella, Algeria; 2018.

[24] N. Bouteraa, S. Benaicha and H. Djourdem, ON the existence and multiplicity of positive radial solutions for nonlinear elliptic equation on bounded annular domains via fixed point index, Maltepe Journal of Mathematics, Volume I, Issue 1,(2019), 30-47.

[25] N. Bouteraa, S. Benaicha and H. Djourdem, Positive solutions for systems of fourth-order two-point boundary value problems with parameter, Journal of Mathematical Sciences and Modeling, 2(1) (2019), 30-38.

[26] N. Bouteraa and S. Benaicha, Existence and multiplicity of positive radial solutions to the Dirichlet problem for the nonlinear elliptic equations on annular domains, Stud. Univ. Babes-Bolyai Math, 65(2020), No. 1, 109-125.

[27] S. Benaicha, N. Bouteraa and H. Djourdem, Triple positive solutions for a class of boundary value problems with integral boundary conditions, Bulletin of Transilvania University of Brasov, Series III : Mathematics, Informatics, Physics, Vol. 13 (62), No. 1 (2020), 51-68.

[28] N. Bouteraa, H. Djourdem and S. Benaicha, Existence of solution for a system of coupled fractional boundary value problem, Proceedings of International Mathematical Sciences, Vol. II, Issue 1 (2020), 48-59.

[29] N. Bouteraa and S. Benaicha, Existence results for second-order nonlinear differential inclusion with nonlocal boundary conditions, Numerical Analysis and Applications, 2021, Vol. 14, No. 1, pp. 3039.

[30] N. Bouteraa M. In, M. A. Akinlar, B. Almohsen, Mild solutions of fractional PDE with noise, Math. Meth. Appl. Sci. 2021;115.

[31] N. Bouteraa, S. Benaicha, Existence Results for Second-Order Nonlinear Differential Inclusion with Nonlocal Boundary Conditions, Numerical Analysis and Applications, 2021, Vol. 14, No. 1, pp. 3039.

[32] N. Bouteraa, S. Benaicha, A study of existence and multiplicity of positive solutions for nonlinear fractional

differential equations with nonlocal boundary conditions, Stud. Univ. Babes-Bolyai Math. 66(2021), No. 2, 361-380.

[33] H. Djourdem, S. Benaicha and N. Bouteraa, Existence and iteration of monotone positive solution for a fourth-order nonlinear boundary value problem, Fundamental Journal of Mathematics and Applications, 1 (2) (2018), 205-211.

[34] H. Djourdem, S. Benaicha, and N. Bouteraa, Two Positive Solutions for a Fourth-Order Three-Point BVP with Sign-Changing Greens Function, Communications in Advanced Mathematical Sciences. Vol. II, No. 1 (2019), 60-68.

[35] H. Djourdem and N. Bouteraa, Mild solution for a stochastic partial differential equation with noise, WSEAS Transactions on Systems, Vol. 19, (2020), 246 256.

[36] R. Ghorbanian, V. Hedayati M. Postolache, and S. H. Rezapour, *On a fractional differential inclusion via a new integral boundary condition*, J. Inequal. Appl. (2014), 20 pages.

[37] M. Inc, N. Bouteraa, M. A. Akinlar, Y. M. Chu, G. W. Weber and B. Almohsen, New positive solutions of nonlinear elliptic PDEs, Applied Sciences, 2020, 10, 4863; doi :10.3390/app10144863, 17 pages.

[38] N. Bouteraa and S. Benaicha, Nonlinear boundary value problems for higher-order ordinary differential equation at resonance, Romanian Journal of Mathematic and Computer Science. 2018. Vol 8, Issue 2 (2018), p. 83-91.

[39] N. Bouteraa, S. Benaicha, A class of third-order boundary value problem with integral condition at resonance, Maltepe Journal of Mathematics, Volume II, Issue 2,(2020), 43-54.

[40] X. Lin, Z. Zhao and Y. Guan, Iterative Technology in a Singular Fractional Boundary Value Problem With q-Difference, Appl. Math. 7 (2016), 91-97.

[41] M. Li, JP. Sun and YH. Zhao, Existence of positive solution for BVP of nonlinear fractional differential equation with integral boundary conditions. Adv Differ Equ 2020, 177 (2020). https://doi.org/10.1186/s13662-020-02618-9

[42] S. H. Rezapour and V. Hedayati, On a Caputo fractional differential inclusion with integral boundary condition for convex-compact and nonconvex-compact valued multifunctions, Kragujevac Journal of Mathematics. **vol.41** (2017), 143-158.

[43] B. AHMAD, SK. NTOUYAS and A. ALSAEDI, Coupled systems of fractional differential inclusions with coupled boundary conditions, Electronic Journal of Differential Equations, Vol. 2019 (2019), No. 69, pp. 121.

[44] H. Covitz and S. B. Jr. Nadler, *Multivalued contraction mappings in generalized metric spaces*, Israel J. Math. **vol.8** (1970), 5-11.

[45] M. H. Annaby, and Z. S. Mansour, *q-Fractional calculus and equations*, Lecture Notes in Mathematics. **vol.2056**, Springer-Verlag, Berlin 2012.

[46] O. Agrawal, *Some generalized fractional calculus operators and their applications in integral equations*, Fract. Cal. Appl. Anal. **Vol.15** (2012), 700-711.

[47] A. A. Kilbas, H. M. Srivastava and J. J. Trijull, *Theory and applications of fractional differential equations*, Elsevier Science B. V, Amsterdam, 2006.

[48] J. P. Aubin and A. Cellina, *Differential Inclusions*, Springer-Verlag, 2012.

[49] K. DEMLING, *Multivalued Differential equations*, Walter De Gryter, Berlin-New-York 1982.

[50] L. GORNIEWICZ, *Topological Fixed Point Theory of Multivalued Map pings, Mathematics and Its Applications*, Vol. 495, Kluwer Academic Publishers, Dordrecht 1999.

[51] S. HU and N. PAPAGEORGIOU, *Handbook of Multivalued Analysis*, **vol.I**, Theory, Kluwer Academic. J. Diff. Equ. No.147, (2013), 1-11.

[52] C. CASTAING and M. VALADIER, *Convex analysis and measurable multifunctions, Lecture Notes in Mathematics*, Springer-Verlage, Berlin-Heidelberg, New-York, 580, 1977.

[53] A. LASOTA and Z. OPIAL, *An application of the Kakutani-Ky Fan theorem in the theory of ordinary differential equations,* Bull. Acad. Pol. Sci. Set. Sci. Math. Astronom. Phy. **vol.13** (1965), 781-786.

[54] J. MUSIELAK, *Introduction to functional analysis,* PWN, Warsaw, 1976, (in Polish).

[55] V. BERINDE and M. PACURAR, *The role of the Pompeiu-Hausdorff metric in fixed point theory,* Creat. Math. Inform. **vol.22** (2013), 35-42.

[56] M. FRIGON and A. GRANAS, *Theoremes d'existence pour des inclusions differentielles sans convexite*, C. R. Acad. Sci. Paris, SerI. **vol.310** (1990), 819-822.

Section 2

Biomedicine

Chapter 4

Agent-Based Modeling and Analysis of Cancer Evolution

Atsushi Niida and Watal M. Iwasaki

Abstract

Before the development of the next-generation sequencing (NGS) technology, carcinogenesis was regarded as a linear evolutionary process, driven by repeated acquisition of multiple driver mutations and Darwinian selection. However, recent cancer genome analyses employing NGS revealed the heterogeneity of mutations in the tumor, which is known as intratumor heterogeneity (ITH) and generated by branching evolution of cancer cells. In this chapter, we introduce a simulation modeling approach useful for understanding cancer evolution and ITH. We first describe agent-based modeling for simulating branching evolution of cancer cells. We next demonstrate how to fit an agent-based model to observational data from cancer genome analyses, employing approximate Bayesian computation (ABC). Finally, we explain how to characterize the dynamics of the simulation model through sensitivity analysis. We not only explain the methodologies, but also introduce exemplifying applications. For example, simulation modeling of cancer evolution demonstrated that ITH in colorectal cancer is generated by neutral evolution, which is caused by a high mutation rate and stem cell hierarchy. For cancer genome analyses, new experimental technologies are actively being developed; these will unveil various aspects of cancer evolution when combined with the simulation modeling approach.

Keywords: cancer, evolution, agent-based model, approximate Bayesian computation, sensitivity analysis

1. Introduction

Cancer is a clump of abnormal cells that originates from normal cells. Normal cells proliferate or stop proliferating depending on their surrounding environment. For example, when skin cells are injured, they proliferate to cover the wound; however, when the wound heals, they stop proliferating. In contrast, cancer cells continue proliferating by ignoring the surrounding environment. Moreover, cancer cells invade surrounding tissues, metastasize to distant organs, and impair functions in the human body.

Malignant transformation from normal to cancer cells generally results from the accumulation of somatic mutations, which are induced by various causes such as aging, ultraviolet rays, cigarette, alcohol, chemical carcinogens, etc. Mutations that contribute to malignant transformation are known as "driver mutations", whereas genes whose function are impaired by driver mutations are named as "driver genes". There are two types of driver genes are categorized: "oncogenes" and "tumor suppressor genes". Oncogenes act as gas pedals for cell proliferation, which are constitutively turned on by driver mutations. Tumor suppressor genes act as brakes to stop cell proliferation, and inhibiting the function of the brakes is necessary for malignant transformation.

Normal cells are transformed into cancer cell when two to 10 driver mutations are acquired. Because these mutations are not induced simultaneously, but rather gradually over a long period of time, this process is known as "multi-stage carcinogenesis" [1]. This process is also regarded as a linear evolutionary process, driven by repeated acquisition of multiple driver mutations and Darwinian selection. Understanding cancer from an evolutionary perspective is important, as therapeutic difficulties against cancer originate from the high evolutionary capacity, which easily endows cancer cells with therapeutic resistance.

Mutations in cancer cells are experimentally detected by DNA sequencing. Next-generation sequencing (NGS) technology, which raised around 2010, enabled cancer genome analysis to comprehensively detect mutations in cancer cells. During the last decade, cancer genome analysis has revolutionized our understanding of cancer [2]. Cancer genome analysis showed that cancer cells harbor a large number of mutations, only a small fraction of which is driver mutations; namely, most mutations in cancer cells are "neutral mutations", which have no selective advantages (also referred to as "passenger mutations" in paired with driver mutations). By sequencing hundreds of tumor samples from different patients with the same cancer type, the repertories of driver genes were also determined across various types of cancer. Moreover, cancer genome analysis has revealed heterogeneity of mutations within one tumor, which is termed intratumor heterogeneity (ITH) [3]. As described above, carcinogenesis was regarded as a linear evolutionary process until the arrival of NGS; however, ITH is actually generated by branching evolution of cancer cells.

However, cancer genome analysis is not sufficient to explain the origin of ITH. To understand the evolutionary principles underlying the generation of ITH, a simulation modeling approach is useful and increasingly employed in the field of cancer research. In this chapter, we introduce such simulation modeling approaches. We first describe agent-based modeling for simulating branching evolution of cancer cells. We next demonstrate how to fit an agent-based model to observational data obtained by cancer genome analyses, employing approximate Bayesian computation (ABC). Finally, we explain how to characterize the dynamics of the simulation models through sensitivity analysis.

2. Agent-based modeling of cancer evolution

To simulate heterogenous cancer evolution, agent-based modeling is widely employed. An agent-based model assumes a set of system constituents, known as independent agents, and specifies rules for the independent behavior of the agents themselves, as well as for the interactions between agents and the agent environment [4]. The agent-based model is a flexible representation of the model, and given the initial conditions and parameters of the system, the behavior of the system can be easily analyzed by computational simulation. For modeling of cancer evolution, if each cell is assumed to be an agent, ITH can be easily represented by the differences in the internal states of each agent. As an example, we explain an agent-based model named as the branching evolutionary process (BEP) model, which was originally introduced by Uchi *et al.* [5] to studying ITH of colorectal cancer (**Figure 1A**). In the BEP model, a cell assumed to be an agent has a genome containing n genes, each of which is represented as a binary value, 0 (wild-type) or

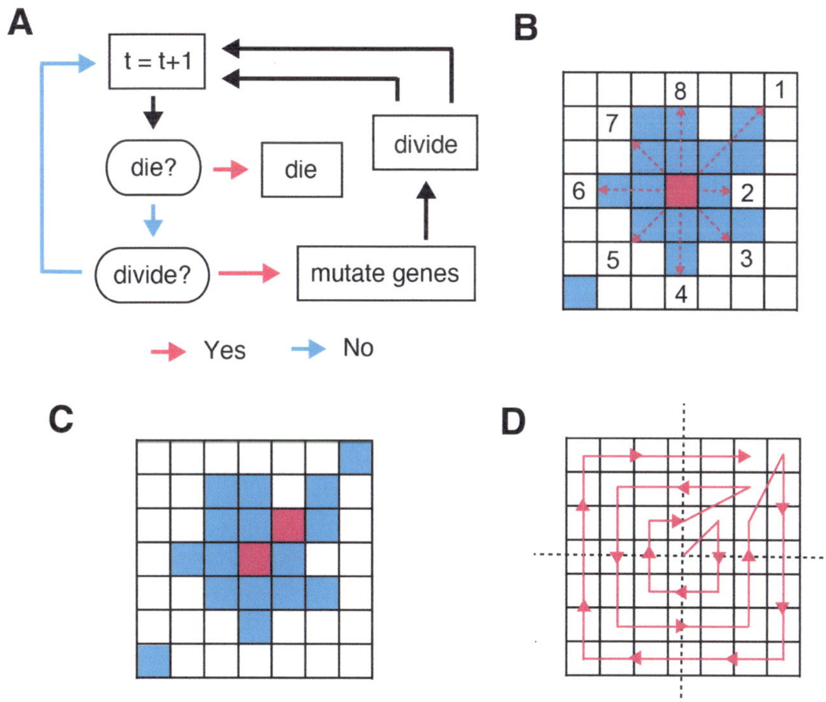

Figure 1.
Illustration of the BEP model. (A) A flowchart of a simulation based on the BEP model. First, a cell is tested for survival and killed with a provability q. Next, the cell is tested for cell division and replicated with a provability p. Before cell division, each gene in the cell is mutated with a provability r. After this process is applied to each cell, the simulation proceeds to the next time step. (B–D) illustration of division operation (see the main text for details). This image originally appeared in [5].

1 (mutated). Thus, the genome is represented as a binary vector **g** with length n. In a unit time step, a cell replicates with a probability p and dies with a probability q. When the cell replicates, a wild-type gene is mutated with a probability r. The first d genes in **g** are considered as driver genes, whose mutations accelerate replication. A normal cell without mutations has a replication probability p_0, and each driver mutation increases p by 10^f-fold; i.e., $p = p_0 \cdot 10^{fk}$, where $k = \sum_{i=1}^{d} g_i$, the number of mutated driver genes. The death probability is fixed as $q = q_0$. Let c and t denote the size of the simulated cell population and number of the time steps, respectively. A simulation is started with c_0 normal cells and the unit time step is repeated while the population size $c \le c_{max}$ and time step $t \le t_{max}$.

The BEP model assumes that a simulated tumor grows in a two-dimensional square lattice where each cell occupies one lattice point. Initially, c_0 cells are initialized as close as possible to the center of the lattice. In a unit time step, along an outward spiral starting from the center, we replicate and kill each cell with probabilities p and q, respectively. When cells replicate, the BEP model places the daughter cell in the neighborhood of the parent cell, assuming a Moore neighborhood (i.e., eight points surrounding a central point). If empty neighbor points exist, we randomly select one of these points. Otherwise, we create an empty point in any of the eight neighboring points as follows. First, for each of the eight directions, we count the number of consecutive occupied points that range from each neighboring point to immediately before the nearest empty cell as indicated in **Figure 1B**. Next, any of the 8 directions is randomly selected proportionally with $1/l_i$, where $l_i (1 \le i \le 8)$ is the count of the consecutive occupied points for each direction. The consecutive occupied points in the selected direction are then shifted by one point so that an empty neighboring point appears as shown in **Figure 1C**. Note that simulation results depend on the order of the division operation in the two-dimensional square lattice. The BEP model first marks cells to be divided and then applies the division operation to the marked cells along an outward spiral starting from the center. In each round on the spiral, the direction is randomly flipped in order to maintain spatial symmetry. An example of such spirals is shown in **Figure 1D**.

Given that a cell without mutations divides according to this rule, after a normal cell acquires its first driver mutation, which accelerates cell division, the proportion of the clone originating from the cell increases in the whole cell population. By repeating these steps, each cell gradually accumulates driver mutations and accompanying passenger mutations, which do not affect the cell division rate, finally forming a tumor with many mutations. Depending on the parameter values during the course of cancer evolution, each cancer cell can accumulate different combinations of mutations to generate different types of ITH. **Figure 2** show an example of snapshots of two-dimensional tumor growth simulated based on the BEP model with an appropriate parameter setting. In this example, driver mutations gradually accumulated in the cells, and a clone with four mutations was selected through Darwinian selection and finally became dominant in the tumor.

The BEP model is a very simple model and has many limitations. Although this BEP model assumes that driver mutations increase the replication probability, it is considered that driver mutations decrease the death probability. The BEP model also assume that each diver mutation has the same effect on the replication probability; however, actual tumors contain different driver mutations of different strengths. Although actual tumors grow in a three-dimensional space, the BEP model assumes tumor growth on the a two-dimensional square lattice; extension to a three-dimensional lattice should be considered as a future improvement. For on-lattice models, various other simulators has been developed for studying tumor growth (off-lattice models which do not assume that tumor growth on the lattice reflects the actual situations more accurately, but are computationally intensive and not commonly used [7]). For example, the pioneering works of agent-based

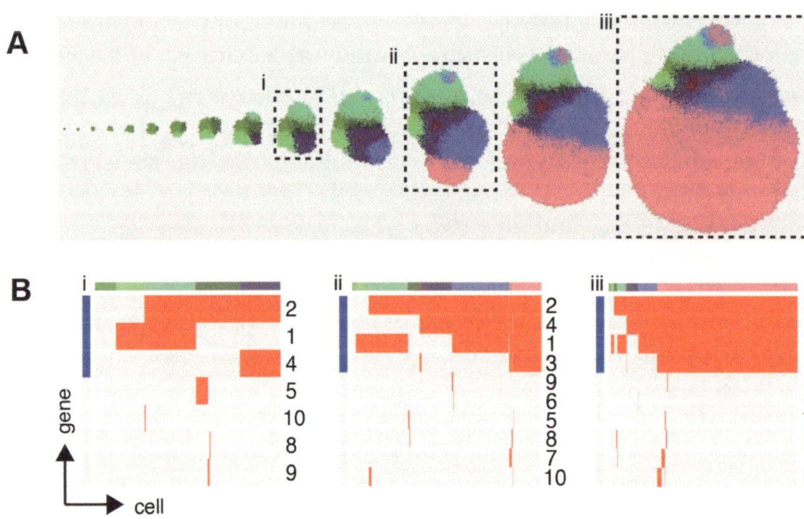

Figure 2.
Visualization of a simulation based on the BEP model. (A) Evolutionary snapshots obtained by simulating two-dimensional tumor growth based on the BEP model with an appropriate parameter setting. The region with the same color represents a clone with the same set of mutated genes. (B) Single-cell mutation profiles at three time points in the simulated tumor growth. Top colored bands represent clones, whereas the blue bands on the left represent driver genes. This image was obtained by modifying a figure that originally appeared in [6].

modeling were performed by Anderson and colleagues [8, 9]. Enderling and colleagues extended the model to incorporate cell differentiation from cancer stem cells where differentiated cells have a limited potential for cell division [10, 11]. Sottoriva *et al.* [12] found that hierarchical organization of cell differentiation affects tumor heterogeneity, which leads to an invasive morphology with finger-like front. Waclaw *et al.* [13] predicted that dispersal and cell turnover limit intratumor heterogeneity.

Each group developed a model different from the others, and thus only limited conditions were considered in each study. To address this issue, Iwasaki and Innan [14] developed a flexible and comprehensive simulation framework named as *tumopp* so that all previous models were included in a single program. This enables researchers to explore the effects of various model settings with simple command-line options. For example, the combined effect of local density on the cell division rate and method of placing new cells had a major impact on ITH. Under the condition with random push and no local competition, all cells undergo cell division at a constant rate regardless of the local density, and new cells are placed while randomly pushing out pre-existing neighbor cells. This behavior creates shuffled patterns of ITH with weak isolation by distance. In contrast, under the conditions of strong resource competition, the division rate of a cell is higher when it has more space (empty sites) in the neighborhood, which generally applies to cells near the surface of the simulated tumor; new cells are placed to fill the empty space without pushing existing cells. This setting tends to create a biased complex shape with clusters of genetically closely related cells, resulting in strong isolation by distance. Thus, it has been demonstrated that various patterns in the shape and heterogeneity of tumors arose depending on the model setting even without Darwinian selection. This suggests a caveat in analyzing ITH data with simulations using limited settings because another setting may predict a different ITH pattern, which could result in a different conclusion.

Moreover, *tumopp* introduced several other factors to relax various assumptions. First, it adopted a gamma function for the waiting time involved in cell division, whereas all previous studies assumed a simple decreasing (i.e., exponential) function implicitly or explicitly. The shape of the gamma distribution can be specified by parameter k, which affects the growth curve and inequality of clones in a tumor. An exponential distribution, which is the most widely used, is included as a special case with $k = 1$, whereas punctual and deterministic cell divisions can be achieved with $k = \infty$. It is reasonable to expect that the true value of k lies somewhere in-between, such as at ~ 10, because cell division is neither a memoryless Poisson process ($k = 1$; equivalent to an exponential distribution) nor perfectly synchronized ($k = \infty$; equivalent to Dirac delta distribution) [15, 16]. Second, a hexagonal lattice has been implemented, which is thought to be biologically more reasonable because the distance to all neighboring cells is identical so that there is only one definition of the neighborhood. A hexagonal lattice should be used for future work when the spatial pattern is of interest to the study. These factors will contribute to simulating the evolutionary processes of cellular populations under more realistic conditions.

3. Fitting the simulation model to observational data

As described in the "Introduction" section, cancer genome analysis demonstrated intratumor heterogeneity and branching evolution of cancer; paticulally, an approach known as multiregion sequencing has been popularly employed for analyzing solid tumors. Here, we introduce a concrete example of a multiregion sequencing study and explain the utility of cancer evolution simulation when combined with multiregion sequencing data.

In multiregion sequencing, multiple samples obtained from physically separate regions within the tumor of a single patient are analyzed (**Figure 3A**), with two categories of somatic single-nucleotide mutations identified: "founder" and "progressor" mutations (**Figure 3B**). Founder mutations are defined as present in all regions, whereas progressor mutations are defined as present in some regions (note that they are also referred to using different terms in different studies, e.g., public/private or trunk/branch mutations). Founder mutations are thought to accumulate during the early phases of cancer evolution. The common ancestor clone acquires all founder mutations, and then branches into subclones, which accumulate progressor mutations and contribute to forming ITH. Through these multiregion mutational profiles, we can infer an evolutionary history of the cancer by constructing a phylogenetic tree (**Figure 3C**).

As a pioneering study, Gerlinger et al. [17] performed multiregion sequencing, revealing extensive ITH and clonal branching evolution in renal cancer. Ther also identified not only founder mutations in some known driver genes such as *VHL*, but also progressor mutations in other known driver genes such as *SETD2* and *BAP1*. Interestingly, in some cases, different mutations in the same driver gene or genes with the same function were acquired independently. This phenomenon known as parallel evolution also indicates that part of the ITH was generated by Darwinian selection.

Uchi et al. [5] also investigated ITH in nine cases of surgically resected late-stage colorectal tumors by multiregion sequencing to identify founder and progressor mutations in each case. **Figure 4** shows the results obtained from one of the nine cases, which contains 20 samples from the primary lesion and one sample from the metastatic lesion. Note that the progressor mutations showed a mutational pattern that was geographically correlated with the sampling locations. Moreover, they found that mutation allele frequencies, which can be approximately regarded as the proportion of cells with mutations in each region, tended to be lower for progressor mutations than for founder mutations. This observation suggests that the founder mutations existed in all the cancer cells while the progressor mutations existed in only a fraction of the cancer cells in each region. Thus, even in each region, extensive ITH may have existed, which was not captured by the resolution of multiregion sequencing. In addition, most mutations in known driver genes such as *APC* and *KRAS* were identified as founder mutations. However, progressor mutations contain few driver mutations and parallel evolution was not confirmed, which contrasts to the findings obtained in renal cancer. These observations suggest that apart from

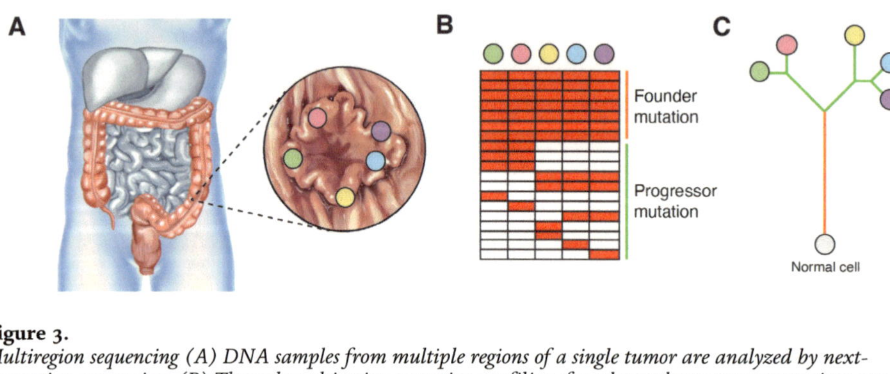

Figure 3.
Multiregion sequencing (A) DNA samples from multiple regions of a single tumor are analyzed by next-generation sequencing. (B) Through multiregion mutation profiling, founder and progressor mutations are identified as common mutations in all regions tested and only restricted regions, respectively. (C) In a phylogenetic tree constructed from the multiregion mutation profile, the trunk and branches correspond to the founder and progressor mutations, respectively. This image originally appeared in [6].

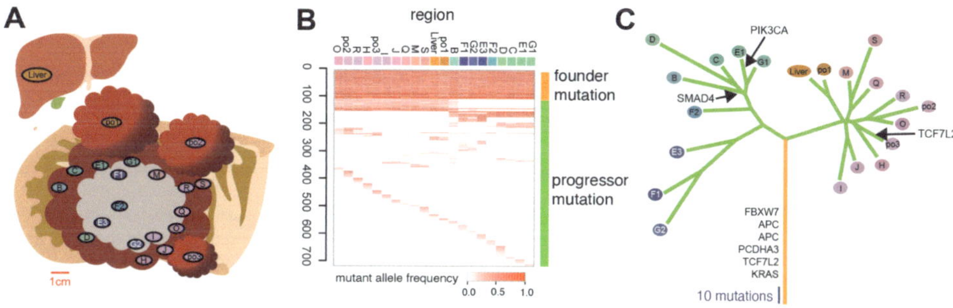

Figure 4.
Multiregion sequencing of colorectal cancer. (A) Schema of the tumor subjected to multiregion sequencing. (B) Multiregion mutation profile. The depth of red represents the mutant allele frequency, whereas the colors of the sample labels were prepared so that the similarities of colors represent those of mutation patterns. (C) Phylogenetic tree constructed from the multiregion mutation profile. The time when mutations in known driver genes of colorectal cancer were acquired is indicated along the tree. This image was obtained by modifying a figure that originally appeared in [5].

Darwinian selection, there are other evolutionary principles generating ITH. To identify these principles, they developed the BEP model as described above; by fitting the BEP model to the multiregion sequencing data, they evaluated the evolutionary principles generating ITH in colorectal cancer.

To fit the simulation model to the observational data, we can employ ABC [18], which constitutes a class of computational methods rooted in Bayesian statistics that can be used to estimate the posterior distributions of model parameters. A common incarnation of Bayes' theorem relates the conditional probability of a specific parameter value θ given data D to the probability of D given θ by the rule, $p(\theta|D) \propto p(D|\theta)p(\theta)$, where $p(\theta|D)$ denotes the posterior, $p(D|\theta)$ the likelihood, and $p(\theta)$ the prior. The prior represents beliefs or knowledge about θ before D is available. To obtain the the posterior, the likelihood function is required. For simple models, an analytical formula for the likelihood function can typically be derived. However, for more complex models, an analytical formula may be elusive or the likelihood function may be computationally very costly to evaluate. Agent-based models also fall into the latter case. ABC methods bypass evaluation of the likelihood function by using summary statistics and simulations, which widen the realm of models for which statistical inference can be considered. ABC has rapidly gained popularity over the last few years, for analizing complex problems arising in biological sciences, e.g., in population genetics, ecology, epidemiology, and systems biology.

In the basic form of ABC, which is known as rejection sampling, we first sample a parameter value (or a combination of parameter values, if there is more than one parameter) from a prescribed prior distribution of the parameter value. Simulated data are then generated from the sampled parameter value. The similarity between the simulated and observational data is evaluated using summary statistics (typically multiple), which is designed to represent the maximum amount of information in the simplest possible form. If the distance of the summary statistics between the simulated and observational data is below a tolerance parameter, the parameter value is accepted and pooled into the posterior probability of the parameter value. Repeating these steps many times, we can approximate the probability distribution. A conceptual overview of the ABC rejection sampling algorithm is presented in **Figure 5**.

In the study of colorectal cancer study by Uchi et al. [5], as summary statistics, they adopted the proportions of founder mutations and unique mutations, which is uniquely observed in each sample, in a multiregion mutation profile.

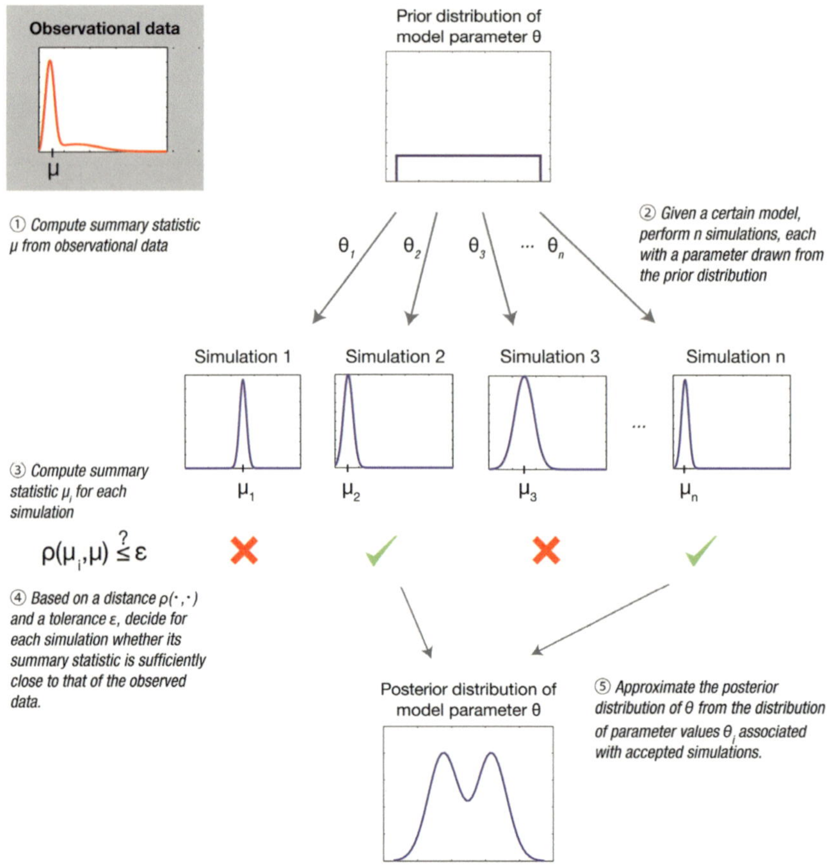

Figure 5.
Conceptual overview of the ABC rejection sampling algorithm. This image originally appeared in [18].

They obtained multiregion mutation profiles for 9 cases with different sample numbers. As the proportions of founder mutations and unique mutations depend on the number of samples, they set the sample number to 5, which is the minimum sample number of the 9 cases, by downsampling the samples in cases containing more than 5 samples. They then estimated the mean of the proportions of founder mutations and unique mutations and used these values as summary statistics values of the observational data (**Figure 6**; note that although we should apply ABC to each of the 9 case separately, they targeted the population mean for simplicity).

For ABC, they generated simulation data while varying 3 parameters, *m* (the mutation rate), *d* (the number of driver genes), and *f* (the effect of driver mutations), which appear to be critical for simulation results (for strategies used to find such parameters, read the next section). In each simulation trial, we simulated multiregion sequencing from a tumor simulated by the BEP model; a multiregion mutation profile was obtained by digging 5 squares out from a simulated tumor and averaging the mutation status of cells in the squares. From the multiregion mutation profile, the proportions of founder mutations and unique mutations were obtained as summery statistics. They performed 50 simulations for each grid point in a three-dimensional rectangular parameter space; namely, they assumed a uniform prior for each of the three parameters. For each grid point in the parameter space, they calculate the proportion of the simulation instances whose statistics fall within 1 standard deviation from the mean of the values observed in the real multiregion

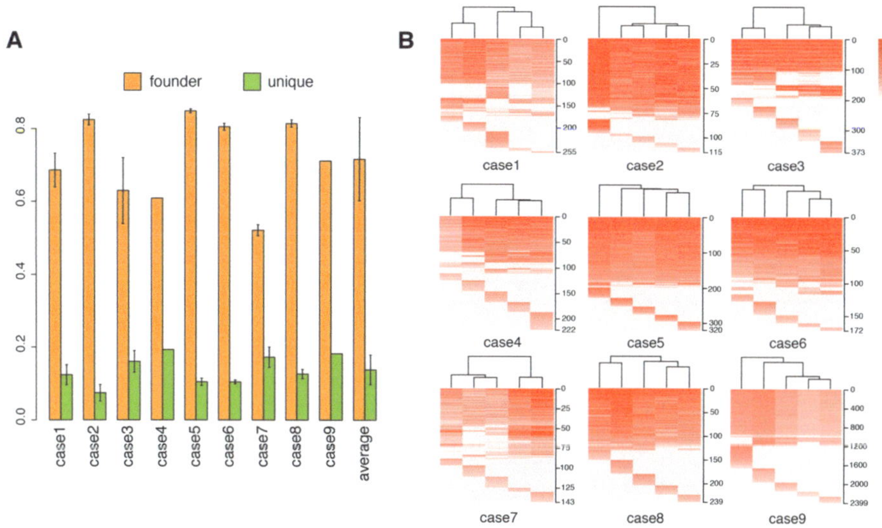

Figure 6.
Fitting the BEP model to multiregion sequencing data by ABC. (A) Observed values of summary statistics in the multiregion sequencing data. After downsampling the samples in cases containing more than 5 samples, the proportions of founder and unique mutations were estimated for 9 cases (case 1–9) and an "average" over the 9 cases was obtained as summary statistics values of the observational data. The error bars at case 1–3 and 5–8 indicate standard deviations for 10 downsampling trials while the error bar at average indicates standard deviations over the 9 cases. (B) Multiregion mutation profiles of 9 colorectal tumors. For the cases except case 4 and 9, representative samples from the downsampling trials were presented as in ***Figure 4B****. This image originally appeared in [5].*

mutation profiles. The distribution of the proportions can be regarded as the posterior and visualized in heat maps (**Figure 7**).

As a result, when cancer evolution was simulated with the assumption of a high mutation rate, we reproduced mutation profiles similar to those obtained by our multiregion sequencing of colorectal cancers (compare **Figure 8A** and **B** with **Figure 4A** and **B**). That is, irrespective of the presence of founder mutations, progressor mutations contributed to the formation of a heterogeneous mutation profile, which was geographically correlated with the sampling locations. Moreover, we also reconstructed local heterogeneity, as illustrated by the finding that progressor mutations existed as mutations with lower allele frequencies in each region. Interestingly, although driver mutations were acquired as founder mutations, progressor mutations contained few driver mutations, and most comprised neutral mutations that did not affect the cell division rate. This suggests that, after the appearance of the common ancestor clone with accumulated driver mutations, extensive ITH was generated by neutral evolution. Moreover, the single-cell mutation profiles of the simulated tumor suggest that the tumor comprises a large number of minute clones with numerous neutral mutations accumulated (**Figure 8C**).

By employing a agent-based model and ABC, Sottoriva et al. [19] also proposed a Big Bang model of human colorectal tumor growth; in their model, tumors grow predominantly as a single expansion producing numerous intermixed subclones that are not subject to stringent selection, which is consistent with the model developed by Uchi *et al.* [5], and both public (clonal) and most detectable private (subclonal) alterations arise early during growth. Hu et al. [20] also employed an agent-based model and ABC to examine the timing of metastasis in colorectal cancer. Multiregion sequencing data containing both primary and metastatic samples were prepared from patients with metastases to the liver or brain. Simultaneously, a spatial agent-based model was developed to simulate tumor growth,

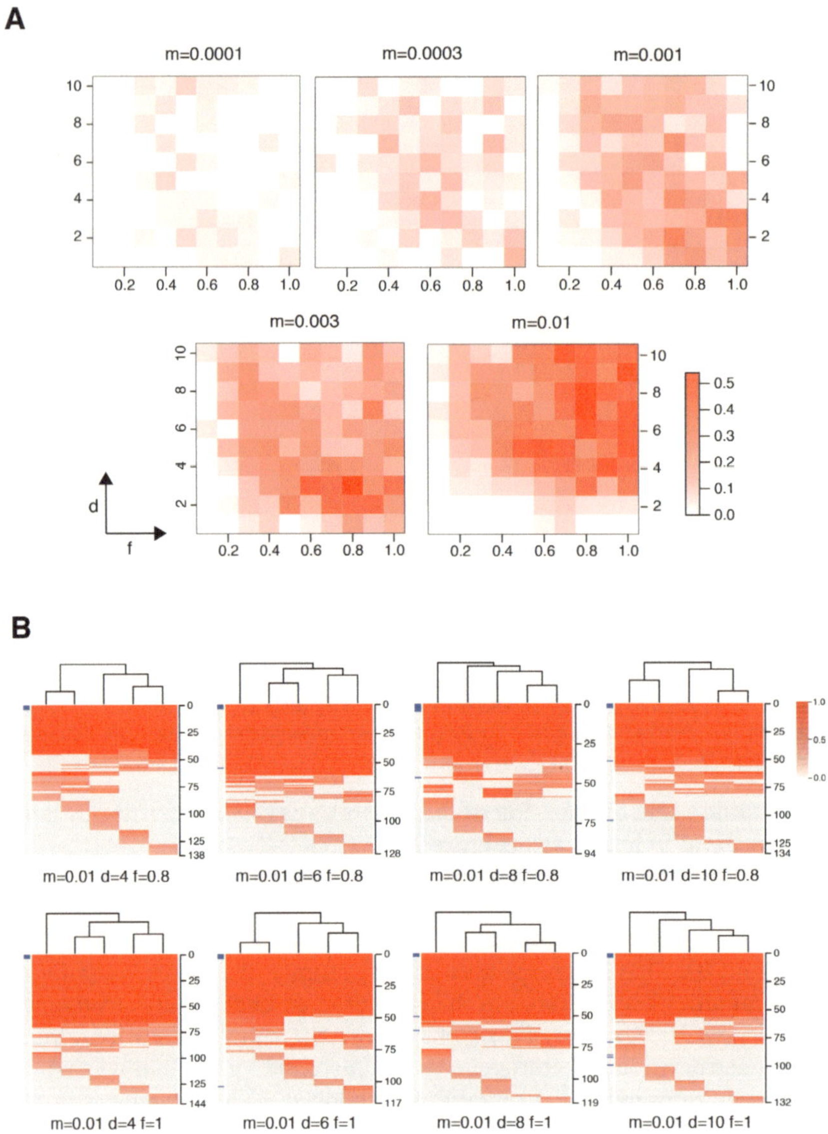

Figure 7.
Fitting the BEP model to multiregion sequencing data by ABC (continued). (A) The proportion of simulation instances fitted to the real data. Multiregion mutation profiles were simulated while varying 3 parameters and, for each parameter settings, the proportion of simulation instances that were judged to be similar to the real data based on summary statistics are visualized as heat maps. (B) Multiregion mutation profiles from the simulations. Representative instances from simulation with indicated parameter settings were presented as in ***Figure 4B****. Left blue bars indicate driver genes. This image originally appeared in [5].*

mutation accumulation, and metastatic dissemination. From multiregion sequencing data of each patient, the time of dissemination, which is a parameter in the agent-based model, was estimated by ABC. The results demonstrated that early disseminated cells commonly (81%, 17 of 21 patients) showed metastases, whereas the carcinoma was clinically undetectable (typically, less than 0.01 cm^3). Collectively, these examples demonstrated that ABC successfully fitted the simulation model of cancer evolution to cancer genome data, providing insight into the mechanisms of cancer evolution.

Although the problem of computational cost generally accompanies ABC, new sampling approaches utilizing Markov chain Monte Carlo and its derivatives [21]

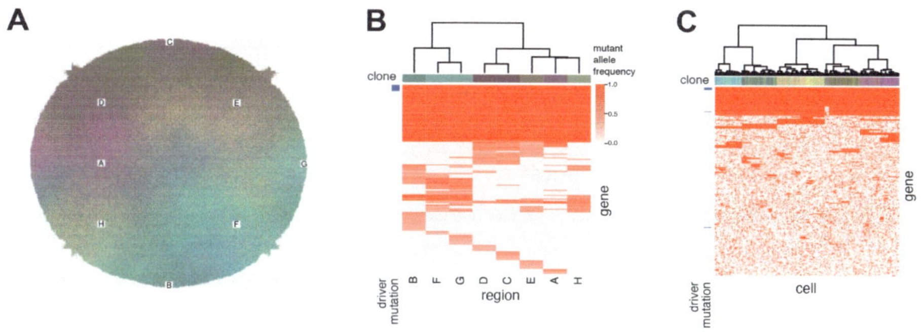

Figure 8.
Computer-simulated tumor with extensive ITH generated by neutral evolution. (A) Tumor simulated based on the BEP model with an assumption of a high mutation rate. (B) Simulated multiregion mutation profile of the simulated tumor. Cell populations in the regions labeled with A–H were extracted from the simulated tumor and their averaged mutation profiles were obtained. (C) Simulated single-cell mutation profile of the simulated tumor. This image was obtained by modifying a figure that originally appeared in [5].

have been developed to overcome this limitation. Moreover, considering the increasing computing power, this problem will potentially be less important. Notably, ABC has many potential pitfalls [18]. For example, setting the tolerance parameter to zero will give accurate results, but typically at a very high computational cost. In practice, therefore, values of greater than zero are used, but this introduces bias. Similarly, sufficient statistics are sometimes not available and other summary statistics are used instead, but this introduces additional bias because of the loss of information. Additionally, prior distributions and choices of parameter ranges are often subject to criticisms, although they are not unique to ABC and apply to all Bayesian methods. Model complexity (i.e., the number of model parameters) is also an important point. If a model is too simple, it can lack predictive power. In contrast, if the model is too complex, there is a risk of overfitting. Moreover, the complex model faces a problem known as the curse of dimensionality, in which the computational cost is severely increased and may, in the worst case, render the computational analysis intractable. When constructing a simulation model, we should follow the Occam's razor principle: i.e., achieve the lowest model complexity that is sufficient to explain the observational data. To determine the optimal model complexity, we can also employ the model selection scheme based on Bayes factor if a choice of summary statistics is appropriate [22].

4. Characterizing the dynamics of the simulation model

In the previous section, we explained how to fit a simulation model to observational data. Another direction for studying a simulation is by characterizing the dynamics of the simulation model without observational data. Namely, we can examine parameter dependance by performing a large number of simulations while varying the parameter values. This approach is known as sensitivity analysis and can provide insights into the modeled system as well as identify parameters that are critical for the system dynamics. In sensitivity analysis, as in ABC, we define a summary statistic Y. A simulation model is then regarded as a function: $Y = F(\mathbf{X})$ where $\mathbf{X} = \{X_1, X_2, \ldots X_k\}$ are model parameters. The aim of sensitivity analysis can also be considered as characterizing the function "F".

So far, a number of approaches have been proposed for sensitivity analysis. For example, one-factor-at-a-time (OFAT) sensitivity analysis is one of the simplest and most common approaches that changes one parameter at a time to determine the effects on a summery statistic [23]. In OFAT sensitivity analysis, we move one parameter, while leaving the other parameters at their baseline (nominal) values, and then return the parameter to its nominal, which is repeated for each of the other parameters. We then plot the relationship between each parameter and a summary statistic to examine the dependency of the summary statistic on the parameter, or the relationship can be measured by partial derivatives or linear regression. In exchange for its simplicity, this approach does not fully explore the input space, as it does not consider the simultaneous variation of multiple parameters. This means that the OFAT approach cannot detect interactions between parameters.

Global sensitivity analysis aims to address this point by sampling a summary statistic over a wide parameter space involving multiple parameters. Sobol's method is a popular approach for estimating the contributions of different combinations of parameters to the variance of the summary statistic while assuming that all parameters are independent [24]. The sensitivity of the summary statistic Y to a parameter X_i is measured by the amount of variance in Y caused by the parameter X_i and can be expressed as a conditional expectation, $\mathrm{Var}(\mathrm{E}_{\mathbf{X}_{\sim i}}(Y|X_i))$, where "*Var*" and "E" denote the variance and expected value operators, respectively, and $\mathbf{X}_{\sim i}$ denotes the set of all input variables except for X_i. This expression essentially measures the contribution X_i alone to uncertainty (variance) in Y (averaged over variations in other variables), and is known as the first-order sensitivity index or main effect index. Importantly, it does not measure the uncertainty caused by interactions with other variables. A further measure, known as the total effect index, gives the total variance in Y caused by X_i and its interactions with any of the other input variables. Both quantities are typically standardized by dividing by $\mathrm{Var}(Y)$. In Sobol's method, we typically attempt full exploration of the parameter space based on a Monte Carlo method to grasp parameter interactions and nonlinear responses.

However, such approaches appears to be insufficient to comprehensively grasp how the parameters judged to be influential control the behaviors of agent-based models. To overcome this point, Niida *et al.* [25] recently developed a new approach to sensitivity analysis for agent-based simulations, named as MASSIVE (Massively parallel Agent-based Simulations and Subsequent Interactive Visualization-based Exploration). MASSIVE overcomes the limitations of existing methods by taking advantage of two currently rising technologies: massively parallel computation and interactive data visualization. MASSIVE employs a full factorial design involving a multiple number of parameters (i.e., test every combination of candidate values of the multiple parameters), which can broadly cover a target parameter space. In addition, when analyzing a stochastic simulation model such as an agent-based model, multiple simulation trials with the same parameter setting are required to examine stochastic effects. To cope with the computational cost problem caused by these features, MASSIVE utilizes a supercomputer, in which agent-based simulations with different parameter settings and the following post-processing step of simulation results are performed in parallel. The massively parallel simulations generate massive results, which then pose a problem for interpretation. This problem was solved by developing a web-based tool that interactively visualizes not only the values of multiple summary statistics, but also output images (e.g., mutation profiles) from simulations with each parameter setting.

Below I explain an example of sensitivity analysis, which was performed by Niida et al. [26] to understand the precise mechanisms underlying neutral evolution induced by a high mutation rate. First, they built an agent-based model, referred to as the "neutral" model, for simulating neutral evolution in cancer. Although the neutral model is similar to the BEP model, the neutral model assumes only neutral mutations and omits spatial information. They also improved the approach used for mutation accumulation in the BEP model. Namely, in the neutral model, they considered only neutral mutations that did not affect cell division and death. In a unit time, a cell divides into two daughter cells with a constant probability g_0 without dying. In each cell division, each of the two daughter cells acquires $k_n \sim \text{Pois}(m_n/2)$ neutral mutations. They assumed that neutral mutations acquired by different division events occur at different genomic positions.The simulation started from one cell without mutations and ended when the population size p reached P or time t reached T.

Through sensitivity analysis based on the MASSIVE method, they confirmed that the mutation rate is the most important factor affecting neutral evolution (**Figure 9**). As a summary statistic for evaluating ITH, they calculated Shannon index 0.05 by the following procedure. After removing mutations with frequencies of less than 0.05, the proportions of different subclones (cell subpopulations with different mutations) were obtained and the Shannon index H was calculated using the following formula: $H = -\sum_{i=1}^{n} p_i \log(p_i)$, where n is the total number of different subclones and p_i is the proportion of each subclone. Based on this definition, a larger Shannon index 0.05 value indicates more extensive ITH. Together with a heat map of the Shannon index 0.05 values, we visualized single-cell mutations profiles obtained for different parameter settings. The mutation profile matrix was obtained by sampling 1,000 cells from a simulated tumor, and visualized after filtering out lower-frequency mutations, such that the maximum number of rows was 300. The rows and columns are reordered by hierarchical clustering and index mutations and samples, respectively. They found that when the mean number of mutations generated by per cell division, m_n, was less than 1, the neutral model just generated sparse mutation profiles with relatively small values of Shannon index 0.05. In contrast, when m_n exceeded 1, the mutation profiles presented extensive ITH, which are characterized by a fractal-like pattern and large values of the ITH score (hereinafter, this type of ITH is referred to as "neutral ITH"). These results suggest that neutral ITH is shaped by neutral mutations that trace the cell lineages in the simulated tumors. Note that the mutation profiles were visualized after filtering out low-frequency mutations. Assuming a high mutation rate, more numerous subclones with different mutations should be observed if mutations existing at lower frequencies are counted. However, the ITH score does not depend on the population size P because low-frequency mutations were filtered out before calculation.

Thus far, several theoretical and computational studies have shown that a stem cell hierarchy can boost neutral evolution in a population of cancer cells [12, 27];

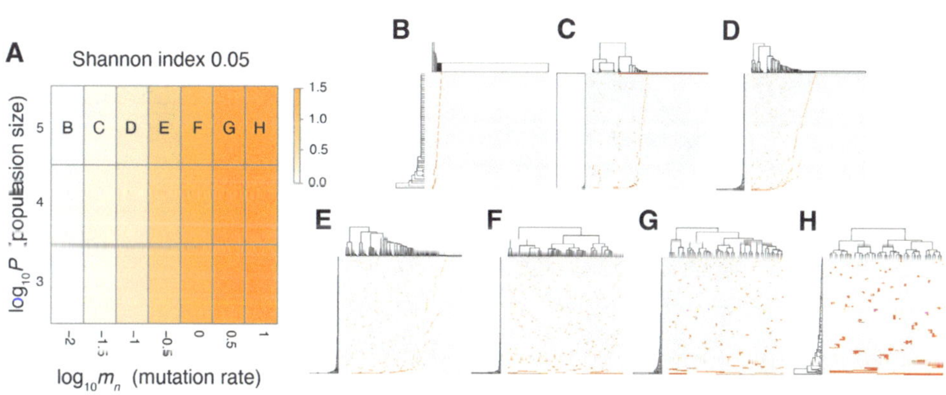

Figure 9.
Sensitivity analysis of the neutral model. (A) Heap map obtained by calculating Shannon index 0.05 *while changing the neutral mutation rate m_n and maximum population size P. (B–H) Single-cell mutations profiles obtained for seven parameter settings, which are indicated on the heat map in A. This image originally appeared in [26].*

based on this, they extended the neutral model to the "neutral-s" model such that it contains a stem cell hierarchy (**Figure 10**). The neutral-s model assumes that two types of cell exist: stem and differentiated. Stem cells divide with a probability g_0 without dying. For each cell division of stem cells, a symmetrical division generating two stem cells occurs with a probability s, whereas an asymmetrical division generating one stem cell and one differentiated cell occurs with a probability $1 - s$. A differentiated cell symmetrically divides to generate two differentiated cells with a probability g_0 but dies with a probability d_0^d. The means of accumulating neutral mutations in the two types of cell is the same as that in the original neutral model, which means that the neutral-s model is equal to the original neutral model when $s = 0$ or $d_0^d = 0$. For convenience, they define $\delta = \log_{10}\left(d_0^d/g_0\right)$ and hereinafter use δ rather than d_0^d.

MASSIVE analysis of the neutral-s model confirmed that incorporation of the stem cell hierarchy boosts neutral evolution (**Figure 11**). To obtain the heat map in **Figure 11A**, the ITH score was measured while d_0^d and δ were changed, whereas $m_n = 0.1$ and $P = 1000$ were maintained as constant. In the heat map, a decrease in s leads to an increase in the ITH score when $\delta \geq 0$ (i.e., $d_0^d \geq g_0$). A smaller value of s means that more differentiated cells are generated per stem cell division, and $\delta \geq 0$ means that the population of differentiated cells cannot grow in total, which is a valid assumption for typical stem cell hierarchy models. That is, this observation indicates that the stem cell hierarchy can induce neutral ITH even with a relatively low mutation rate setting (i.e., $m_n = 0.1$), with which the original neutral model cannot generate neutral ITH.

The underlying mechanism boosting neutral evolution can be explained as follows. Only stem cells were considered for an approximation, as differentiated cells do not contribute to tumor growth with $\delta \geq 0$. While one cell grows to a population of P cells, let cell divisions synchronously occur across x generations during the clonal expansion. Then, $(1+s)^x = P$ holds because the mean number of stem cells generated per cell division is estimated as $1+s$. Solving the equation for x gives $x = \log P/\log(1+s)$; that is, it can be estimated that during clonal expansion, each of the P cells experiences $\log P/\log(1+s)$ cell divisions and accumulates $m_n \log P/2\log(1+s)$ mutations on average. They confirmed that the expected mutation count based on this formula fit well with the values observed in their simulation (data not shown). These arguments mean that a tumor with a stem cell hierarchy accumulates more mutations until reaching a fixed population size than

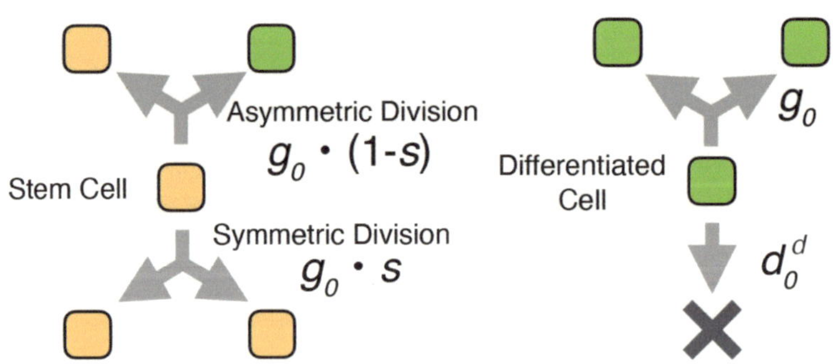

Figure 10.
Schema of the neutral-s model. Stem cells divide with a probability g_0 without dying. For each cell division of stem cells, a symmetrical division generating two stem cells occurs with probability s, whereas an asymmetrical division generating one stem cell and one differentiated cell occurs with probability $1 - s$. A differentiated cell symmetrically divides to generate two differentiated cells with probability g_0 but dies with probability d_0^d. This image originally appeared in [26].

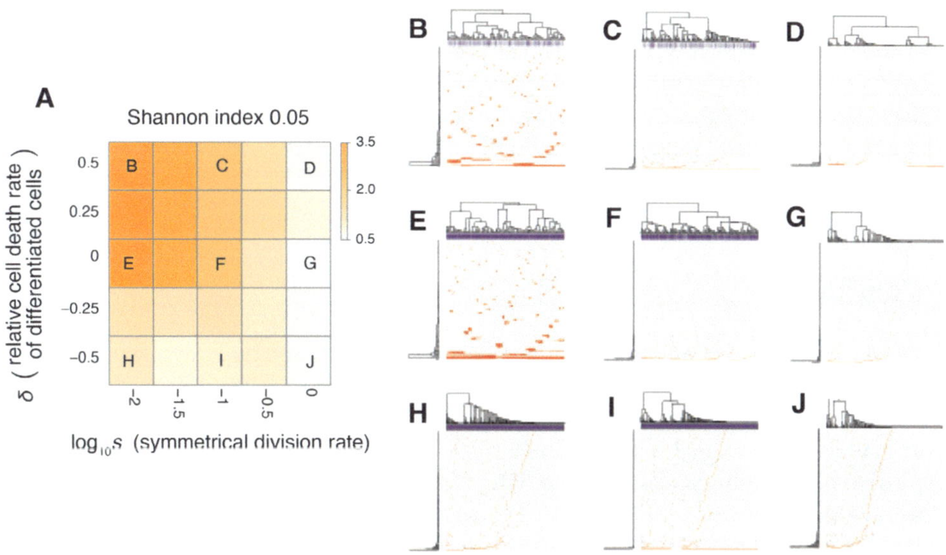

Figure 11.
Sensitivity analysis of the neutral-s model. (A) Heat map obtained by calculating Shannon index 0.05 while changing the relative death rate of differentiated cells $\delta = \log_{10}\left(d_0^d/g_0\right)$ *and symmetrical division rate s. The neutral mutation rate* m_n *and maximum population size P set to* 10^{-1} *and* 10^5*, respectively. (B–J) Single-cell mutation profiles obtained for nine parameter settings, indicated on the heat map presented in A. This image originally appeared in [26].*

does a tumor without a stem cell hierarchy. That is, a stem cell hierarchy increases the apparent mutation rate by $\log 2/\log(1+s)$-fold, which induces neutral evolution even with relatively low mutation rate settings.

Recent genomic analysis demonstrated that multiple evolutionary modes exists in cancer systems. For example, as described above, ITH in renal cancer is generated by Darwinian selection, which is in contrast to neutral evolution in colorectal cancer. Moreover, by multiregion sequencing of early-stage colorectal tumors, Saito *et al.* [26] showed that ITH is shaped by Darwinian selection in the early phase of colorectal cancer evolution, which means that a temporal shift of the evolutionary principle shaping ITH occurs during colorectal tumorigenesis. Employing agent-based modeling and MASSIVE analysis, Niida *et al.* [26] also constructed a model that explain this evolutionary shift. Darwinian ITH in an early-stage tumor is reproduced by the assumption of multiple driver mutations of relatively weak strength. At some point, growth of the early colorectal tumor slows because resource limitations, which is reproduced by introducing the carrying capacity into the simulation model. When they assumed that an explosive mutation that negates the carrying capacity was obtained with a small probability, a clone acquiring the explosive mutation overcame the resource limitation and expanded as late-stage tumors, in which ITH was generated neutral evolution. Another simulation study by West *et al.* [28] proposed that spatial constraints and limited cellular mixing play important roles in a similar Darwinian-neutral shift.

Sensitivity analysis also provides insight into metastatic tumor progression, which is poorly understood despite its clinical importance. Evaluation of genomic divergence between paired metastatic and primary tumors (M-P divergence) from multiregion sequencing is a good starting point for addressing this problem. Sun and Nikolakopoulos [29] extended *tumopp* [14] to simulate paired primary and metastatic tumors, and explored factors affecting M-P divergence by sensitivity analysis. As a result, they found that M-P divergence depends not only on the metastatic dissemination time, but also on the evolutionary dynamics and detectability of seedling cell lineages in a primary tumor. It was concluded that investigating tumor growth dynamics in detail is important, particularly when researchers interpret heterogeneity among longitudinal samples to infer the evolutionary timeline of cancer progression. Collectively, these examples demonstrated that agent-based modeling combined with sensitivity analysis is a useful tool for studying cancer evolutionary dynamics.

5. Conclusion

In this chapter, we introduced agent-based modeling of cancer evolution along with methodologies for data fitting and sensitivity analysis. Although there is a long history of theoretical science in the field of cancer research, this approach has been overshadowed by experimental science until recently. However, with a recent explosive increase in cancer genome data, there is now an increasing need to integrate experimental and theoretical science. As an example, this chapter introduced methods for modeling and analyzing the evolutionary processes generating ITH, which is experimentally observed by multiregion sequencing. We also presented exemplifying applications: e.g., agent-based simulation modeling and analysis successfully demonstrated that ITH in colorectal cancer is generated by neutral evolution, which is caused by a high mutation rate and stem cell hierarchy. For cancer genome analyses, new experimental technologies are actively being developed. For example, single-cell sequencing technologies can profile IHT at the ultimate resolution [30] while liquid biopsy technologies, such as the sequencing of circulating tumor DNA, enables us to non-invasively track cancer evolution during treatment [31]. These technologies will unveil more various aspects of cancer evolution when combined with the approach introduced in this chapter. This chapter also exemplified how simulation modeling helps to solve scientific problems raised by new experimental technologies. We hope that this chapter will provides readers with some hints to solve their own problems using simulation modeling.

Acknowledgements

This work was supported by the JSPS KAKENHI (19K12214) and AMED (JP21cm0106504).

Author details

Atsushi Niida[1*] and Watal M. Iwasaki[2]

1 The Institute of Medical Science, The University of Tokyo, Tokyo, Japan

2 Graduate School of Life Sciences, Tohoku University, Sendai, Japan

*Address all correspondence to: aniida@ims.u-tokyo.ac.jp

References

[1] Eric R Fearon and Bert Vogelstein. A genetic model for colorectal tumorigenesis. *Cell*, 61(5):759–767, 1990.

[2] Bert Vogelstein, Nickolas Papadopoulos, Victor E Velculescu, Shibin Zhou, Luis A Diaz, and Kenneth W Kinzler. Cancer genome landscapes. *Science*, 339 (6127):1546–1558, 2013.

[3] Nicholas McGranahan and Charles Swanton. Biological and therapeutic impact of intratumor heterogeneity in cancer evolution. *Cancer Cell*, 27(1):15–26, 2015.

[4] Charles M Macal and Michael J North. Tutorial on agent-based modeling and simulation. In *Proceedings of the Winter Simulation Conference, 2005.*, pages 14–pp. IEEE, 2005.

[5] Ryutaro Uchi, Yusuke Takahashi, Atsushi Niida, Teppei Shimamura, Hidenari Hirata, Keishi Sugimachi, Genta Sawada, Takeshi Iwaya, Junji Kurashige, Yoshiaki Shinden, et al. Integrated multiregional analysis proposing a new model of colorectal cancer evolution. *PLOS Genetics*, 12(2): e1005778, 2016.

[6] Atsushi Niida, Satoshi Nagayama, Satoru Miyano, and Koshi Mimori. Understanding intratumor heterogeneity by combining genome analysis and mathematical modeling. *Cancer Science*, 109(4):884–892, 2018.

[7] PADM Van Liedekerke, A Buttenschön, and D Drasdo. Off-lattice agent-based models for cell and tumor growth: numerical methods, implementation, and applications. In *Numerical methods and advanced simulation in biomechanics and biological processes*, pages 245–267. Elsevier, 2018.

[8] Alexander RA Anderson, Alissa M Weaver, Peter T Cummings, and Vito Quaranta. Tumor morphology and phenotypic evolution driven by selective pressure from the microenvironment. *Cell*, 127(5):905–915, 2006.

[9] Alexander RA Anderson, Katarzyna A Rejniak, Philip Gerlee, and Vito Quaranta. Microenvironment driven invasion: a multiscale multimodel investigation. *Journal of Mathematical Biology*, 58(4):579–624, 2009.

[10] Heiko Enderling, Lynn Hlatky, and Philip Hahnfeldt. Migration rules: tumors are conglomerates of self-metastases. *British Journal of Cancer*, 100 (12):1917–1925, 2009.

[11] Jan Poleszczuk, Philip Hahnfeldt, and Heiko Enderling. Evolution and phenotypic selection of cancer stem cells. *PLOS Computational Biology*, 11 (3):e1004025, 2015.

[12] Andrea Sottoriva, Joost JC Verhoeff, Tijana Borovski, Shannon K McWeeney, Lev Naumov, Jan Paul Medema, Peter MA Sloot, and Louis Vermeulen. Cancer stem cell tumor model reveals invasive morphology and increased phenotypical heterogeneity. *Cancer Research*, 70(1): 46–56, 2010.

[13] Bartlomiej Waclaw, Ivana Bozic, Meredith E Pittman, Ralph H Hruban, Bert Vogelstein, and Martin A Nowak. A spatial model predicts that dispersal and cell turnover limit intratumour heterogeneity. *Nature*, 525 (7568):261–264, 2015.

[14] Watal M Iwasaki and Hideki Innan. Simulation framework for generating intratumor heterogeneity patterns in a cancer cell population. *PLOS One*, 12(9): e0184229, 2017.

[15] Camilla Hurwitz and LJ Tolmach. Time-lapse cinemicrographic studies of x-irradiated hela s3 cells: I. cell progression and cell disintegration. *Biophysical Journal*, 9(4):607–633, 1969.

[16] Darren R Tyson, Shawn P Garbett, Peter L Frick, and Vito Quaranta. Fractional proliferation: a method to deconvolve cell population dynamics from single-cell data. *Nature Methods*, 9 (9):923–928, 2012.

[17] Marco Gerlinger, Andrew J Rowan, Stuart Horswell, James Larkin, David Endesfelder, Eva Gronroos, Pierre Martinez, Nicholas Matthews, Aengus Stewart, Patrick Tarpey, et al. Intratumor heterogeneity and branched evolution revealed by multiregion sequencing. *New England Journal of Medicine*, 366:883–892, 2012.

[18] Mikael Sunnåker, Alberto Giovanni Busetto, Elina Numminen, Jukka Corander, Matthieu Foll, and Christophe Dessimoz. Approximate bayesian computation. *PLOS Computational Biology*, 9(1):e1002803, 2013.

[19] Andrea Sottoriva, Haeyoun Kang, Zhicheng Ma, Trevor A Graham, Matthew P Salomon, Junsong Zhao, Paul Marjoram, Kimberly Siegmund, Michael F Press, Darryl Shibata, et al. A big bang model of human colorectal tumor growth. *Nature Genetics*, 47(3): 209–216, 2015.

[20] Zheng Hu, Jie Ding, Zhicheng Ma, Ruping Sun, Jose A Seoane, J Scott Shaffer, Carlos J Suarez, Anna S Berghoff, Chiara Cremolini, Alfredo Falcone, et al. Quantitative evidence for early metastatic seeding in colorectal cancer. *Nature Genetics*, 51(7):1113–1122, 2019.

[21] SA Sisson and Y Fan. Abc samplers. *Handbook of Approximate Bayesian Computation*, pages 87–123, 2018.

[22] Jean-Michel Marin, Natesh S Pillai, Christian P Robert, and Judith Rousseau. Relevant statistics for bayesian model choice. *Journal of the Royal Statistical Society: Series B: Statistical Methodology*, pages 833–859, 2014.

[23] Veronica Czitrom. One-factor-at-a-time versus designed experiments. *The American Statistician*, 53(2):126–131, 1999.

[24] Ilya M Sobol. Global sensitivity indices for nonlinear mathematical models and their monte carlo estimates. *Mathematics and computers in simulation*, 55(1–3):271–280, 2001.

[25] Atsushi Niida, Takanori Hasegawa, and Satoru Miyano. Sensitivity analysis of agent-based simulation utilizing massively parallel computation and interactive data visualization. *PLOS one*, 14(3):e0210678, 2019.

[26] Atsushi Niida, Takanori Hasegawa, Hideki Innan, Tatsuhiro Shibata, Koshi Mimori, and Satoru Miyano. A unified simulation model for understanding the diversity of cancer evolution. *PeerJ*, 8: e8842, 2020.

[27] Ricard V Solé, Carlos Rodrguez-Caso, Thomas S Deisboeck, and Joan Saldaña. Cancer stem cells as the engine of unstable tumor progression. *Journal of Theoretical Biology*, 253(4):629–637, 2008.

[28] Jeffrey West, Ryan O Schenck, Chandler Gatenbee, Mark Robertson-Tessi, and Alexander RA Anderson. Normal tissue architecture determines the evolutionary course of cancer. *Nature Communications*, 12(1):1–9, 2021.

[29] Ruping Sun and Athanasios N Nikolakopoulos. Elements and evolutionary determinants of genomic divergence between paired primary and metastatic tumors. *PLOS Computational Biology*, 17(3):e1008838, 2021.

[30] Bora Lim, Yiyun Lin, and Nicholas Navin. Advancing cancer research and medicine with single-cell genomics. *Cancer Cell*, 37(4):456–470, 2020.

[31] David W Cescon, Scott V Bratman, Steven M Chan, and Lillian L Siu. Circulating tumor dna and liquid biopsy in oncology. *Nature Cancer*, 1(3):276–290, 2020.

Chapter 5

AI Modeling to Combat COVID-19 Using CT Scan Imaging Algorithms and Simulations: A Study

Naser Zaeri

Abstract

The coronavirus disease 2019 (COVID-19) outbreak has been designated as a worldwide pandemic by World Health Organization (WHO) and raised an international call for global health emergency. In this regard, recent advancements of technologies in the field of artificial intelligence and machine learning provide opportunities for researchers and scientists to step in this battlefield and convert the related data into a meaningful knowledge through computational-based models, for the task of containment the virus, diagnosis and providing treatment. In this study, we will provide recent developments and practical implementations of artificial intelligence modeling and machine learning algorithms proposed by researchers and practitioners during the pandemic period which suggest serious potential in compliant solutions for investigating diagnosis and decision making using computerized tomography (CT) scan imaging. We will review the modern algorithms in CT scan imaging modeling that may be used for detection, quantification, and tracking of Coronavirus and study how they can differentiate Coronavirus patients from those who do not have the disease.

Keywords: Artificial intelligence, COVID-19, CT algorithms, modeling

1. Introduction

Today, the world is facing one of its most dangerous risks, if not the most one throughout the century. It is a pandemic that is draining the whole world's resources and threatening the development of human civilization. The COVID-19 pandemic continues to have a devastating effect on the health and well-being of global population, caused by the infection of individuals by the severe acute respiratory syndrome coronavirus 2 (SARS-CoV-2) [1]. On the 30th of January 2020, the WHO declared the SARS-CoV-2 outbreak a public health emergency of international concern. On March 11th, WHO characterized COVID-19 as a pandemic. At the time of writing this manuscript (May 29, 2021), the number of infected people has surpassed 169,118,995 confirmed cases and more than 3,519,175 deaths in 223 countries [2]. The World Trade Organization has announced that the world has effectively entered a recession period. The world's economy and many countries' economies are

IntechOpen

in danger of collapsing. Schools in many countries are closed and students around the world are forced to stay at home [3, 4].

One of the early challenges that emerged at the beginning of the pandemic is the detection of COVID-19 cases. The most important method used for detecting COVID-19 cases is polymerase chain reaction (PCR) testing, that can detect SARSCoV-2 RNA from respiratory specimens [5]. Though PCR testing is the standard, it is a time-consuming, laborious, and complicated manual process that is in short supply [6]. Accurate and rapid diagnosis of COVID-19 suspected cases plays a crucial role in timely quarantine and medical treatment. This limitation of human expert-based diagnosis has provided a strong motivation for the use of computer simulation and modeling to improve the speed and accuracy of the detection process [7, 8]. Another related issue is the manual contouring of lung lesions which tends to be a tedious and time-consuming work, and could lead to subsequent assessment discrepancies in case of inconsistent delineation. Thus, a fast auto-contouring tool for COVID-19 infection is needed in the onsite applications for quantitative disease assessment [9].

Since the early days of this catastrophic crisis, there has been an upsurge in the exploration and use of artificial intelligence (AI), computer simulation, and data modeling and analytic tools, in a multitude of areas. AI and machine learning (ML) have demonstrated great performance in various medical fields and have proven their vital role in complicated therapeutic scenes. These systems have shown high level of accuracy in different applications, such as lung disease classification, breast cancer, skin lesion classification, identifying diabetic retinopathy, and Alzheimer [10–12].

Scientists and healthcare professionals have realized the importance of AI and imaging technologies in slowing the spread of COVID-19 at preliminary stages, and containing the virus at later stages. Currently, many AI and computer modeling systems are used in disease diagnosis, examining, identifying, and treating patients. AI-based simulations have also been employed for evaluating disease progression, economic downturn and recovery, contingency planning, demand sensing, supply chain disruptions, workforce planning, as well as for management decision-making on site openings [13]. For example, AI-based simulations were critical in integrating multiple decision-making domains (e.g., COVID-19 disease progression, government interventions, people behavior, demand sensing, supply disruptions etc.) [14].

In this paper, we provide an extensive review and a deep study on how AI and ML can help the world to deliver efficient responses and combat the COVID-19 pandemic using CT scan imaging. More specifically, we will focus on the modern algorithms in CT scan imaging that may be used for detection, quantification, and tracking of Coronavirus and study how they can differentiate Coronavirus patients from those who do not have the disease. We provide recent theoretical developments, technological advancements, and practical implementations of AI algorithms and ML techniques that uses CT imaging to suggest possible solutions in investigating diagnosis, severity level, prediction, tracking, treatments and other decision making scenarios related to COVID-19. In this regard, we explore a vast number of important studies that have been performed by various academic and research communities from numerous disciplines during the period of pandemic since the early days of 2020 up to the very recent days (May 2021). Before we further proceed, we note that many of the articles cited are still preprints at the time of writing this manuscript. Given the fast-moving nature of the crisis, we endeavored to be comprehensive of coverage. We understand that the full scientific rigor for many articles should still be assessed by the scientific community through peer-reviewed evaluation and other quality control mechanisms. However, the whole story is a striking dilemma and a big challenge to the global scientific communities.

Researchers, physicians, technical-background individuals, and academics are putting all their efforts to come up with solutions and cures to this fatal disease. All of these efforts have emerged during a very short period of time, and a lot are yet to emerge in the coming few months, and possibly years.

1.1 A view of AI and ML in healthcare

AI is becoming one of the highest priorities for healthcare decision makers, governments, investors and innovators. An increasing number of governments have set out targets for AI in healthcare, in countries as diverse as the United States, China, Finland, Germany, and the UK, and many are investing heavily in AI-related research. The private sector is also playing a significant role, with venture capital funding for the top 50 firms in healthcare-related AI reaching $8.5 billion [15]. Though the US dominates the list of firms with highest venture capital funding in healthcare AI to date, and has the most related research studies and trials, China is emerging as the fastest growing country in this field. Europe, meanwhile, benefits from the vast depot of health data collected by national health systems and has significant strengths in terms of the number of research studies, established clusters of innovation and collaborations related to AI [16].

AI applications based on imaging, are already in use in specialties such as radiology, oncology, cardiology, neurology, pathology and ophthalmology. It is expected that more AI solutions would support the shift from hospital-based to home-based care, such as remote monitoring, AI-powered alerting systems, and virtual assistants [17, 18]. Also, AI is anticipated to be embedded more extensively in clinical workflows through the intensive engagement of professional bodies and providers. Moreover, AI solutions are expected to emerge in clinical practices based on evidence from clinical trials, with increasing focus on improved and scaled clinical decision-support tools [19]. Advances in AI mean that algorithms can generate layers of abstract features that enable computers to recognize complicated concepts (such as a diagnosis). This enables them to learn discriminative features automatically and approximate highly complex relationships [20, 21].

2. Convolutional neural network

Neural networks have been successfully applied to many real-world problems. The most general type of neural network is Multilayer Perceptron (MLP). While MLPs can be used to effectively classify small images, they are impractical for large images. The reason for this can be explained by the fact that the implementation of a MLP would result in a huge output vector of weights for each class (size of millions). MLPs not only are computationally expensive to train (both in terms of time and memory usage), but they also have high variance due to the large number of weights [22, 23].

Convolutional Neural Networks (CNNs) have been driving the heart of computer vision in recent years. The key concept of CNNs is to find local features from an input (usually an image) at higher layers and combine them into more multifaceted features at lower layers [24, 25]. CNNs are very good in extracting patterns in the input image, such as lines, gradients, circles, or even eyes and faces. It is this property that makes CNNs so powerful for computer vision. Unlike earlier computer vision algorithms, CNNs can operate directly on a raw image and do not need any preprocessing. In the medical field, CNNs are used to improve image quality in low-light images from a high-speed video endoscopy [26] and is also applied to recognize the nature of pulmonary nodules via CT images and the identification of pediatric pneumonia via chest X-ray images [27].

A CNN comprises several layers where each neuron of a subsequent higher layer connects to a subset of neurons in the previous lower layer. This permits the receptive field of a neuron of a higher layer to cover a greater part of images compared to that of a lower layer. The higher layer is capable to learn more abstract features of images than the lower layer by considering the spatial relationships between different receptive fields. It should be noted that CNNs significantly reduce the number of weights, and in turn reduce the variance. Like MLPs, CNNs use fully connected (FC) layers and non-linearities, but they introduce two new types of layers: convolutional and pooling layers. A convolutional layer takes a $W \times H \times D$ dimensional input "I" and convolves it with a $w \times h \times D$ dimensional filter (or kernel) G. The weights of the filter can be hand designed, but in the context of machine learning they are automatically tuned, just like the way weights of an FC layer are tuned. In line with convolutional layers where reducing the number of weights in neural networks reduces the variance, pooling layers directly reduce the number of neurons in neural networks. The sole purpose of a pooling layer is to downsample (also known as pool, gather, consolidate) the previous layer, by sliding a fixed window across a layer and choosing one value that effectively "represents" all of the units captured by the window. There are two common implementations of pooling. In max-pooling, the representative value just becomes the largest of all the units in the window, while in average-pooling, the representative value is the average of all the units in the window. In practice, pooling layers are stridden across the image with the stride equal to the size of the pooling layer. None of these properties actually involve any weights, unlike fully connected and convolutional layers.

3. CT diagnosis algorithms

Medical imaging is a useful supplement to reverse transcription polymerase chain reaction (RT-PCR) testing for the confirmation of COVID-19. Researchers have found that CT images of COVID-19 patients exhibit typical imaging characteristics. During the last year, studies have shown that typical chest CT patterns of COVID-19 viral pneumonia include multifocal bilateral peripheral ground-glass areas associated with subsegmental patchy consolidations, mostly subpleural, and predominantly involving lower lung lobes and posterior segments [28–32]. In more detail, chest CT images of COVID-19 patients could be evaluated using the following characteristics [33–40]:

- presence of ground-glass opacities (GGOs)
- laterality of GGO and consolidation
- presence of nodules
- presence of pleural effusion
- presence of thoracic lymphadenopathy
- degree of involvement of each lung lobe, in addition to the overall extent of lung involvement measured
- presence of underlying lung disease such as emphysema or fibrosis
- bilateral distribution

- number of lobes affected where either ground-glass or consolidative opacities are present
- interlobular septal thickening
- presence of cavitation
- bronchial wall thickening
- air bronchogram
- perilesional vessel diameter
- lymphadenopathy
- pleural pericardial effusion

GGO, which is defined as hazy increased lung attenuation with preservation of bronchial and vascular margins [41], is the most common early finding of COVID-19 on chest CT. Besides GGO, bilateral patchy shadowing is one of the most common radiologic findings on chest CT [42]. In another study containing 51 COVID-19 patients, Song et al. [43] found that disease progression can be determined by lesions with consolidation. Multiple lesions and crazy-paving pattern are also common in COVID-19 patients. The diagnosis of chest CT depending on visual diagnosis of radiologists suffers from some problems [44]. For example, chest CT contains hundreds of slices, which takes a long time to diagnose. Also, it was found that the chest CT images of some COVID-19 patients share some similar manifestations with other types of pneumonia. This could add extra challenges to inexperienced radiologists, considering that COVID-19 is a new lung disease.

During the last year, AI methods and ML techniques have played a very important role in COVID-19 diagnosis in applications utilizing the CT imaging. The aim of AI techniques and ML methods was always to extract the distinguished features of COVID-19 presented in the different types of images. In this section, we review and thoroughly discuss the major works and articles that have addressed AI and ML in COVID-19 diagnosis using CT imagery. AI and deep learning methods have shown great ability to address the aforementioned problems by detecting this disease and distinguishing it from community acquired pneumonia (CAP) and other non-pneumonic lung diseases using chest CT. We explore important studies that have been performed by various academic and research communities from numerous disciplines which focus on detecting, quantifying, and tracking of Coronavirus and study how they can differentiate Coronavirus patients from those who do not have the disease.

Li et al. [45] developed a 3D deep learning framework for the detection of COVID-19, referred to as COVID-19 detection neural network (COVNet). The proposed CNN consists of a RestNet50 as the backbone, which takes a series of CT slices as the input and generates features for the corresponding slices. In more detail, it extracts visual features from volumetric chest CT scans both in 2D local and 3D global representation. The extracted features from all slices are then combined by a max-pooling operation. CAP and other non-pneumonia CT scans were included to test the robustness of the proposed model. The final feature map is fed to a fully connected layer and softmax activation function to generate a probability score for each type (COVID-19, CAP, and non-pneumonia), and produce a classification prediction. The CT scans are performed using different manufacturers with standard imaging protocols. Each volumetric scan contains 1094 CT slices with a varying slice-thickness from 0.5 mm to 3 mm. The reconstruction matrix is 512x512 pixels with in-plane pixel spatial resolution from 0.29x0.29 mm^2 to 0.98x0.98 mm^2. The CT scans are preprocessed and the lung region is extracted as the region of interest (ROI) using a U-net based segmentation method. Then, the image is passed to the COVNet for the predictions, as shown in **Figure 1**.

The authors have tested the system on datasets collected from six hospitals between August 2016 and February 2020. The collected datasets consisted of 4356 chest CT scans from 3322 patients. Diagnostic performance was assessed by the area under the receiver operating characteristic curve (AUC), sensitivity and specificity. The COVID-19 cases were affirmed as positive by RT-PCR and were obtained from Dec 31, 2019 to Feb 17, 2020. The most shared symptoms were fever (81%) and cough (66%). Moreover, the patients were 49±15 years old and there are slightly more male patients than female (1838 vs. 1484). CT scans with multiple reconstruction kernels at the same imaging session or acquired at multiple time points were included. The final dataset consisted of 1296 (30%) scans for COVID-19, 1735 (40%) for CAP and 1325 (30%) for non-pneumonia.

For each patient, one or multiple CT scans at several time points during the course of the disease were acquired (Average CT scans per patient was 1.8, with a range from 1 to 6). The per-scan sensitivity and specificity for detecting COVID-19 in the independent test set was 114 of 127 (90% [95% confidence interval: 83%, 94%]) and 294 of 307 (96% [95% confidence interval: 93%, 98%]), respectively, with an AUC of 0.96. The details of their tests are given in **Table 1**.

In another study, a weakly-supervised deep learning-based software system was developed by Zheng et al. [46] using 3D CT volumes to detect COVID-19. The authors have searched unenhanced chest CT scans of patients with suspected COVID-19 from the picture archiving and communication system of radiology department (Union Hospital, Tongji Medical College, Huazhong University of

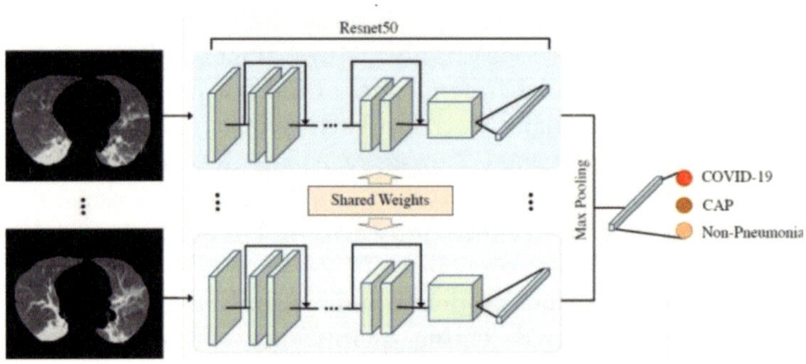

Figure 1.
COVID-19 detection neural network (COVNet) architecture [45].

	Sensitivity %	Specificity %	AUC
COVID-19	90 (114 of 127)	96 (294 of 307)	0.96
CAP	87 (152 of 175)	92 (239 of 259)	0.95
Non-Pneumonia	94 (124 of 132)	96 (291 of 302)	0.98

Note: Values in the parentheses are the numbers for the percentage calculations.

Table 1.
The performance of COVNet as per [45].

DOI: http://dx.doi.org/10.5772/intechopen.99442

Science and Technology). 540 patients (age of 42.5 ± 16.1 years; range 3–81 years) were enrolled into the study, including 313 patients (age, 50.7 ± 14.7 years; range 8–81 years) with clinical diagnosed COVID-19 (COVID-positive group) and 227 patients (age of 31.2 ± 10.0 years; range, 3–69 years) without COVID-19 (COVID-negative group). As shown in **Figure 2**, the system takes a CT volume and its 3D lung mask as input, where the 3D lung mask is generated by a pre-trained U-Net. The proposed system is divided into three stages. The first stage consists of a 3D convolution with a kernel size of 5 × 7 × 7, a batchnorm layer and a pooling layer. The second stage is composed of two 3D residual blocks. In each one of the residual block, a 3D feature map is handed into both a 3D convolution with a batchnorm layer and a shorter connection containing a 3D convolution. The third stage is a progressive classifier, which contains three 3D convolution layers and a fully-connected layer with the softmax activation function. As described in **Figure 3**, a U-Net is trained for lung region segmentation on the labeled training set using the ground-truth lung masks generated by an unsupervised learning method. Then, the pre-trained U-Net is used to test all CT volumes to obtain the lung masks. The lung mask is concatenated with CT volume and serves as the input of the system. The authors have used the spatially global pooling layer and the temporally global pooling layer to technically handle the weakly-supervised COVID-19 detection problem.

Furthermore, Gozes et al. [47] presented a system that exploits 2D and 3D deep learning models. **Figure 4** shows a block diagram of the developed system. The system is comprised of several components and analyzes the CT case at two distinct levels: *Subsystem A and Subsystem B.* Subsystem A provides a 3D analysis of the case volume for nodules and focal opacities using existing, previously developed algorithms, where Subsystem B provides newly developed 2D analysis of each slice of

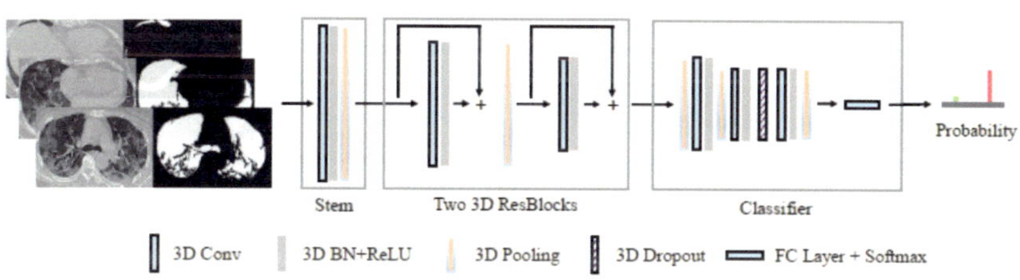

Figure 2.
The architecture proposed in [46]. The network takes a CT volume with its 3D lung mask as the input and outputs the probabilities of COVID-19 positive/negative.

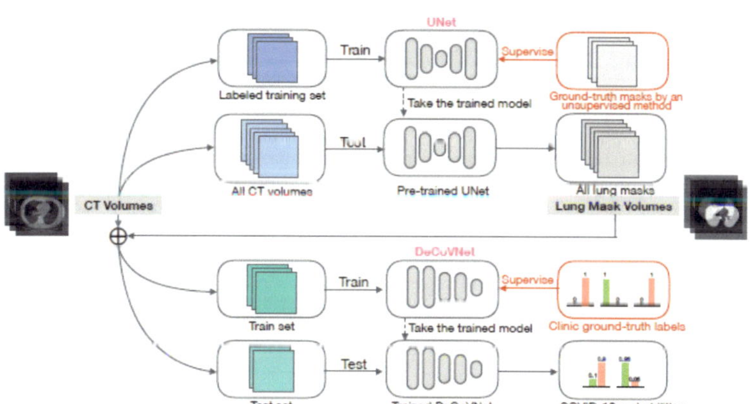

Figure 3.
Training and testing procedures [46].

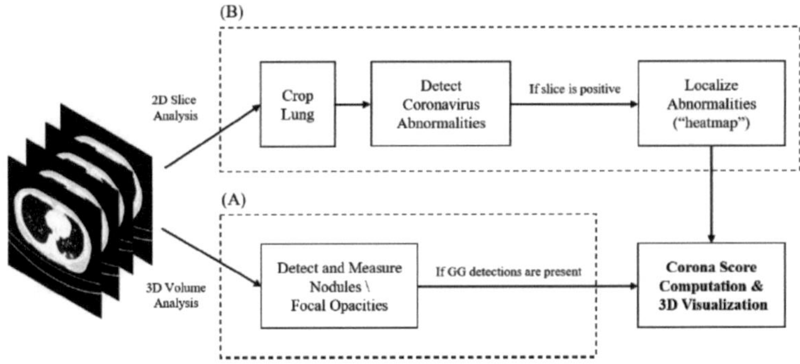

Figure 4.
System block diagram [47].

the case to detect and localize larger-sized diffuse opacities including ground glass infiltrates which have been clinically described as representative of Coronavirus. As argued by the authors, working in the 2D space has several advantages for deep learning based algorithms in limited data scenarios. These include an increase in the training samples (with many slices per single case), the ability to use pre-trained networks that are common in 2D space, and an easier annotation for segmentation purposes.

For Subsystem A, the authors used commercial off-the-shelf software that detects nodules and small opacities within a 3D lung volume. This software was developed as a solution for lung pathology detection and provides quantitative measurements (including volumetric measurements, axial measurements, calcification detection and texture characterization). For Subsystem B, the first step is the *lung crop stage*, where the lung region of interest is extracted using a lung segmentation module. In the following step, Coronavirus related abnormalities are detected using Resnet-50, which is a 2D CNN architecture that consists of 50 layers. In the classification stage, the authors calculated the ratio of positive detected slices out of the total slices of the lung (*positive ratio*). A positive case-decision is made if the positive ratio exceeds a pre-defined threshold. The system was tested on 157 patients from China and U.S. The sensitivity and the specificity of the system were 98.2% and 92.2%, respectively. **Figure 5** shows a patient case visualization.

The authors have also proposed a *Corona score* which is a volumetric measurement of the opacities burden. The corona score is computed by a volumetric summation of the network-activation maps. The system output enables quantitative measurements for smaller opacities (volume, diameter) and visualization of the larger opacities in a slice-based "heat map" or a 3D volume display. The authors claim that the score is robust to slice thickness and pixel spacing as it includes pixel volume. For patient-specific monitoring of disease progression, they suggested the *Relative Corona score* in which they normalize the corona score by the score computed at the first time point. The suggested "Corona score" measures the progression of disease over time. An example of such an implementation is shown in **Figure 6** which demonstrates a tracking over time of a specific opacity in a Coronavirus patient (red box). In this example, a patient was imaged in time points where the first CT scan was obtained few days following the first signs of the virus (fever, cough). This case involves multiple opacities and shows an overview of the patient recovery process with its corresponding Corona score over time.

Barstugan et al. [48] presented a classification system consisting of five different feature extraction methods followed by support vector machine (SVM). The feature extraction methods were Gray Level Co-occurrence Matrix (GLCM), Local

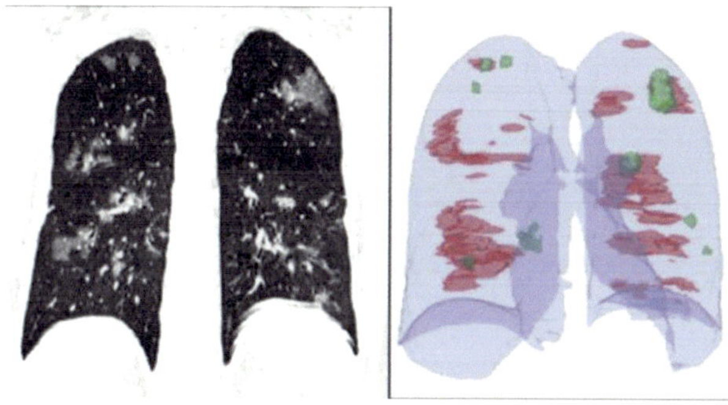

Figure 5.
Patient case visualization. Left: Coronal view; right: Automatically generated 3D volume map of focal opacities (green) and larger diffuse opacities (red) [47].

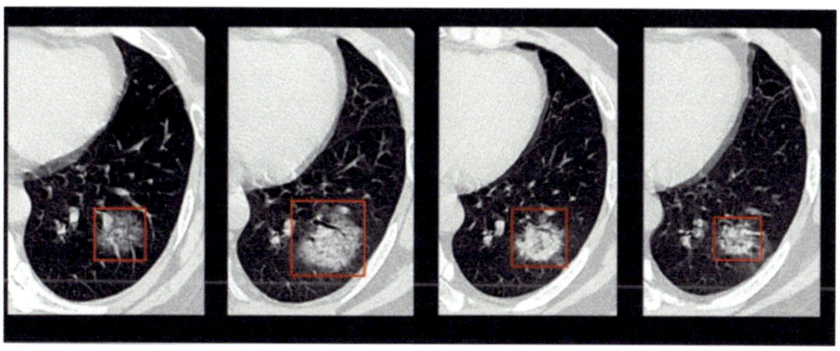

Figure 6.
Multi time point tracking of disease progression [47].

Directional Patterns (LDP), Gray Level Run Length Matrix (GLRLM), Gray Level Size Zone Matrix (GLSZM), and Discrete Wavelet Transform (DWT). To test the proposed system, four different datasets were formed by taking patches of size 16x16, 32x32, 48x48 and 64x64 from 150 CT images belonging to 53 infected cases, from the "Societa Italiana di Radiologia Medica e Interventistica". The samples of datasets were labeled as Coronavirus/non-Coronavirus (infected/non-infected). **Table 2** shows the four different subsets created from patch regions. The authors have implemented 2-fold, 5-fold and 10-fold cross-validations during the classification process. Sensitivity, specificity, accuracy, precision, and F-score metrics were used to evaluate the classification performance. **Figure 7** shows patch regions and patch samples from the four different subsets.

Caruso et al. [49] investigated chest CT features of patients with COVID-19 in Rome, Italy, and compared the diagnostic performance of CT with that of RT-PCR. All chest CT examinations were performed with patients in the supine position on a 128-slice CT scanner. Radiologists in consensus with thoracic imaging experience evaluated the images using a clinically available dedicated application (Thoracic VCAR, GE Medical Systems), defining patients as having positive CT findings when a diagnosis of viral pneumonia was reported. The study comprised 158 participants, of them fever was witnessed in 97 (61%) and cough and dyspnea were observed in 88 (56%) and 52 (33%), respectively. Of these patients, 62 (39%) had positive RT-PCR results and 102 (64%) had positive CT findings. Sensitivity, specificity, and accuracy of CT for COVID-19 pneumonia were 97% (60 of 62 participants), 56% (54 of 96 participants), and 72% (114 of 158 participants), respectively.

Table 3 details the CT features in participants with COVID-19 infection confirmed with RT-PCR as reported in [49]. The results presented in [49] agree with

Subset	Patch Dimension	Number of Non-Coronavirus Patches	Number of Coronavirus Patches
Subset 1	16 × 16	5912	6940
Subset 2	32 × 32	942	1122
Subset 3	48 × 48	255	306
Subset 4	64 × 64	76	107

Table 2.
Four different subsets created from patch regions [48].

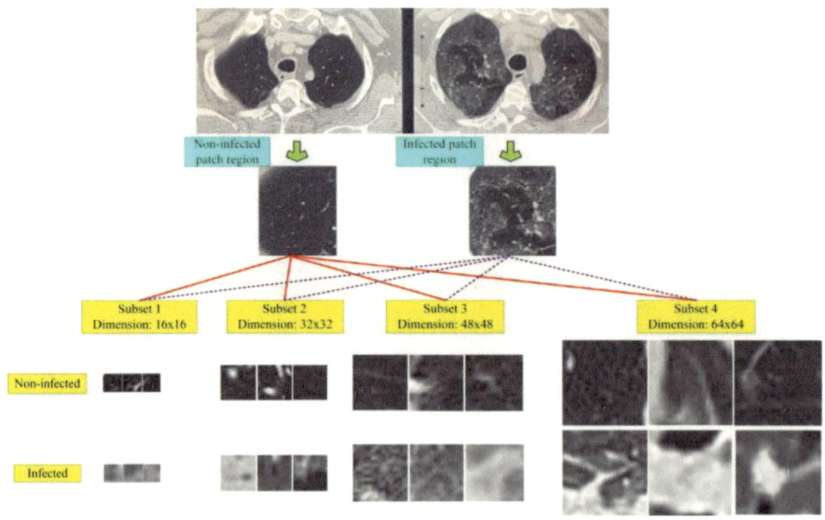

Figure 7.
Patch regions and patch samples from the four different subsets. Sample images for infected and non-infected situations for all subsets are shown as well [48].

the study performed by Salehi et al. [50] of 919 patients, despite some differences. However, the population in [50] varies from the population examined in [49]. Also, Chung et al. [51] analyzed a small population consisting of 21 patients and found a very low frequency of crazy paving pattern compared with [49] (19% vs. 39%).

Furthermore, Xu et al. [52] established a model to distinguish COVID-19 from influenza-A viral pneumonia (IAVP) and healthy cases through pulmonary CT images. The authors have discussed that the RT-PCR detection of viral RNA from sputum or nasopharyngeal swab have a relatively low positive rate in the early stage. They argued that the manifestations of COVID-19 as seen through CT imaging show individual characteristics that differ from those of other types of viral pneumonia such as IAVP. The suggested model consists of multiple CNNs, where the candidate infection regions are segmented out from the pulmonary CT image set. Then, these separated images are categorized into the COVID-19, IAVP, and irrelevant to infection groups, together with the corresponding confidence scores, using a location-attention classification model. Finally, the infection type and overall confidence score for each CT case are calculated using the Noisy-OR Bayesian function.

Figure 8 shows the whole process. As described in the figure, the CT images are first preprocessed to excavate the effective pulmonary regions. Then, a 3D CNN segmentation model is used to segment multiple candidate image cubes. After that, an image classification model is used to classify all the image patches into three

CT Feature	No. of Participants (n = 58)	Percentage
GGO	58	100
Multilobe involvement (≥2 lobes)	54	93
Bilateral distribution	53	91
Posterior involvement	54	93
GGO location (peripheral)	52	89
Subsegmentel vessel enlargement (>3 mm)	52	89
Consolidation	42	72
Subsegmental	32	55
Segmental	10	17
Lymphadenopathy	34	58
Bronchiectasis	24	41
Air bronchogram	21	36
Pulmonary nodules surrounded by GGO	10	17
Interlobular septal thickening	8	13
Halo sign	7	12
Pericardial effusion	3	5
Pleural effusion	2	3
Bronchial wall thickening	1	1
Cavitation	0	0

Table 3.
CT features in participants with COVID-19 infection confirmed with RT-PCR [49].

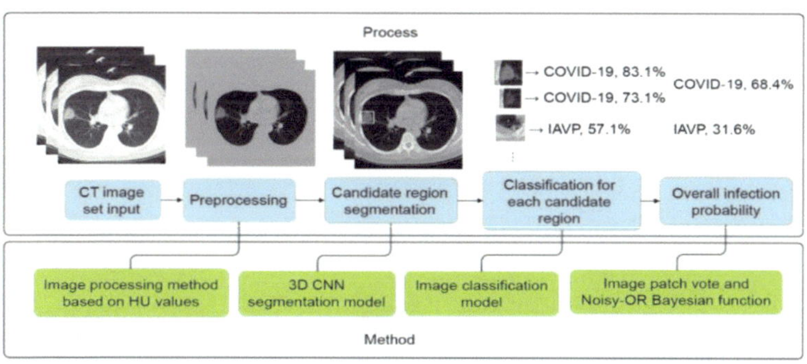

Figure 8.
The process flow chart of [52].

kinds: COVID-19, IAVP, and irrelevant to infection. Image patches from the same group "vote" for the type and confidence score of this candidate as a whole. Finally, the Noisy-OR Bayesian function is used to calculate the overall analysis report for one CT sample. It is worth mentioning that the model uses a V-Net as the backbone feature extraction part. The authors have further discussed how the variable 3D structures of the lesion regions can aggravate the results. For example, when the border between a healthy region and the infected one becomes blurred and indistinct, it will be difficult to label pixel-level masks for lesion regions of pneumonia. As such, the model uses the RPN structure [52] to capture the region of interest with 3D bounding boxes instead of pixel-level segmented masks.

To evaluate the system, two classification models were used, as shown in **Figure 9**. The first one was the ResNet model and the other was designed based on the first network structure by concatenating the location-attention mechanism in the full-connection layer to improve the overall accuracy rate. The resultant model was added to the first full-connection layer to enhance the influence of this factor on the whole network. The output of the convolution layer was flattened to a 256-dimensional feature vector and then converted into a 16-dimensional feature vector using a full-connection network. The overall accuracy rate was 86.7% in terms of all the CT cases taken together.

Belfiore et al. [53] presented a practice of a good tool for radiologists (Thoracic VCAR) that can be used in COVID-19 diagnosis. Thoracic VCAR offers quantitative measurements of the lung involvement. Further, it can generate a clear, fast and concise report that communicates vital medical information to referring physicians. In the post-processing phase, the software can recognize the ground glass and differentiate it from consolidation and quantifies them as a percentage with respect to the healthy parenchyma. This information is useful for evaluating regression or progression disease in response to drug therapy as well as evaluating the effectiveness of pronation maneuvers for alveolar recruitment in ICU patients. The authors in [53] have discussed the importance of such high-resolution CT (HRCT) technique in investigating the patients with suspicion COVID-19 pneumonia. They have argued that the HRCT is a very accurate technique in identifying pathognomic findings of interstitial pneumonia as ground glass areas, crazy paving, nodules and consolidations, mono- or bilateral, patchy or multifocal, central and/or peripheral distribution, declivous or nondeclivous. As per the discussion, during the follow-up, HRCT examination can quantify the course of the disease and evaluate the effectiveness of the experimental trial and the patient's prognosis.

In [54], Mei et al. have also used AI algorithms to integrate chest CT findings with clinical symptoms, exposure history and laboratory testing to rapidly diagnose patients who are positive for COVID-19. Among a total of 905 patients

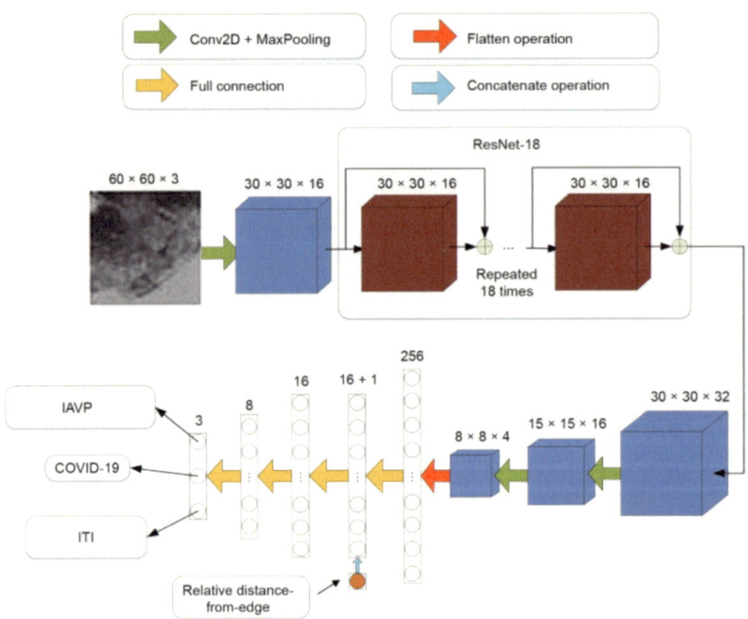

Figure 9.
The network structure of ResNet-18-based classification model [52].

tested by real-time RT–PCR test, 419 (46.3%) tested positive for SARS-CoV-2. In this study, the dataset included patients aged from 1 to 91 years (with mean of 40.7 year and standard deviation of 6.5 years) where 488 of the patients were men and 417 were women. All scans were acquired using a standard chest CT protocol and were reconstructed using the multiple kernels and displayed with a lung window. Clinical information included travel and exposure history, leukocyte counts (including absolute neutrophil number, percentage neutrophils, absolute lymphocyte number and percentage lymphocytes), symptomatology (presence of fever, cough and sputum), patient age and patient sex. More specifically, the authors developed a CNN to learn the imaging characteristics of patients on the initial CT scan. They used multilayer perceptron classifiers to classify patients with COVID-19 according to the radiological data and clinical information.

Of the 134 positive cases in the test set, 90 were correctly categorized by both the joint model and the senior thoracic radiologist and 33 were classified differently. Of the 33 patients, 23 were correctly classified as positive by the joint model, but were misclassified by the senior thoracic radiologist. Ten patients were classified as negative by the joint model, but correctly diagnosed by the senior thoracic radiologist. Eleven patients were misclassified by both the joint model and the senior thoracic radiologist. Of the 145 patients negative for COVID-19 in the test set, 113 were correctly classified by both the joint model and the senior thoracic radiologist. Thirty-two out of 145 were classified differently by the joint model and the senior thoracic radiologist. Seven were correctly classified as negative by the joint model, but were diagnosed as positive by the senior thoracic radiologist. Twenty-three were classified as positive by the joint model, but correctly diagnosed as negative by the senior thoracic radiologist. Two patients were misclassified by both the joint model and the senior thoracic radiologist. As discussed in [54], patient's age, presence of exposure to SARS-CoV-2, presence of fever, cough, cough with sputum, and white blood cell counts are significant features associated with SARS-CoV-2 status. However, it should be pointed out that difficulties on model training have been witnessed due to the limited sample size.

Moreover, Fei et al. [55] developed a deep learning-based system for automatic segmentation of lung and infection sites using chest CT. Likewise, Xiaowei et al. [56] distinguished COVID-19 pneumonia and Influenza-A viral pneumonia from healthy cases. Further, Shuai et al. [57] developed a system to extract the graphical features in order to provide a clinical diagnosis before pathogenic testing and thus save critical time. Also, Zheng et al. [58] developed a model for automatic detection using 3D CT volumes. Bai et al. [59] established and evaluated an AI system for differentiating COVID-19 and other pneumonia from chest CT to assess radiologist performance. As they have discussed, distinguishing COVID-19 from normal lung or other lung diseases, such as cancer from chest CT, may be straightforward. However, a major difficulty in controlling the current pandemic is making out subtle radiologic differences between COVID-19 and pneumonia of other origins. A total of 521 patients with positive RT-PCR results for COVID-19 and abnormal chest CT findings were retrospectively identified from 10 hospitals. A total of 665 patients with non–COVID-19 pneumonia and definite evidence of pneumonia from chest CT were retrospectively selected from three hospitals.

Further, the authors have performed data augmentation dynamically during training and included flips, scaling, rotations, random brightness and contrast manipulations, random noise, and blurring. Training was performed for 20 epochs, where each epoch was defined as 16000 slices. A classification model was trained to distinguish between slices with and those without pneumonia-like findings (both COVID-19 and non–COVID-19). In more technical details, the EfficientNet B4 architecture was used for the pneumonia classification task. Each slice was stacked to three channels as the input of EfficientNet that used pretrained weights on ImageNet. EfficientNets with dense top fully connected layers were used. There were four fully connected layers of 256, 128, 64, and 32 neurons, respectively. Also, a fully connected layer with 16 neurons with batch normalization and a classification layer with sigmoid activation were added at the end of EfficientNet. Then, the slices were pooled using a two-layer fully connected neural network to make predictions at the patient level. **Figure 10** shows the proposed classification neural network model, while **Figure 11** demonstrates the model's flowchart.

Kumar et al. [60] proposed a framework that collects a big amount of data from various hospitals and trains a deep learning model over a decentralized network using the most recent information related to COVID-19 patients based on CT slices. The authors suggested the integration of blockchain and federated-learning technology that allow the collection of data from different hospitals without the leakage of data; a step that adds the necessary privacy to the model. They employed Google's Inception V3 network for feature extraction and tested various learning models (VGG, DenseNet, AlexNet, MobileNet, ResNet, and Capsule Network) in order

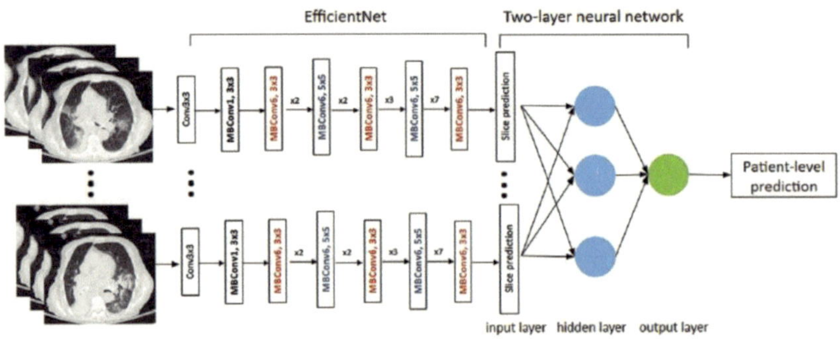

Figure 10.
Classification neural network model proposed by [59].

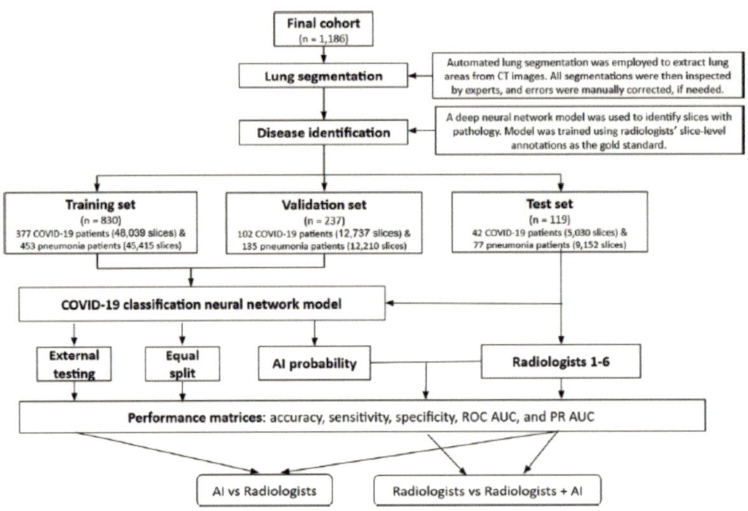

Figure 11.
The flowchart showing the AI model used to distinguish COVID-19 from non–COVID-19 pneumonia. (PR AUC = precision recall area under curve, ROC AUC = receiver operator characteristics area under the curve) [59].

to recognize the patterns from lung screening. They found that Capsule network achieved the best performance when compared to other learning models. **Figure 12** shows the suggested model in [60].

The Capsule network contains four layers: i) Convolutional layer, ii) Hidden layer, iii) PrimaryCaps layer, and iv) DigitCaps layer. A capsule is made when input features are in the lower layer. Each layer of the Capsule Network contains many capsules. To train it, the activation layer represents the parameters of the entity and computes the length of the Capsule network to re-compute the scores of the feature part. The capsule acts as a neuron. Capsule networks tend to describe an image at a component level and associate a vector with each component. The probability of the existence of a component is represented by the vectors lengths.

In federated learning, the hospitals keep their data private and share only the weights and gradients while blockchain technology is used to distribute the data securely among the hospitals. Federated learning was proposed by McMahan et al. [61] to learn from the shared model while protecting the privacy of data. In this context, the federated learning is used to secure data and aggregate the parameters from multiple organizations. As argued by the authors, since the volume of data is big, placing them on the blockchain directly with its limited storage space will be very expensive and resource-intensive. As such, a special data manipulation is needed. So, the hospital needs to store a transaction in the block to verify the ownership. The hospital data include the data type and size. It is noteworthy that federated learning does not affect the accuracy but it adds the privacy while sharing the data. Some selected 3D samples from the dataset are shown in **Figure 13**. The authors have claimed that the system sensitivity is 0.96, and its precision is 0.83. However, its specificity was not very attractive.

A simple 2D deep learning framework was developed in [62] to diagnose COVID-19 pneumonia based on a single chest CT image using transfer learning. For training and testing, the authors collected 3993 chest CT images of patients with COVID-19 pneumonia, other pneumonia and nonpneumonia diseases. These CT images were split into a training set and a testing set at a ratio of 8:2. After a simple preprocessing stage, three channels (256 × 256 × 3 pixels) were arranged in the input layer and fed into the pretrained model layers. In the pre-trained model layers, the authors included one of these four models (VGG16, ResNet-50, Inception-v3, and Xception). Each model comprises two parts: a convolutional base and a classifier. The convolutional base is composed of a stack of convolutional and pooling layers to generate features from the images. The role of the classifier is to categorize the image based on the extracted features. The activations from the pretrained model layers were fed into the additional layers. In the additional layers, the activations were first flattened and connected to two fully connected layers: one consisted of 32 nodes, and the other consisted of three nodes. Subsequently, the activations from the second fully connected layer were fed into a SoftMax layer, which provided the probability for each of the

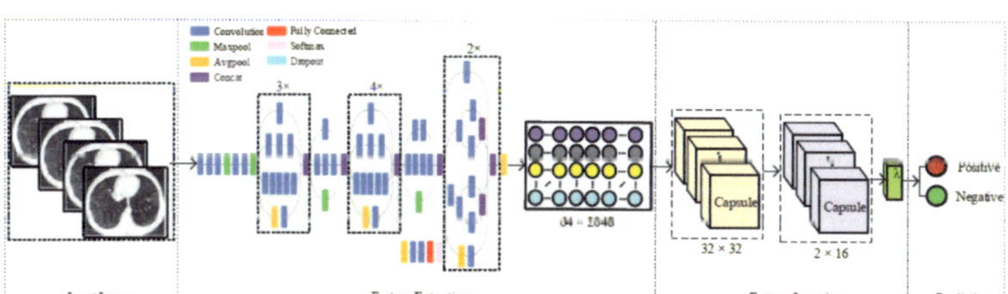

Figure 12.
COVID-19 model suggested by [60].

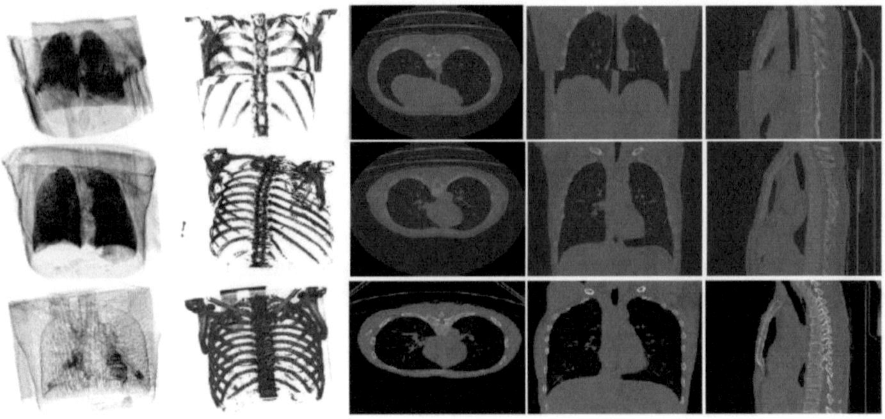

Figure 13.
Selected samples from [60].

classes (COVID-19, other pneumonia, and nonpneumonia). However, the study has several limitations as well. First, the testing dataset was obtained from the same sources as the training data set. This may raise issues of generalizability and overfitting of the models. Indeed, the authors have mentioned that the detection accuracy decreased when datasets from other published papers were used.

Song et al. [63] first extracted the main regions of the lungs and filled the blank of lung segmentation with the lung itself to avoid noise caused by different lung contours. Then, they extracted the top-K details in the CT images and obtained image-level predictions. Finally, the image-level predictions were combined to attain patient-level diagnoses. In the testing set, the model achieved an AUC of 0.95 and sensitivity of 0.96. In [64], Jin et al. built a method to accelerate the diagnosis speed. This model was trained using 312 images. Yet, it achieved a comparable performance with experienced radiologists. Among 1255 independent testing cases, the proposed deep-learning model achieved an accuracy of 94.98%, an AUC of 97.91%, a sensitivity of 94.06% and a specificity of 95.47%.

Zheng et al. [65] used U-Net to segment the lung area automatically, and then used 3DResNet for classification. As they have discussed, infectious areas can be distributed in many locations in the lungs, and automatic infectious area detection may not guarantee very high precision. Consequently, using the whole lung for classification is more convenient in practice. In [66], 3506 patients (468 with COVID-19, 1551 with CAP, and 1303 with non-pneumonia) were used to train and test another deep-learning model. The authors first used U-net to extract the whole lung region as an ROI. Afterwards, 2D RestNet50 was used for classifying COVID-19. Since each CT scanning includes multiple 2D image slices, the features in the last layer of ResNet50 were max pooled and combined for prediction. The model achieved an AUC of 0.96 in classifying COVID-19 from CAP and other pneumonia. Moreover, Shi et al. [67] included 1658 patients with COVID-19 and 1027 patients with CAP for classification. They first used VBNet to segment the infected areas, bilateral lungs, 5 lung lobes, and 18 lung pulmonary areas. Then, hand-crafted features such as location specific features, infection size, and radiomic features were extracted, and least absolute shrinkage and selection operator (LASSO) was used for feature selection. The method reached sensitivity of 0.9, specificity of 0.8, and accuracy of 0.88.

Further, Dong et al. [68] reviewed the use of various imaging characteristics and computing models that have been applied for the management of COVID-19. Specifically, they have quantitatively analyzed the use of imaging data for detection and treatment by means of CT, positron emission tomography - CT (PET/CT), lung ultrasound, and magnetic resonance imaging (MRI). PET is a sensitive but invasive imaging method that plays an important role in evaluating inflammatory and infectious pulmonary diseases, monitoring disease progression and treatment effect, and improving patient management. It is worth mentioning that lung ultrasound is a non-invasive, radiation-free, and portable imaging method that allows for an initial bedside screening of low-risk patients, diagnosis of suspected cases in the emergency room setting, prognostic stratification, and monitoring of the changes in pneumonia [69, 70].

Also, Jin et al. [71] presented their experience in building and deploying an AI system that analyzes CT images and detects COVID-19 pneumonia features. They obtained the image samples from five different hospitals with 11 different models of CT equipment to increase the model's generalization ability. The combined "segmentation - classification" model pipeline, can highlight the lesion regions in addition to the screening result. The model pipeline is divided into two stages: 3D segmentation and classification. The pipeline leverages a model library that contains different segmentation models such as FCN-8 s, U-Net, V-Net, and 3D U-Net++, as well as the classification models like dual path network (DPN-92), Inception-v3, ResNet-50, and Attention ResNet-50. As for the training set, in addition to the positive cases, they assembled a set of negative images of inflammatory and neoplastic pulmonary diseases, such as lobar pneumonia, lobster pneumonia, and old lesions. Their aim was enabling the model to learn different COVID-19 features from various resources. Using 1136 training cases (723 positives for COVID-19), they were able to achieve a sensitivity of 0.974 and a specificity of 0.922 on the test set. Further, the system achieved an AUC of 0.991. According to the authors, the system is in use in 16 hospitals and has a daily capacity of over 1300 screenings. Similarly, Jin et al. [72] performed an extensive statistical analysis on CT images diagnosed by COVID-19.

They evaluated the system on a large dataset with more than 10000 CT volumes from COVID-19, influenza-A/B, non-viral CAP and non-pneumonia subjects. **Figure 14** shows the workflow of the suggested system. The system consists of five key parts: (1) lung segmentation network, (2) slice diagnosis network, (3) COVID-infectious slice locating network, (4) visualization module for interpreting the vital region, and (5) image phenotype analysis module for features explanation. CT volumes were divided into different cohorts. The authors claimed that the system achieved an AUC of 97.81% on a test set of 3199 scans.

Jin et al. [73] drafted a guideline according to the guidelines methodology and general rules of WHO in relation to CT imaging. This guideline includes the epidemiological characteristics, disease screening, diagnosis, treatment, and nosocomial infection prevention. In this regard, the authors have discussed that the imaging findings vary with the patient's age, immunity status, disease stage at the time of scanning, underlying diseases, and drug interventions. The imaging features of lesions show: (1) dominant distribution (mainly subpleural, along the bronchial vascular bundles), (2) quantity (often more than three or more lesions, occasional single or double lesions), (3) shape (patchy, large block, nodular, lumpy, honeycomb-like or grid-like, cord-like, etc.), (4) density (mostly uneven, a paving stones-like change mixed with ground glass density and interlobular septal thickening, consolidation and thickened bronchial wall, etc.), and (5) concomitant signs variations (air-bronchogram, rare pleural effusion and mediastinal lymph nodes enlargement, etc.).

In addition, Chen et al. [74] constructed a system based on deep learning for detecting COVID-19 pneumonia from high resolution CT. For model development and validation, 46096 anonymous images from 106 admitted patients, including 51

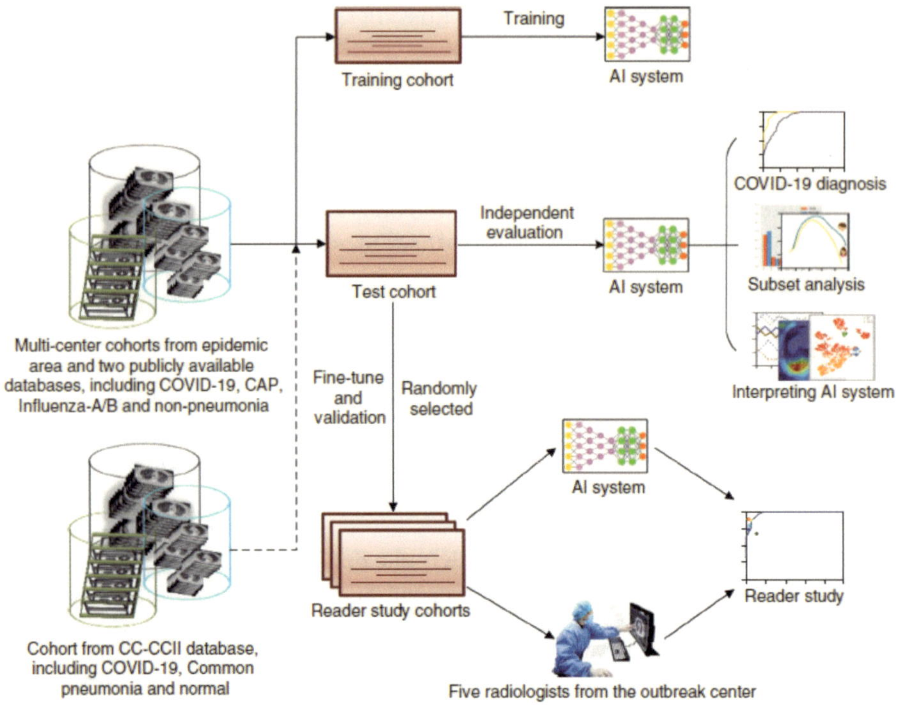

Figure 14.
The workflow of the AI system suggested in [72].

patients of laboratory confirmed COVID-19 pneumonia and 55 control patients of other diseases in Renmin Hospital of Wuhan University were retrospectively collected and processed. Twenty-seven consecutive patients who underwent CT scans were prospectively collected to evaluate and compare the efficiency of radiologists against COVID-19 pneumonia with that of the model. The authors have first filtered the images where 35355 images were selected and split into training and testing datasets. In more detail, the authors implemented UNet++ being a well-known architecture for medical image segmentation. They trained UNet++ to extract valid areas in CT images using 289 randomly selected CT images and tested it on other 600 randomly selected CT images. The training images were labeled with the smallest rectangle containing all valid areas. With the raw CT scan images taken as the input, and the labeled map from the expert as the output, UNet++ was used to train in an image-to-image manner. The model successfully extracted valid areas in 600 images from the testing set with an accuracy of 100%. Based on system performance, the authors constructed a cloud-based platform to provide a worldwide assistance for detecting COVID-19 pneumonia [75].

In [76], Vinod and Prabaharan have elaborated a methodology that helps identifying COVID-19 infected people among the normal individuals by utilizing CT scan. The image diagnosis tool utilizes decision tree classifier for finding Coronavirus infected person. The percentage accuracy of an image was analyzed in terms of precision, recall score and F1 score. Moreover, Gieraerts et al. [77] hypothesized that the use of semi-automated AI may allow for more accurate patient detection. They assessed COVID-19 patients who underwent chest CT by conventional visual and AI-based quantification of lung injury. They also studied the impact of chest CT variability in determining the potential response to novel antiviral therapies. In their study, 250 consecutive patients with clinical suspicion of COVID-19 pneumonia were tested with both RT-PCR and CT within a 2-hour interval of hospital admission. Epidemiological, demographic, clinical, and laboratory data at admission were obtained from the electronic patient management system.

In Zhang et al. [78], 4695 manually annotated CT slices were used to build seven classes, including background, lung field, consolidation, ground-glass opacity, pulmonary fibrosis, interstitial thickening, and pleural effusion. After a comparison between different semantic segmentation approaches, the authors selected DeepLabv3 as the segmentation detection backbone. The diagnostic system was based on a neural network fed by the lung-lesion maps. The results showed a COVID-19 diagnostic accuracy of 92.49% when tested on 260 subjects. In Bai et al. [79], a direct classification of COVID-19 specific pneumonia versus other etiologies was performed using an EfficientNet B5 network followed by a two-layer fully connected network to pool the information from multiple slices and provide a patient-level diagnosis. This system yielded 96% accuracy on a testing set of 119 subjects compared to an average accuracy of 85% for six radiologists.

Also, Ying et al. [80] used 2D slices including lung regions segmented by OpenCV. Fifteen slices of complete lungs were derived from each 3D chest CT images, and each 2D slice was used as the input to the system. A pretrained ResNet-50 was used and the Feature Pyramid Network (FPN) was added to extract the top-K details from each image. An attention module was coupled to learn the important details. Chest CT images from 88 patients with COVID-19, 101 patients with bacterial pneumonia, and 86 healthy persons were used. The model achieved an accuracy of 86% for pneumonia classification (COVID-19 or bacterial pneumonia), and an accuracy of 94% for pneumonia diagnosis (COVID-19 or healthy). Wang et al. [81] used 1065 chest CT scan images of COVID-19 patients to build a classifier using InceptionNet. They reported an accuracy of 89.5%, a specificity of 0.88, and a sensitivity of 0.87. In [82], different deep learning approaches (VGG16, InceptionResNetV2, ResNet50, VGG19, MobilenetV2, and NasNetMobile) have been modified and tested on 400 CT scan images. The results have shown that NasNetMobile outperformed all other models in terms of accuracy (81.5% –95.2%). On the other hand, Mucahid et al. [83] used classical feature extraction techniques for COVID-19 detection. For example, they have implemented gray level co-occurrence matrices (GLCM), local directional pattern (LDP), gray-level run length matrix (GLRLM), and discrete wavelet transform (DWT). They reported an accuracy of 99.68% in the best configuration settings.

Modegh et al. [84] proposed a system to distinguish healthy people, patients with COVID-19, and patients with other pneumonia diseases from axial lung CT-scan images. The general workflow for the proposed model is shown in **Figure 15**. The Ground Glass Opacity Axial (GGOA) CT-scan images are preprocessed and the lobes of the lungs are detected and extracted from the axial slices. The images of the left and right lobes of all the slices are then fed into two deep CNNs, one for calculating the probability of being diseased versus healthy, and the other for calculating the probability of diagnosis to be COVID-19 versus other diseases. In addition, the system detects the infected areas in the lung images. At the end, the probabilities assigned to the lobes are aggregated to make a final decision.

Figure 16 shows the model used for calculating the probability of each slice lobe being infected. The model was evaluated on a dataset of 3359 samples from 6 different medical centers and achieved sensitivities of 97.8% and 98.2%, and specificities of 87% and 81% in distinguishing normal cases from the diseased and COVID-19 from other diseases, respectively. Authors in [85] examined the effect of generalizability of the deep learning models, given the heterogeneous factors in training datasets such as patient demographics and pre-existing clinical conditions. The examination was done by evaluating the classification models trained to identify COVID-19 positive patients on 3D CT datasets from different

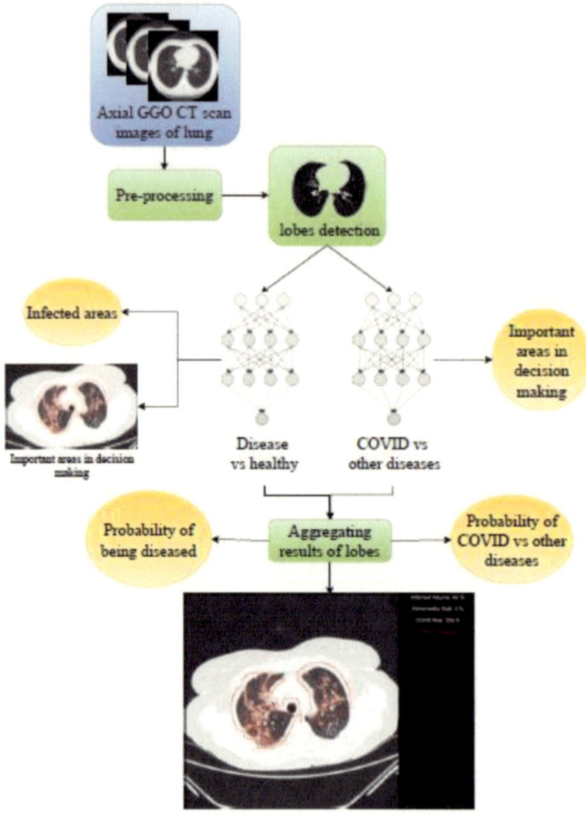

Figure 15.
The general workflow for the interpretable COVID-19 detection proposed in [84].

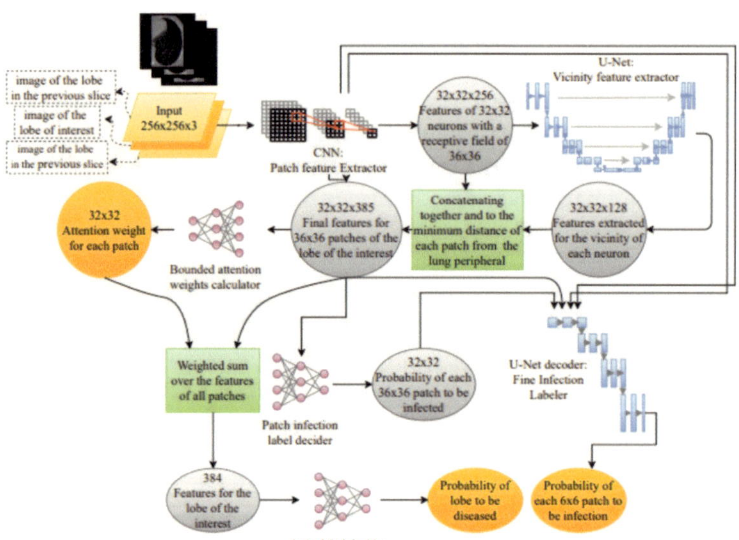

Figure 16.
The deep model used for calculating the probability of each slice lobe [84].

countries: UT Southwestern (UTSW), CC-CCII Dataset (China), COVID-CTset (Iran), and MosMedData (Russia). The data were divided into two classes: COVID-19 positive and COVID-19 negative patients. The models trained on a single dataset achieved accuracy/AUC values of 0.87/0.826 (UTSW), 0.97/0.988 (CC-CCCI), and 0.86/0.873 (COVID-CTset) when evaluated on their own dataset.

In addition, Shah et al. [86] developed a deep learning network (CTnet-10) for COVID-19 classification. The model is fed with an input image of size 128 × 128 × 3. It passes through two convolutional blocks of dimensions 126 × 126 × 32, 124 × 124 × 32 respectively. Then it passes through a max-pooling of dimension 62 × 62 × 32 followed by two convolutional layers of dimensions 60 × 60 × 32, 58 × 58 × 32 respectively. Then, it is passed through a pooling layer of dimension 29 × 29 × 32, a flattened layer of 26912 neurons, and dropout layers of 256 neurons. After that, it is passed through a dense layer of a single neuron, where the CT scan image is classified as COVID-19 positive or negative. The system achieved an accuracy of 82.1%. The CTnet-10 model architecture is shown in **Figure 17**.

VB-Net, a deep learning network, was developed by Shan et al. [87] to quantify longitudinal changes in the follow-up CT scans of COVID-19 patients, and to explore the quantitative lesion distribution. VB-Net is a modified 3D CNN that consists of two paths. The first is a contracting path including down-sampling and convolution operations to extract global image features. The second is an expansive path including up-sampling and convolution operations to integrate fine-grained image features. Compared with V-Net, the VB-Net is much faster. The system not only performs auto-contouring of infection regions, but also accurately estimates their shapes, volumes and percentage of infection (POI) in CT scans. In addition, it measures the severity of COVID-19 and the distribution of infection within the lung. The accurate segmentation provides quantitative information that is necessary

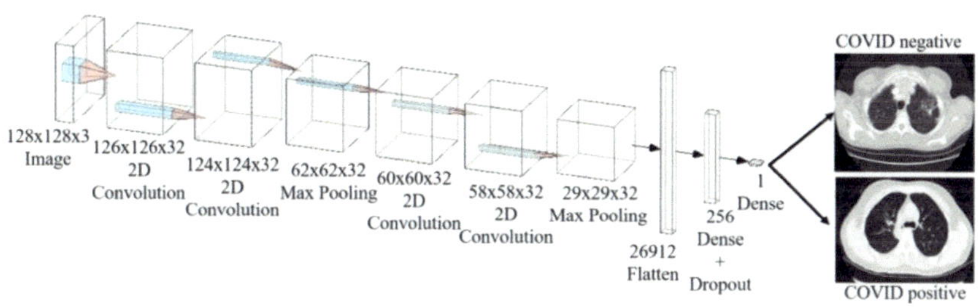

Figure 17.
CTnet-10 model architecture [86].

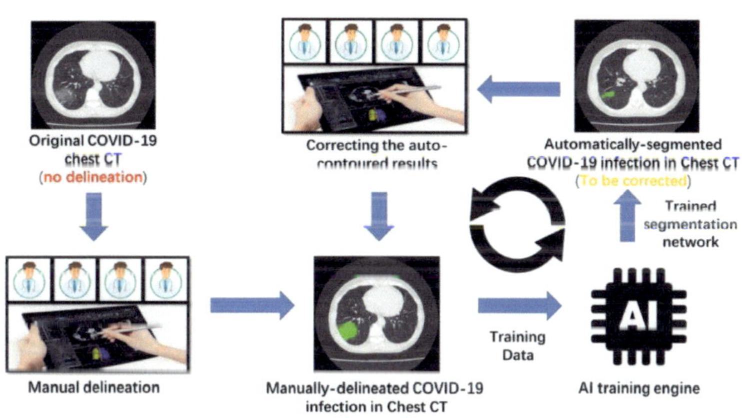

Figure 18.
The human-in-the-loop workflow [87].

to track disease progression and analyze longitude changes of COVID-19 during the entire treatment period. After segmentation, various metrics are computed to quantify the infection, including the volumes of infection in the whole lung, and the volumes of infection in each lobe and each bronchopulmonary segment.

The system was trained using 249 COVID- 19 patients, and validated using 300 new COVID-19 patients. To accelerate the manual delineation of CT images for training, a human-in-the-loop (HITL) strategy (shown in **Figure 18**) was adopted to assist radiologists to refine automatic annotation of each case. To evaluate the performance of the system, the Dice similarity coefficient, the differences of volume and the POI were calculated between the automatic and the manual

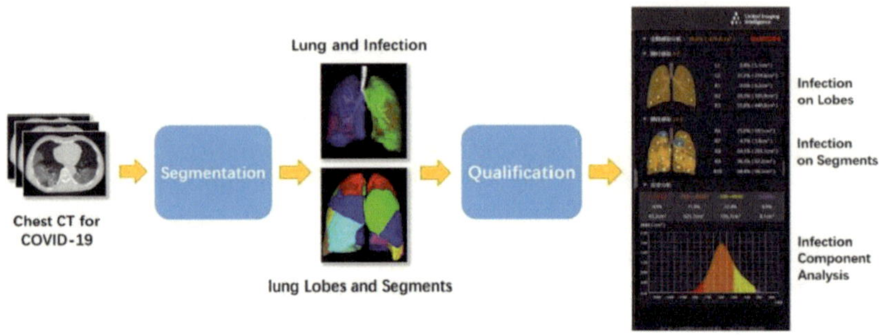

Figure 19.
The proposed pipeline for quantifying COVID-19 infection [87].

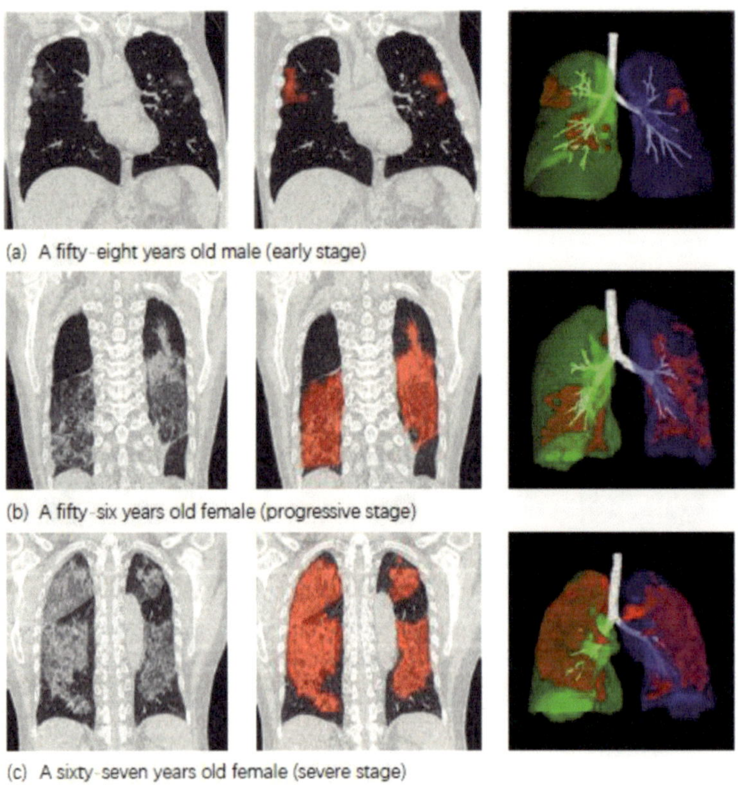

Figure 20.
Typical infection segmentation results of CT scans for three COVID-19 patients. Rows 1–3: Early, progressive and severe stages. Columns 1–3: CT image, CT images overlaid with segmentation, and 3D surface rendering of segmented infections [87].

segmentation results on the validation set. The system yielded a Dice similarity coefficient of 91.6% and a mean POI estimation error of 0.3% for the whole lung on the validation dataset. Moreover, the proposed HITL strategy reduced the delineation time to 4 minutes after 3 iterations of model updating, compared to the cases of fully manual delineation that often take 1 to 5 hours. **Figure 19** shows the pipeline for quantifying COVID-19 infection, whereas **Figure 20** shows typical infection segmentation results of CT scans for three COVID-19 patients.

4. Discussion

The specific nature of COVID-19 pandemic requires strong coordination of connected data, people, and systems to facilitate worldwide collaboration in fighting against it. From this study, we notice that healthcare stakeholders are not using the same systems, data formats, or standards. This can obstruct the ability to identify the possible trends of solutions to the pandemic related challenges and develop interventions among the associated efforts. Public health researchers, epidemiologists, and government officials need to be connected via integrated systems with connected data to understand the evolving pandemic better and make collective decisions on addressing this crisis. An important key action that needs to be taken to ensure best possible fight of the current (or even future) pandemic(s) is to catalyze scaling up the implementation of AI and ML in health sector.

Our study specifically suggests the following important issues that need to be addressed intensely and more efficiently:

- Due to the low contrast of the infection regions in some images and large variation of both shape and position across different patients, delineating the infection regions from the chest images is very challenging [88, 89]. Researchers are challenged to investigate AI techniques that may help in this direction.

- Although CT provides rich pathological information, it was noticed that in some cases only qualitative evaluation has been provided and precise changes across follow-up CT scans are often ignored. Actually, contouring infection regions in the Chest CT is necessary for quantitative assessment [90, 91]. As such, more investigation is required in this area.

- Quantifying imaging metrics and correlating them with syndromes, epidemiology, and treatment responses is essential and could further reveal insights about imaging markers and findings toward improved diagnosis and treatment for COVID-19 [92, 93].

- Some segmentation models were trained using imperfect ground-truth data. This could be improved by using 3D segmentation networks and adopting precise ground-truth annotated by experts [94, 95].

- The images in some datasets were acquired from different devices. This situation makes the classification process a kind of challenging. This could be explained by the fact that some gray-levels in one image represent certain Coronavirus infected levels, and the same gray-levels in another image may represent different levels (may lead to different decisions) [96, 97].

- Some systems were developed to quantify infections only, and it may not be applicable for quantifying other pneumonia, for example bacterial pneumonia [98, 99].

- Many datasets were collected in one center, which may not be representative of all COVID-19 patients in other geographic areas. The generalization of the deep learning system needs to be further validated on multi-center datasets [100].

- Experimental evidence is presented on datasets of hundreds (maybe thousands). However, the need is to go to real world settings, in which databases consist of hundreds of thousands and even more cases, with large variability [101].

5. Conclusion

The COVID-19 is a disease that has spread all over the world. This work attempted to provide a detailed study on how the AI and ML can help in various domains related to COVID-19, specifically in the area of disease diagnosis using CT imagery. In pursuing so, we considered, examined, discussed, and analyzed comprehensive studies and detailed researches proposed by intellectuals and researchers from various scientific communities and international academic institutions. Deep learning techniques and algorithms have shown immense appearances and implementations in different domains and COVID-19 related applications.

AI solutions have the potential to detect and analyze any abnormalities in health conditions in general, and related to COVID-19 in particular. The study has demonstrated that AI solutions can assist in differentiating Coronavirus patients from those who do not have the disease and can provide support in tracking disease progression. AI technology can potentially support radiologists in the triage, quantification, and trend analysis of data. For example, if the developed technique suggests a significantly increased likelihood of disease, then the case can be flagged for further review by a radiologist or clinician for possible treatment/quarantine. Moreover, AI technology can provide a consistent method for rapid evaluation of high volumes of diagnostic that can reliably exclude images which are negative for findings associated with COVID-19. This decreases the volume of cases passing through to the radiologist without overlooking positive cases. Using AI solutions, progression and regression of positive findings could be monitored more quantitatively and regularly. This could lead to more effective identification and containment of early cases. The study also discovered that a critical existing impediment to effective AI implementation is the lack of COVID-19-related clinical data that can be maintained and processed into easily accessible databases. Integrating COVID-19-related clinical data with existing biobanks, as well as pre-existing patient data, could help bioinformaticians and computational scientists develop a faster and more practical way to useful data-mining.

It is our hypothesis that AI and ML tools can leverage the ability to modify and adapt existing models and combine them with initial clinical understanding to address COVID-19 challenges and the new emerging strains or mutations of the virus. Researchers and scientists are hoping to speed up the development of extremely precise and useful AI, ML, and deep learning technologies to combat COVID-19. If our societies could not reach the best expected AI solutions during this pandemic, we strongly anticipate that AI technology will be of greater help with the next pandemic.

Acknowledgements

The author would like to express his gratitude and grateful appreciation to the Kuwait Foundation for the Advancement of Sciences (KFAS) for financially supporting this project. The project was fully funded by KFAS under project code: PN20-13NH-03.

Author details

Naser Zaeri
Faculty of Computer Studies, Arab Open University, Kuwait

*Address all correspondence to: n.zaeri@aou.edu.kw

References

[1] Sousa, R. T., O. Marques, Iwens I. G. Sene, Anderson S. Soares and L. L. G. D. Oliveira. "Comparative performance analysis of machine learning classifiers and dimensionality reduction algorithms in detection of childhood pneumonia." (2013)

[2] https://www.who.int/emergencies/diseases/novel-coronavirus-2019, 2021

[3] B. Wang, R. Li, Z. Lu, Y. Huang, Does comorbidity increase the risk of patients with COVID-19: evidence from meta-analysis, Aging (Albany NY) 12 (7) (2020) 6049

[4] S. Szymkowski. COVID-19 Shut Down 93% of All US Auto Production. Roadshow, 2020. [Online]. Available: https://www.cnet.com/roadshow/news/covid-19-shut-down-us-autoproductio%n-coronavirus/

[5] Ng et al. Imaging profile of the covid-19 infection: Radiologic findings and literature review. Radiology: Cardiothoracic Imaging, 2(1), 2020

[6] Pham, Q.-V.; Nguyen, D.C.; Huynh-The, T.; Hwang, W.-J.; Pathirana, P.N. Artificial intelligence (AI) and big data for coronavirus (COVID-19) pandemic. 2020, 2020040383

[7] A. Bernheim et al., "Chest CT Findings in Coronavirus Disease-19 (COVID-19): Relationship to Duration of Infection," Radiology, pp. 200463-200463, 2020-Feb-20 2020, doi: 10.1148/radiol.2020200463.

[8] S. Ardabili, *A. Mosavi*, S. S. Band and A. R. Varkonyi-Koczy, "Coronavirus Disease (COVID-19) Global Prediction Using Hybrid Artificial Intelligence Method of ANN Trained with Grey Wolf Optimizer," 2020 IEEE 3rd International Conference and Workshop in Óbuda on Electrical and Power Engineering (CANDO-EPE), Budapest, Hungary, 2020, pp. 000251-000254, doi: 10.1109/CANDO-EPE51100.2020.9337757

[9] Tang, L, Tian, C, Meng, Y, Xu, K., "Longitudinal evaluation for COVID-19 chest CT disease progression based on Tchebichef moments," International Journal of Imaging Systems and Technology, pp. 1– 8, 2021, https://doi.org/10.1002/ima.22583

[10] Reza Mohammadi, Iman Shokatian, Mohammad Salehi, Hossein Arabi, Isaac Shiri, Habib Zaidi, "Deep learning-based auto-segmentation of organs at risk in high-dose rate brachytherapy of cervical cancer," Radiotherapy and Oncology, Volume 159, 2021, Pages 231-240, ISSN 0167-8140, https://doi.org/10.1016/j.radonc.2021.03.030

[11] D. -P. Fan et al., "Inf-Net: Automatic COVID-19 Lung Infection Segmentation From CT Images," in IEEE Transactions on Medical Imaging, vol. 39, no. 8, pp. 2626-2637, 2020, doi: 10.1109/TMI.2020.2996645

[12] 7.Y. Feng et al., "COVID-19 with different severities: A multicenter study of clinical features", Amer. J. Respir. Crit. Care Med., vol. 201, no. 11, pp. 1380-1388, 2020

[13] Anand Rao, Global Leader, Artificial Intelligence, PwC and Kay Firth-Butterfield, Head, Artificial Intelligence and Machine Learning, World Economic Forum, "3 ways COVID-19 is transforming advanced analytics and AI," 23 Jul 2020

[14] Zhongxiang Chen, Jun Yang and Binxiang Dai, "Forecast Possible Risk for COVID-19 Epidemic Dissemination under Current Control Strategies in Japan," Int. J. Environ. Res. Public Health 2020, 17, 3872; doi:10.3390/ijerph17113872

[15] EIT-a body of the European Union "Transforming healthcare with AI: The impact on the workforce and organisations,", 2020

[16] Holzinger, A. et al., "What do we need to build explainable AI systems for the medical domain?", arXiv:1712. 09923, 2017.

[17] Mohammad (Behdad) Jamshidi et al., "Artificial Intelligence and COVID-19: Deep Learning Approaches for Diagnosis and Treatment," IEEE Special Section On Emerging Deep Learning Theories And Methods For Biomedical Engineering, June 24, 2020

[18] S. Dodge and L. Karam, "Understanding how image quality affects deep neural networks," International Conference on Quality of Multimedia Experience (QoMEX), 2016 http://image-net.org/challenges/LSVRC/2010/results; http://image-net.org/challenges/LSVRC/2017/results.

[19] Daniel Nelson, "Baidu Beats Out Google And Microsoft, Creates New Technique For Language Understanding", Unite. AI, 28 December 2019, https://www.unite.ai/baidu-beats-out-google-and-microsoft-creates-new-technique-for-languageunderstanding/.

[20] "Can science be automated?" ScienceDaily, April 2019, https://www.sciencedaily.com/releases/2019/04/190418105730.htm

[21] Hinton, G., Deep learning—a technology with the potential to transform health care. Jama, 2018, 320(11), pp.1101-1102; Gottesman, O., et al., "Guidelines for reinforcement learning in healthcare". Nature Medicine, 2019, 25(1), pp.16-18

[22] Soroush Nasiriany, Garrett Thomas, William Wang, Alex Yang, Jennifer Listgarten, Anant Sahai, "A Comprehensive Guide to Machine Learning," Department of Electrical Engineering and Computer Sciences University of California, Berkeley, 2019

[23] Garrett Thomas, "Mathematics for Machine Learning" Department of Electrical Engineering and Computer Sciences, University of California, Berkeley, 2018

[24] J. Wan , D. Wang , S.C. Hoi , P. Wu , J. Zhu , Y. Zhang , J Li , Deep learning for con- tent-based image retrieval: a comprehensive study, in: Proceedings of the 22nd ACM international conference on Multimedia, 2014 Nov 3, pp. 157-166

[25] M.A . Wani , F.A . Bhat , S. Afzal , A .I Khan , Advances in Deep Learning, Springer, 2020

[26] Nicolas Coudray, Paolo Santiago Ocampo, Theodore Sakellaropoulos, Navneet Narula, Matija Snuderl, David Fenyo, Andre L Moreira, Narges Razavian, and Aristotelis Tsirigos. Classification and mutation prediction from non–small cell lung cancer histopathology images using deep learning. Nature medicine, 24(10):1559-1567, 2018

[27] Kaiming He, Xiangyu Zhang, Shaoqing Ren, and Jian Sun. Delving deep into rectifiers: Surpassing human-level performance on imagenet classification. In Proceedings of the IEEE international conference on computer vision, pages 1026-1034, 2015

[28] Ng MY, Lee EY, Yang J, et al. Imaging Profile of the COVID-19 Infection: Radiologic Findings and Literature Review. Radiol Cardiothorac Imaging 2020;2(1):e200034

[29] Pan F, Ye T, Sun P, et al. Time Course of Lung Changes On Chest CT During Recovery From 2019 Novel Coronavirus (COVID 19) Pneumonia. Radiology 2020 Feb 13:200370

[30] Chung M, Bernheim A, Mei X, et al. CT Imaging Features of 2019 Novel

Coronavirus (2019-nCoV). Radiology 2020;295(1):202-207.

[31] Song F, Shi N, Shan F, et al. Emerging 2019 Novel Coronavirus (2019-nCoV) Pneumonia. Radiology 2020;295(1):210-217.

[32] Pan Y, Guan H, Zhou S, et al. Initial CT findings and temporal changes in patients with the novel coronavirus pneumonia (2019-nCoV): a study of 63 patients in Wuhan, China. Eur Radiol 2020

[33] Bernheim A, Mei X, Huang M, et al. Chest CT Findings in Coronavirus Disease- 19 (COVID-19): Relationship to Duration of Infection. Radiology 2020 Feb 20:200463

[34] Bai HX, Hsieh B, Xiong Z, et al. Performance of radiologists in differentiating COVID- 19 from viral pneumonia on chest CT. Radiology 2020 Mar 10:200823

[35] Peng, Q.-Y., Wang, X.-T. & Zhang, L.-N. Findings of lung ultrasonography of novel corona virus pneumonia during the 2019-2020 epidemic. Intensive Care Med 1-2 (2020) doi:10.1007/s00134-020-05996-6

[36] Huang, Y. et al. A Preliminary Study on the Ultrasonic Manifestations of Peripulmonary Lesions of Non-Critical Novel Coronavirus Pneumonia (COVID-19). https://papers.ssrn.com/abstract=3544750 (2020) doi:10.2139/ssrn.3544750

[37] Bertalmio, M., Bertozzi, A. L. & Sapiro, G. Navier-stokes, fluid dynamics, and image and video inpainting. in Proceedings of the 2001 IEEE Computer Society Conference on Computer Vision and Pattern Recognition. CVPR 2001 vol. 1, 2001

[38] Buda, N., Segura-Grau, E., Cylwik, J. & Wełnicki, M. Lung ultrasound in the diagnosis of COVID-19 infection - A case series and review of the literature. Advances in Medical Sciences 65, 378-385 (2020)

[39] Brahier, T. et al. Lung ultrasonography for risk stratification in patients with COVID-19: a prospective observational cohort study. Clinical Infectious Diseases (2020) doi:10.1093/cid/ciaa1408

[40] Karagöz, A., Saglam, C., Demirbas, H. B., Korkut, S. & Ünlüer, E. E. Accuracy of Bedside Lung Ultrasound as a Rapid Triage Tool for Suspected Covid-19 Cases. Ultrasound Quarterly 36, 339-344 (2020)

[41] H. Zhang , G. Chen , X. Li , Resource management in cloud computing with optimal pricing policies, Comput. Syst. Sci. Eng. 34 (4) (2019) 249-254

[42] M.J. Van Der Donckt , D. Weyns , M.U. Iftikhar , R.K. Singh , Cost-benefit analysis at runtime for self-adaptive systems applied to an internet of things application., in: Proceedings of the ENASE, 2018, pp. 478-490

[43] D. Gupta , O. Kayode , S. Bhatt , M. Gupta , A.S. Tosun , Learner's Dilemma: IoT devices training strategies in collaborative deep learning, IEEE 6th World Forum Internet Things (WF-IoT) (2020)

[44] Bassetti, M., Kollef, M. H. & Timsit, J. F. Bacterial and fungal superinfections in critically ill patients with COVID-19. Intensive Care Med. 46, 2071-2074 (2020)

[45] Lin Li,1b, Lixin Qin, Zeguo Xu, Youbing Yin, Xin Wang, Bin Kong, Junjie Bai, Yi Lu, Zhenghan Fang, Qi Song, Kunlin Cao, Daliang Liu, Guisheng Wang, Qizhong Xu, Xisheng Fang, Shiqin Zhang, Juan Xia, Jun Xia, "Artificial Intelligence Distinguishes COVID-19 from Community Acquired Pneumonia on Chest CT", Radiology

[46] Chuangsheng Zheng, Xianbo Deng, Qiang Fu1, Qiang Zhou, Jiapei Feng, Hui Ma, Wenyu Liu, Xinggang Wang, "Deep Learning-based Detection for COVID-19 from Chest CT using Weak Label," medRxiv 2020.03.12.20027185; doi: https://doi.org/10.1101/2020.03.12.20027185

[47] O. Gozes, M. Frid-Adar, H. Greenspan, P. D. Browning, H. Zhang, W. Ji, et al., "Rapid AI development cycle for the coronavirus (covid-19) pandemic: Initial results for automated detection & patient monitoring using deep learning ct image analysis," arXiv:2003.05037, 2020

[48] Barstugan, M., Ozkaya, U., and Ozturk, S., "Coronavirus (COVID-19) Classification using CT Images by Machine Learning Methods", arXiv e-prints, 2020

[49] Damiano Caruso, Marta Zerunian, Michela Polici, Francesco Pucciarelli, Tiziano Polidori, Carlotta Rucci, Gisella Guido, Benedetta Bracci, Chiara De Dominicis, Andrea Laghi, "Chest CT Features of COVID-19 in Rome, Italy," Radiology: Volume 296: Number 2—August 2020

[50] Salehi S, Abedi A, Balakrishnan S, Gholamrezanezhad A. Coronavirus Disease 2019 (COVID-19): A Systematic Review of Imaging Findings in 919 Patients. AJR Am J Roentgenol 2020 Mar 14:1-7

[51] Chung M, Bernheim A, Mei X, et al. CT Imaging Features of 2019 Novel Coronavirus (2019-nCoV). Radiology 2020;295(1):202-207.

[52] Xiaowei Xu, Xiangao Jiang, Chunlian Mac, Peng Dud, Xukun Li, Shuangzhi Lv, Liang Yu, Qin Ni, Yanfei Chen, Junwei Su, Guanjing Lang, Yongtao Li, Hong Zhao, Jun Liu, Kaijin Xu , Lingxiang Ruan, Jifang Sheng, Yunqing Qiu, Wei Wua, Tingbo Liang, Lanjuan Li, "A Deep Learning System to Screen Novel Coronavirus Disease 2019 Pneumonia," Engineering, 2020

[53] Maria Paola Belfiore, Fabrizio Urraro, Roberta Grassi, Giuliana Giacobbe, Gianluigi Patelli, Salvatore Cappabianca, Alfonso Reginelli, "Artificial intelligence to codify lung CT in Covid-19 patients," La radiologia medica (2020) 125:500-504, https://doi.org/10.1007/s11547-020-01195-x

[54] Xueyan Mei et al., "Artificial intelligence–enabled rapid diagnosis of patients with COVID-19," Nature Medicine, VOL 26, August 2020, pp. 1224-1228, www.nature.com/naturemedicine

[55] Kuruvilla, J., Gunavathi, K.: Lung cancer classification using neural networks for ct images. Computer methods and programs in biomedicine 113(1), 202-209 (2014)

[56] Brunese, L., Mercaldo, F., Reginelli, A. & Santone, A. Explainable Deep Learning for Pulmonary Disease and Coronavirus COVID-19 Detection from X-rays. Computer Methods and Programs in Biomedicine 196, 105608 (2020)

[57] Wikramaratna, P. S., Paton, R. S., Ghafari, M. & Lourenço, J. Estimating the false-negative test probability of SARS-CoV-2 by RT-PCR. medRxiv 2020.04.05.20053355 (2020) doi:10.1101/2020.04.05.20053355

[58] Born, J. et al. POCOVID-Net: Automatic Detection of COVID-19 From a New Lung Ultrasound Imaging Dataset (POCUS). arXiv:2004.12084 [cs, eess] (2020)

[59] Harrison X. Bai, Robin Wang, Zeng Xiong, Ben Hsieh, Ken Chang, Kasey Halsey, Thi My Linh Tran, Ji Whae Choi, Dong-Cui Wang, Lin-Bo Shi, Ji Mei, Xiao-Long Jiang, Ian Pan, Qiu-Hua Zeng, Ping-Feng Hu, Yi-Hui Li, Fei-Xian Fu, Raymond Y. Huang, Ronnie Sebro,

Qi-Zhi Yu, Michael K. Atalay, Wei-Hua Liao, "Artificial Intelligence Augmentation of Radiologist Performance in Distinguishing COVID-19 from Pneumonia of Other Origin at Chest CT," Radiology 2020; 296:E156–E165, https://doi.org/10.1148/radiol.2020201491, Volume 296: Number 3—September 2020

[60] Rajesh Kumar, Abdullah Aman Khan, Sinmin Zhang, WenYong Wang, Yousif Abuidris, Waqas Amin , and Jay Kumar, "Blockchain-Federated-Learning and Deep Learning Models for COVID-19 detection using CT Imaging," JOURNAL OF LATEX CLASS FILES, VOL. 14, NO. 8, AUGUST 2020

[61] Bourcier, J.-E. et al. Performance comparison of lung ultrasound and chest x-ray for the diagnosis of pneumonia in the ED. The American Journal of Emergency Medicine 32, 115-118 (2014)

[62] Hoon Ko, Heewon Chung, Wu Seong Kang, Kyung Won Kim, Youngbin Shin, Seung Ji Kang, Jae Hoon Lee, Young Jun Kim, Nan Yeol Kim, Hyunseok Jung, Jinseok Lee, "COVID-19 Pneumonia Diagnosis Using a Simple 2D Deep Learning Framework With a Single Chest CT Image: Model Development and Validation," JOURNAL OF MEDICAL INTERNET RESEARCH, 2020, vol. 22, iss. 6, e19569

[63] S. Ardabili, *A. Mosavi*, S. S. Band and A. R. Varkonyi-Koczy, "Coronavirus Disease (COVID-19) Global Prediction Using Hybrid Artificial Intelligence Method of ANN Trained with Grey Wolf Optimizer," 2020 IEEE 3rd International Conference and Workshop in Óbuda on Electrical and Power Engineering (CANDO-EPE), Budapest, Hungary, 2020, pp. 000251-000254, doi: 10.1109/CANDO-EPE51100.2020.9337757

[64] F. Gao, K. Deng and C. Hu, "Construction of TCM Health Management Model for Patients with Convalescence of Coronavirus Disease Based on Artificial Intelligence," 2020 International Conference on Big Data and Informatization Education (ICBDIE), Zhangjiajie, China, 2020, pp. 417-420, doi: 10.1109/ICBDIE50010.2020.00104

[65] S. Tabik et al., "COVIDGR Dataset and COVID-SDNet Methodology for Predicting COVID-19 Based on Chest X-Ray Images," in IEEE Journal of Biomedical and Health Informatics, vol. 24, no. 12, pp. 3595-3605, Dec. 2020, doi: 10.1109/JBHI.2020.3037127

[66] D. -P. Fan et al., "Inf-Net: Automatic COVID-19 Lung Infection Segmentation From CT Images," in IEEE Transactions on Medical Imaging, vol. 39, no. 8, pp. 2626-2637, Aug. 2020, doi: 10.1109/TMI.2020.2996645

[67] F. Shi et al., "Review of Artificial Intelligence Techniques in Imaging Data Acquisition, Segmentation, and Diagnosis for COVID-19," in IEEE Reviews in Biomedical Engineering, vol. 14, pp. 4-15, 2021, doi: 10.1109/RBME.2020.2987975

[68] Di Dong, Zhenchao Tang, Shuo Wang, Hui Hui, Lixin Gong, Yao Lu, Zhong Xue, Hongen liao, Fang Chen, Fan Yang, Ronghua Jin, Kun Wang, Zhenyu Liu, Jingwei Wei, Wei Mu, Hui Zhang, Jingying Jiang, Jie Tian, Hongjun Li, "The role of imaging in the detection and management of COVID-19: a review," IEEE DOI 10.1109/RBME.2020.2990959

[69] Beovic, B. et al. Antibiotic use in patients with COVID-19: A 'snapshot' Infectious Diseases International Research Initiative (ID-IRI) survey. J. Antimicrob. Chemother. 75, 3386-3390 (2020)

[70] Tabassum, N., Zhang, H. & Stebbing, J. Repurposing Fostamatinib to combat SARS-CoV-2-induced acute

lung injury. Cell Reports Med. 1, 100145 (2020)

[71] Shuo Jin et al., "AI-assisted CT imaging analysis for COVID-19 screening: Building and deploying a medical AI system in four weeks," medRxiv preprint doi: https://doi.org/10.1101/2020.03.19.20039354

[72] Cheng Jin, Weixiang Chen, Yukun Cao, Zhanwei Xu, Zimeng Tan, Xin Zhang, Lei Deng, Chuansheng Zheng, Jie Zhou, Heshui Shi, Jianjiang Feng, "Development and evaluation of an artificial intelligence system for COVID-19 diagnosis,", NATURE COMMUNICATIONS, (2020) 11:5088, https://doi.org/10.1038/s41467-020-18685-1, www.nature.com/naturecommunications

[73] Jin et al., "A rapid advice guideline for the diagnosis and treatment of 2019 novel coronavirus (2019-nCoV) infected pneumonia (standard version);"Military Medical Research (2020) 7:4 https://doi.org/10.1186/s40779-020-0233-6

[74] Jun Chen et al. "Deep learning-based model for detecting 2019 novel coronavirus pneumonia on high-resolution computed tomography: a prospective study," medRxiv preprint doi: https://doi.org/10.1101/2020.02.25.20021568

[75] Chen J, Wu L, Zhang J, Zhang L, Gong D, Zhao Y, Chen Q, Huang S, Yang M, Yang X, Hu S, Wang Y, Hu X, Zheng B, Zhang K, Wu H, Dong Z, Xu Y, Zhu Y, Chen X, Zhang M, Yu L, Cheng F, Yu H, Open-access website available at: http://121.40.75.149/znyx-ncov/index

[76] Dasari Naga Vinod, S.R.S. Prabaharan, "Data science and the role of Artificial Intelligence in achieving the fast diagnosis of Covid-19," Chaos, Solitons and Fractals 140 (2020) 110182

[77] Christopher Gieraerts, Anthony Dangis, Lode Janssen, Annick Demeyere, Yves De Bruecker, Nele De Brucker, Annelies van Den Bergh, Tine Lauwerier, André Heremans, Eric Frans, Michaël Laurent, Bavo Ector, John Roosen, Annick Smismans, Johan Frans, Marc Gillis, Rolf Symons, "Prognostic Value and Reproducibility of AI-assisted Analysis of Lung Involvement in COVID-19 on Low-Dose Submillisievert Chest CT: Sample Size Implications for Clinical Trials," Radiology: Cardiothoracic Imaging, 2020

[78] Wu J, Feng CL, Xian XY, Qiang J, et al (2020) Novel Coronavirus Pneumonia (COVID-19) CT Distribution and Sign Features. Zhonghua Jie He He Hu Xi Za ZhiPMID: 32125131 DOI: 10.3760 / cma.j.cn112147-20200217-00106

[79] Bernheim A, Mei X, Huang M, et al (2020) Chest CT Findings in Coronavirus Disease-19 (CO-VID-19): Relationship to Duration of Infection. Radiology https://doi.org/10.1148/radiol. 2020200463

[80] S. Ying, S. Zheng, L. Li, X. Zhang, X. Zhang, Z. Huang, et al., "Deep learning enables accurate diagnosis of novel Coronavirus (COVID-19) with CT images.," MedRxiv, 2020

[81] ShuaiWang, Bo Kang, Jinlu Ma, Xianjun Zeng, Mingming Xiao, Jia Guo, Mengjiao Cai, Jingyi Yang, Yaodong Li, Xiangfei Meng, et al. 2020. Adeep learning algorithm using CT images to screen for Corona Virus Disease (COVID-19). MedRxiv (2020).

[82] Manjurul Ahsan, Kishor Datta Gupta, Mohammad Maminur Islam, Sajib Sen, Lutfar Rahman, Mohammad Shakhawat Hossain, "Study of Different Deep Learning Approach With Explainable AI For Screening Patients With Covid-19 Symptoms: Using Ct Scan and Chest X-Ray Image Dataset," arXiv:2007.12525v1 [eess.IV] 24 Jul 2020

[83] Mucahid Barstugan, Umut Ozkaya, and Saban Ozturk. 2020. Coronavirus (covid-19) classification using ct images by machine learning methods. arXiv preprint arXiv:2003.09424 (2020).

[84] Rassa Ghavami Modegh et al., "Accurate and Rapid Diagnosis of COVID-19 Pneumonia with Batch Effect Removal of Chest CT-Scans and Interpretable Artificial Intelligence," arXiv:2011.11736v2, 2021

[85] Dan Nguyen, Fernando Kay, Jun Tan, Yulong Yan, Yee Seng Ng, Puneeth Iyengar, Ron Peshock, Steve Jiang, "Deep learning–based COVID-19 pneumonia classification using chest CT images: model generalizability," 2021

[86] Vruddhi Shah, Rinkal Keniya, Akanksha Shridharani, Manav Punjabi, Jainam Shah, Ninad Mehendale, "Diagnosis of COVID-19 using CT scan images and deep learning techniques," Emergency Radiology, https://doi.org/10.1007/s10140-020-01886-y, 2021

[87] Fei Shan, Yaozong Gao, Jun Wang, Weiya Shi, Nannan Shi, Miaofei Han, Zhong Xue, Dinggang Shen, Yuxin Shi, "Abnormal lung quantification in chest CT images of COVID-19 patients with deep learning and its application to severity prediction," International Journal of Medical Physics Research and Practice, 2020, https://doi.org/10.1002/mp.14609

[88] C. Zheng, X. Deng, Q. Fu, Q. Zhou, J. Feng, H. Ma, et al., "Deep learning-based detection for COVID-19 from chest CT using weak label," MedRxiv, 2020

[89] L. Huang, R. Han, T. Ai, P. Yu, H. Kang, Q. Tao, et al., "Serial quantitative chest CT assessment of COVID-19: Deep-Learning Approach," Radiology: Cardiothoracic Imaging, vol. 2, p. e200075, 2020

[90] Lionel Roques, Etienne Klein, Julien Papa, Antoine Sar and Samuel Soubeyrand, "Using early data to estimate the actual infection fatality ratio from COVID-19 in France ," Biology doi: 10.3390/biology9050097

[91] Athanasios S. Fokas, Nikolaos Dikaios, George A. Kastis, "COVID-19: Predictive Mathematical Models for the Number of Deaths in South Korea, Italy, Spain, France, UK, Germany, and USA," doi: https://doi.org/10.1101/2020.05.08.20095489

[92] I. Apostolopoulos, S. Aznaouridis, and M. Tzani. Extracting possibly representative covid-19 biomarkers from x-ray images with deep learning approach and image data related to pulmonary diseases. arXiv preprint arXiv:2004.00338, 2020

[93] Song, P., Wang, L., Zhou, Y., He, J., Zhu, B., Wang, F., Tang, L., and Eisenberg, M. (2020). An Epidemiological Forecast Model and Software Assessing Interventions on COVID-19 Epidemic in China. medRxiv, (https://doi.org/10.1101/2020.02.29.20029421)

[94] Zhou, Z., Siddiquee, M. M. R., Tajbakhsh, N. & Liang, J. UNet++: A nested U-Net architecture for medical image segmentation. In Deep Learning in Medical Image Analysis and Multimodal Learning for Clinical Decision Support, 3-11 (Springer, 2018).

[95] Long, J., Shelhamer, E. & Darrell, T. Fully convolutional networks for semantic segmentation. In Proceedings of the IEEE Conference on Computer Vision and Pattern Recognition, 3431-3440 (2015).

[96] Tang Z, Zhao W, Xie X, Zhong Z, Shi F, Liu J, et al. Severity assessment of coronavirus disease 2019 (COVID-19) using quantitative features from chest CT images. arXiv. (2020) 2003.11988. Available online at: https://arxiv.org/ abs/2003.11988 (accessed May 10, 2020)

[97] Y. Jiang, H. Chen, M. Loew and H. Ko, "COVID-19 CT Image Synthesis With a Conditional Generative Adversarial Network," in IEEE Journal of Biomedical and Health Informatics, vol. 25, no. 2, pp. 441-452, Feb. 2021, doi: 10.1109/JBHI.2020.3042523

[98] Khalid El Asnaoui, Youness Chawki, and Ali Idri. 2020. Automated methods for detection and classification pneumonia based on x-ray images using deep learning. arXiv preprint arXiv:2003.14363 (2020)

[99] Li K, Wu J, Wu F, Guo D, Chen L, Fang Z, Li C. The Clinical and Chest CT Features Associated With Severe and Critical COVID-19 Pneumonia. Invest Radiol 2020;55(6):327-331. doi: 10.1097/RLI.0000000000000672

[100] Jocelyn Zhu, Beiyi Shen, Almas Abbasi, Mahsa Hoshmand-Kochi, Haifang Li, Tim Q. Duong, "Deep transfer learning artificial intelligence accurately stages COVID-19 lung disease severity on portable chest radiographs," PLOS ONE, https://doi.org/10.1371/journal.pone.0236621, 2020

[101] Jasjit S. Suri et al., "COVID-19 pathways for brain and heart injury in comorbidity patients: A role of medical imaging and artificial intelligence-based COVID severity classification: A review," Computers in Biology and Medicine 124 (2020) 103960

Section 3

Systems of Systems

Chapter 6

Systems-of-Systems MS&A for Complex Systems, Gaming and Decision for Space Systems

Tien M. Nguyen

Abstract

This chapter discusses advanced modeling, simulation and analysis (MS&A) approaches for supporting complex space system, gaming and decision support system (DSS) using systems-of-systems perspective. The systems-of-systems MS&A approaches presented here also address capability-based approach for supporting US defense acquisition life cycle with a laser focus on the pre-award acquisition phase and combined game theory and wargaming for acquiring complex defense space systems. The chapter also provides an overview of existing models and tools for the design, analysis and development of the government reference system architecture solution and corresponding acquisition strategy in a complex defense systems-of-systems environment. Although, the proposed MS&A approaches presented here are focused on defense space systems, but the approaches are flexible and robust that can be extended to any civilian and commercial applications.

Keywords: System-of-Systems, Systems-of-Systems, Space System, Airborne System, Pre-Milestone A, Pre-Award Phase, Modeling Simulation and Analysis, Game Theory, Decision Support Systems, War-gaming, Acquisition, Family of Systems

1. Introduction

In the past twenty years, US DoD has been undergoing three major transformations concerning the way to (i) fight, (ii) conduct business, and (iii) collaborate with allied countries. These transformations have led to significant changes in US DoD acquisition process, moving away from requirement-based to capability-based acquisition, the adaptation of the Joint Capability Integration & Development System (JCIDS) and Systems-of-Systems perspective in the design and build of future space systems [1–6]. **Figure 1** illustrates the key differences between the requirement-based and capability-based approaches. The red dotted line shown in **Figure 1** denotes the area of responsibility between the US Government (USG)[1] and its selected contractor. For requirement-based, the US DoD is responsible for (i) defining the reference system architecture, architecture performance attributes (APAs) and associated key performance parameters (KPP), and architecture trade-space, and (ii) developing Level-A specifications (spec) that can potentially

[1] Practically, USG team refers pre-Milestone A acquisition activities as the pre-award phase in the US Department of Defense (DoD) acquisition life cycle.

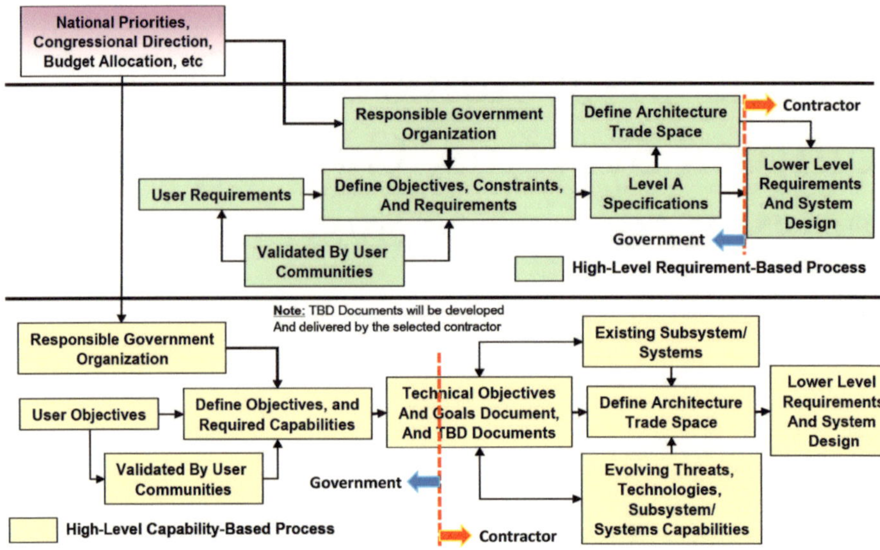

Figure 1.
Description of requirement-based and capability-based approaches.

achieving optimum KPP within the defined trade-space. A selected defense contractor is responsible for using the Level A spec to derive lower level requirements (subsystems and components), design and build the system.

Unlike requirement-based approach, the capability-based approach requires USG to define user objectives and provide required capabilities for meeting warfighter needs. As shown in **Figure 1**, USG is also responsible for providing technical objectives and goals documents along with the Initial Capability Document (ICD)[2] that presents APAs, required capabilities, threshold, and objective criteria for meeting the required capabilities. On the other hand, a selected contractor is responsible for defining the architecture trade space and developing appropriate "TBD" documents for the derivation of Level A spec using the USG's ICD and inputs (e.g., evolving adversary threats, existing US DoD systems' capabilities, etc). The "TBD" documents shown in **Figure 1** are dependent on the acquisition phase. The "TBD" document can be a Capability Development Document (CDD) or a Revised-CDD[3]. Like requirement-based approach, the selected contractor is also responsible for the (i) flow-down of Level A spec to lower level requirements, and (ii) design and build of the systems. For some defense system acquisition programs, USG also provides Technical Requirement Document (TRD) or System Requirement Document (SRD) along with the ICD to help the selected contractor concentrates on specific operational use cases, APAs and KPPs.

Designing a system for operation in complex Systems-of-Systems environment requires a good understanding of the types of systems-of-systems that the designed system would be operated in. There are three types of Systems-of-Systems, namely, Type 1: A family of System-of-Systems that provides similar core services, e.g., communication services - But each system provides different core service types, e.g., non-secure FDMA vs. secure TDMA communication services; Type 2: An integration of many families of System-of-Systems, when combined, this type of system provides unique Systems-of-Systems capabilities at the enterprise level (i.e., integrated level) - An example of this complex system is a combination of a family

[2] Per JCIDS process, the required system capabilities are usually provided in ICD.

[3] Formerly known as Capability Production Document.

of communications Systems with a family of Global Position Satellite systems; and Type 3: An integration of many heterogenous, independent but interrelated types of systems with each system providing distinctive core services.

As pointed out in [7], most of current professional papers, technical reports and System-of-Systems standards considered integration of (i) many systems of the same type of systems together, which is identical to Type 1, and (ii) many different types of systems as a system consisted of many systems and referred to as System-of-Systems, which is identical to Type 3. In this chapter, we focus our discussion on Type 2, since existing System-of-Systems engineering standards and current MS&A approaches can be directly applied to Type 1 and Type 3 but not Type 2.

This chapter presents advanced concepts on systems-of-systems MS&A approaches to support capability-based acquisition of defense space systems. The MS&A frameworks and processes presented here are emphasized on the systems-of-systems architecture trade support phase of the US DoD defense acquisition life cycle. In addition, it addresses systems-of-systems MS&A frameworks, processes, models and tools using game theoretical modeling and DSS for developing optimum acquisition strategy to acquire complex systems. The complex systems discussed in this chapter are mainly focused on defense space systems, but they can be extended to any systems-of-systems for civilian and commercial applications.

The chapter is organized as follow: (i) Section 2 presents an advanced capability-based MS&A framework, including processes and required MS&A tools, to support US DoD acquisition life cycle from the system architecture design phase to sustainment phase; (ii) Section 3 describes a MS&A approach supporting architecture design and analysis of complex systems using systems-of-systems perspective; (iii) Section 4 provides a MS&A approach for acquisition strategy development and optimization supporting pre-award phase; (iv) Section 5 describes existing available systems-of-systems MS&A models and tools supporting the pre-award phase of the acquisition life cycle; and (v) Section 6 concludes the chapter with a conclusion and way-forward.

2. MS&A approach supporting defense acquisition life cycle

Existing US DoD acquiEsition life cycle employs capability-based acquisition and has three key milestones, namely, Milestone A, Milestone B and Milestone C, which correspond to (i) analysis of alternative (AoA) and technology development phase, (ii) system and prototype development and demonstration phase, and (iii) produce and deploy phase, respectively [4]. For US Air Force and Space Force, the MS&A domains[4] for supporting the three key milestones and associated three phases can be defined as operation, training, test and evaluation, acquisition, analysis, education, experimentation and war-gaming exercise. Based on the identified MS&A domains, **Figure 2** proposes an advanced MS&A framework and associated processes supporting US DoD defense acquisition life cycle. The proposed framework defines the (i) input in terms of systems-of-systems CONOPS meeting warfighter needs and stakeholders' inputs and requirements, (ii) output in terms of data products, and (ii) related MS&A components to support the DoD acquisition life cycle. Following is a high-level description of the proposed framework, including four phases with required MS&A Models and tools and related processes supporting US DoD acquisition life cycle [4]: (i) Phase 1: systems-of-systems architecture trade study supporting pre and post-Milestone A – Addresses required MS&A tools for architecture refinement, evolution and spiral development planning, capability gap analysis, system solution development, convert capabilities to system

[4] Depending on the warfighter's needs, the M&S domains can be different.

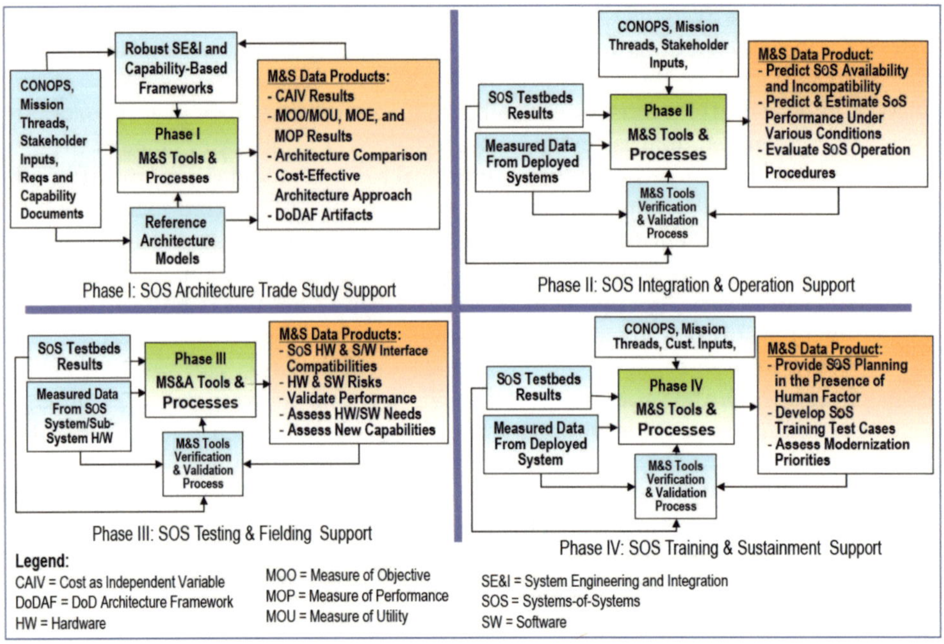

Figure 2.
MS&A framework supporting US DoD defense acquisition life cycle.

requirements – This chapter focuses on systems-of-systems architecture design and analysis for the development of cost effective acquisition strategy and acquisition of optimum reference architecture solutions; (ii) Phase 2: systems-of-systems integration and operation support at pre-Milestone B – Addresses MS&A models and tools for assessing and evaluating incompatibility testing and operation; (iii) Phase 3: systems-of-systems testing and fielding support at post Milestone B – Addresses MS&A models and tools for evaluating hardware and software testing and fielding; and (iv) Phase 4: training and sustainment at pre and post Milestone C – Addresses models and tools for supporting on/off-line training and sustainment evaluation.

The proposed M&SA framework, including processes, models and tools, presented in **Figure 2** can provide support the US DoD acquisition life cycle. The framework allows for USG stakeholders to incorporate their needs using systems-of-systems perspective taking into account (i) Warfighter's needs (CONOPS and mission threads), (ii) M&SA domains (e.g., operation, training, test and evaluation, acquisition, analysis, education, experimentation and war-gaming), (iii) JCIDS analyses and US DoD Defense of Acquisition Guide (DAG) process [2–4], (iv) DoD Joint Tactical Architecture (JTA) MS&A standards, (v) USG Stakeholder M&S strategic plan (e.g., [8]), and (vi) USG Stakeholder's goals, scopes, objectives, Statement of Objectives (SOO), and System Engineering Plan (SEP).

3. MS&A approach supporting architecture design and analysis

The proposed systems-of-systems MS&A approach presented in this section addresses the system architecture design and analysis for Phase 1 at pre and post Milestone A with an emphasis on achieving integrated capabilities at the enterprise level. As discussed in Section 1, US DoD has been using capability-based approach for developing government reference system architecture (GRA) solutions that would be used for generating optimum acquisition strategy to select appropriate contractor(s) for system acquisition. To avoid potential stovepipe GRA solutions, the capability-based approach allows the (i) USG team to develop desired GRA solution(s) in terms of high level required capabilities that are independent of technologies, and (ii) selected contractor to decide what technology enablers (TEs)[5] should be used to meet the required capabilities dictated by the GRA. Based on the choices of TEs, the selected contractor defines the system trade space and derives the Level-A specification. From the USG perspective, at pre-Milestone A, the system architecture trade space is not well-defined[6] since the choices of TEs are not available for the system architect to perform architecture design trade study making the search for an optimum GRA solution becomes very challenging. This section discusses how MS&A models and tools should be developed to support the system architecture trade study at (i) pre-Milestone A, where the USG team is responsible for the trade study to generate a reference architecture solution for the development of optimum acquisition strategy, and (ii) post-Milestone A, where a contractor (or multiple contractors) is (are) selected to work with USG team to refine the reference solution and develop associated CDD at Pre-Milestone B.

Practically, at pre-Milestone A, a contractor has not been selected[7] and the USG team obtains required inputs from warfighter needs and associated stakeholders' requirements in terms of desired systems-of-systems Enterprise (SOSE) CONOPS and associated mission threads for the desired system to be acquired. It is assumed at this stage that the Capability-Base Analysis (CBA) was completed and that the capability gaps were identified and associated potential capability solutions for the identified gaps were documented in the preliminary Initial Capability[8] Document (ICD). From USG's perspective, the USG team's objective is two-fold, namely, (i) to develop an optimum reference architecture solution meeting warfighter needs along with affordable cost and deployment schedule, and (ii) to finalize the ICD for post-Milestone A. The goal of the MS&A models and tools for this phase is to support USG team achieving these objectives. The key challenge for Pre-Milestone A is the lack of a clearly defined system architecture design trade space due to the intent[9] of capability-based approach. **Figure 3** presents a MS&A approach to address this challenge and support key Milestone A activities, including pre- and post-Milestone A activities. As shown in **Figure 3**, Sections 3.1 and 3.2 discuss systems-of-systems MS&A approaches for pre-Milestone A and post-Milestone A, respectively.

3.1 MS&A approach for pre-milestone A

This section emphasizes on systems-of-systems MS&A approach for the pre-award phase at pre-Milestone A. As mentioned earlier, at pre-Milestone A, a contractor has not been selected, and the USG team is responsible for developing reference system architecture with associated program and technical risks. **Figure 3** shows that there are four key pre-Milestone A activities requiring MS&A support, including: (i) SOSE CONOPS assessment for identifying SOSE architecture solutions and generating corresponding alternative system architecture solutions, (ii) System architecture assessment and trade study for selecting optimum system architecture solutions, (iii) Acquisition strategy development and optimization, and (iv) Pre-award risk

[5] TE is a specific technology solution meeting a required capability alone or in combination with other TEs.

[6] ICD captured CBA results identifying desired capabilities, where the system trade space is not defined until the contract is awarded.

[7] Pre-Milestone A is the pre-award phase, and post-Milestone A is the post-award phase.

[8] Capability is defined as an ability that a system has, which fulfills a warfighter need. As examples, the abilities to manage satellite trajectories and disseminate mission data to users at video streaming.

[9] A contractor will be selected to define the trade space for the system design and build.

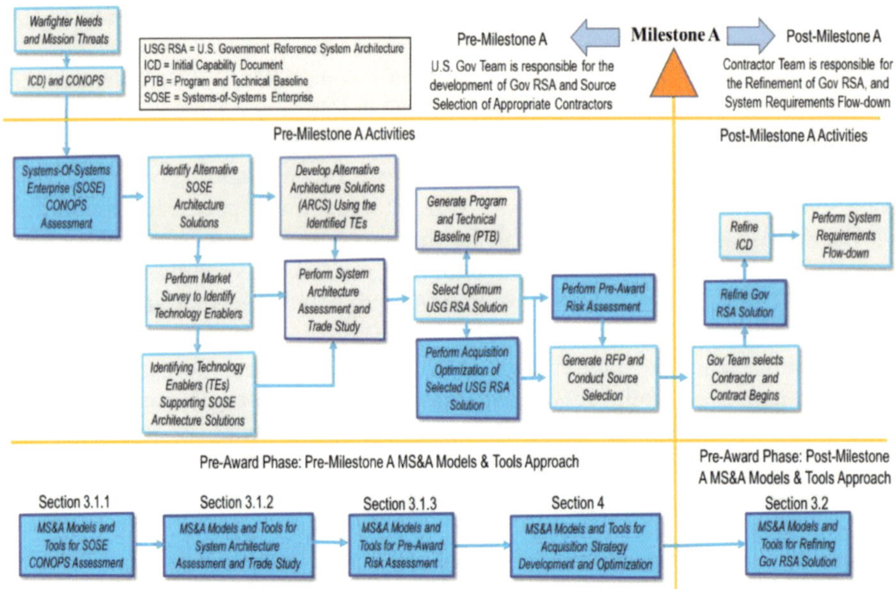

Figure 3.
MS&A framework and processes supporting milestone A.

assessment. This section discusses MS&A models and tools for supporting these four key activities. Sections 3.1.1, 3.1.2, and 3.1.3 describe SOSE MS&A approaches for pre-Milestone A activities, including SOSE CONOPS assessment, system architecture assessment, and pre-award risk assessment, respectively. As shown in **Figure 3**, the SOSE MS&A approach for the acquisition strategy development and optimization will be addressed in Section 4.

3.1.1 Approach for SOSE CONOPS assessment in pre-milestone A

The MS&A approach for SOSE CONOPS assessment discussed in this subsection is derived from [9, 10] with an emphasis on the design and build of a new space system that can be deployed in a complex space SOSE. The complex space SOSE can be assumed to have three families of systems (FOSs), namely, FOS of communications satellites, FOS of sensing satellites and FOS of position-navigation-and-timing (PNT) satellites. This section describes a MS&A approach to design and build of a new space system in this complex space SOSE environment.

The proposed MS&A approach employs advanced orbital mathematical and complex space systems simulation models for the assessment of a pre-defined SOSE CONOPS to identify the alternative systems-of-systems architecture solutions meeting warfighter and stakeholders needs [9]. This approach allows the system architecture solution to be optimized within a selected set of alternative systems-of-systems architectures using appropriate APAs and KPPs. Recently, USG has been using the "Resilience" attribute for assessing and optimizing SOSE CONOPS performance [11–14]. The Resilience attribute encompasses avoidance, robustness, reconstitution and recovery. Practically, Resilience Capacity (RC) metric is defined as the system resilience against an adversary threat, and RC is a value that represents a fraction of system capability that is retained after the recovery and reconstitution steps. Mathematically, RC is a function of:

- Avoidance - R_{AV}: is a measure of how likely it is that the threat can be fully avoided,

- Robustness - R_{RO}: is a measure of how much capability is preserved should avoidance failed,

- Recovery - R_{RV}: is a measure of the lost capability can be recovered, and perhaps how quickly it can be recovered for a specific mission, and

- Reconstitution - R_{RC}: is a measure of the total capability can be replaced, and perhaps how quickly it can be replaced.

Mathematically, RC can be expressed as follow [14]:

$$\begin{aligned} RC = & R_{AV} + (1-R_{AV})R_{RO} + (1-R_{AV})(1-R_{RO})R_{RV} \\ & + (1-R_{AV})(1-R_{RO})(1-R_{RV})R_{RC} \end{aligned} \tag{1}$$

For defense space applications, the most pronounce threat is the radio frequency interference (RFI) threats from both friendly and unfriendly sources. Thus, R_{AV}, R_{RO}, R_{RV}, and R_{RC} can be defined in terms of the SOSE architecture[10] performance as follow:

- R_{AV} = % of time SOSE is free from any RFI threats and the required SOSE network nodes relate to sufficient Link Margin (LM), i.e., no drops of communications links due to insufficient LM in the presence of RFI,

- R_{RO} = Mean SOSE Network Score when RFI present and/or no connectivity drops due to insufficient LM

- R_{RV} = Mean Network Score when band switching increases SOSE Network Score

- R_{RC} = Mean increase in SOSE Network Score due to optimal assisting satellite.

SOSE network score is used to assess and evaluate the SOSE network states. The SOSE network score is calculated by the number of communication pairings (e.g., Ground Terminal 1 connected to Satellite 1) possible in the current state divide by the number of pairings possible in an ideal State. It is the probability two arbitrary SOSE network nodes can communicate or connect to each other. Thus, the SOSE network score is defined as:

$$SOSE\ Network\_Score = \frac{\sum_{i=l}^{N} \binom{l}{2}}{\binom{N}{2}} \tag{2}$$

where l is the number of fragmented network i, N is the total number of fragmented networks, and $\binom{l}{2}$ is the Binomial coefficient.

RC and SOSE network score models are used to evaluate and assess SOSE communications LM and SOSE network availability [9, 10]. In addition to RC model,

[10] SOSE architecture consists of families of space systems (FOS) and FOS are connected by communications datalinks. A datalink connects two system nodes, and the nodes are connected when the communications datalink maintains a specified link margin (LM).

[9, 10] also recommended two additional models that are very useful in the SOSE CONOPS assessment, namely, Resilience Assessment Index (RAI) and Spectrum Resiliency Assessment Index (SRAI). RAI Model is used to generate a "Heat-Map" for identifying areas impacted by RFI threats and associated reconstitution's quality (R_{RC}). SRAI Model generates a "Heat-Map" to show the likelihood that a space system can access to the allocated frequency-band in the presence of RFI events. A description of RAI and SRAI models is provided in [10].

Figure 4 describes an advanced MS&A approach with desired simulation models for SOSE CONOPS assessment. Based on the warfighter needs and related stakeholders, SOSE CONOPS can be developed to address warfighter needs using required SOSE databases. The required SOSE databases include practical operational systems that can impact the pre-defined SOSE CONOPS. The pre-defined SOSE CONOPS focuses on defense space systems and defense space enterprise, and the operational space systems can impact the defense space enterprise operations, including civilian space and commercial space enterprises. in addition to RAI-RFI, SRAI and RC models, additional mathematical and simulations models are required to perform SOSE CONOPS assessment, including SOSE orbital analysis and simulation, dynamic LM calculation, satellite performance, and avoidance models.

SOSE orbital-analysis-and-simulation model is used to simulate the RFI threats and dynamics of space systems of interest providing accurate SOSE network nodes and associated positions and network nodes connectivity. The dynamic-LM-calculation model simulates and evaluates link margins of SOSE communications links among SOSE network nodes and calculates network score. The satellite-performance model simulates and evaluates satellite system performance, including signal-to-noise ratio (SNR) calculation and processing time estimation for assessing the recovery time from the threat. The avoidance model simulates space system threat avoidance techniques including antenna beam nulling and adaptive modulation-and-coding techniques to assess and evaluate if the RFI threats can be avoided **Figure 4**.

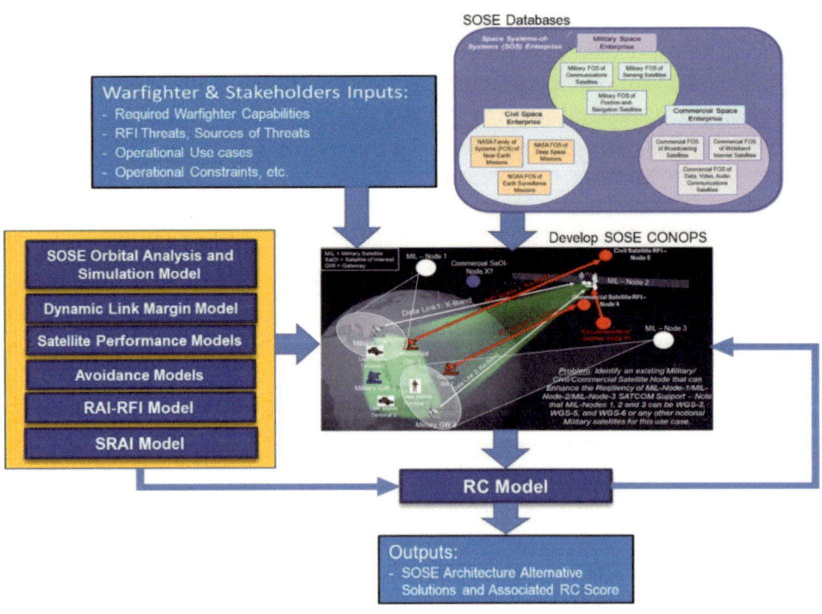

Figure 4.
MS&A approach for SOSE CONOPS assessment.

As described in **Figure 4**, the inputs to the SOSE CONOPS MS&A models and tools are required warfighter capabilities, RFI threats and sources of threats, operational use cases and operational constraints. The MS&A output is a set of optimum (or the best) alternative architectures based on a pre-defined SOSE CONOPS. Note that the USG team will adjudicate of what is the "best" or "optimum" set of alternative system architecture solutions based on warfighter and stakeholder needs and the SOSE network score for each operational use case associated with the pre-defined SOSE CONOPS.

3.1.2 MS&A approach using multi-criteria decision support system for system architecture assessment in pre-milestone A

To address the system requirements trade space challenges, the proposed system architecture assessment approach should be based on the required warfighter capabilities and market survey results to identify desired TEs for providing the required capabilities. The MS&A approach is derived from [15, 16], where system architecture assessment is based on the technical performance, market, cost, and schedule risks. Technical performance risk is referred to as technology risk and is quantified using Technology Readiness Level (TRL), while market risk is related to market uncertainty and is quantified by Manufacturing Readiness Level (MRL). Rough Order Magnitude (ROM) Cost, TRL and MRL data are collected from a market survey for cost and schedule risk assessment. **Figure 5** illustrates a recent advanced MS&A approach for system architecture assessment and an example of an architecture solution output [15, 16]. The approach uses game theory combined with the war-gaming concept to assess and optimize the system architecture solutions using the market survey results. The approach requires input from warfighter and associated stakeholders along with a set of "optimum" alternative architectural solutions obtained from SOSE CONOPS assessment described in Section 3.1.1, and a pre-defined Payoff-and-Cost Function (PCF). The outputs are (i) optimum architecture solution, (ii) associated technology and market risks, and (iii) predicted related schedule and cost risks. The selected optimum architecture solution is captured in terms of selected TEs and DoD Architecture Framework (DoDAF) views, including Capability View-1 (CV-1) and CV-2.

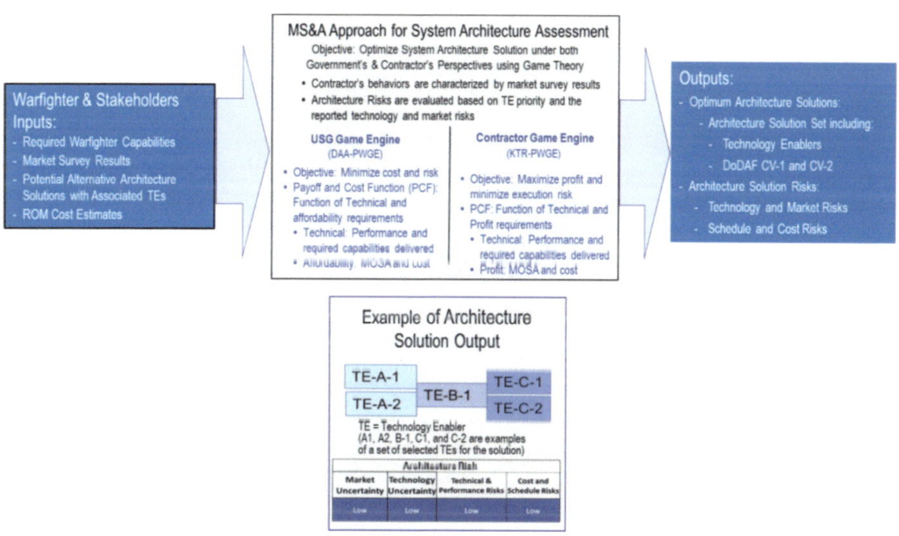

Figure 5.
MS&A approach for SOSE system architecture assessment.

The MS&A approach requires systems-of-systems analysts to develop the USG game engine (a.k.a. DAA-PWGE) and Contractor game engine (a.k.a. KTR-PWGE) for assessing and optimizing the architecture solutions under USG perspective and contractor perspective, respectively [15, 16]. The objective of the USG game engine is to minimize cost and technical risk using an appropriate PCF for trading off the affordability and technical requirements. The objective of the contractor game engine is to maximize profit and minimize execution risk using an appropriate PCF for trading off the profit and execution risk. The game engines can play pure game or mixed game depending or survey results. Pure game is used when the contractors are surer of their risk assessments, and there are no "belief" and "weighting" functions are needed for assessing the TE risks. Mixed game is used when contractors are more uncertain of their risk assessments, and hence "belief" and "weighting" functions are needed to characterize the TE risks. For this case, TEs are weighted based on their priorities using either a uniform or triangular distribution. The games are static Bayesian games with the goal to reach Nash equilibrium, where the games have stable solutions to game theoretic problem involving multiple players in which no individual player can improve their payoff by a unilateral change in behavior. The objective of MS&A models and tools is to selects the best architectural solution and associated architecture solution type for risk assessment. Classification of architecture solution type depends on the system and associated systems-of-systems requirements and associated market and technology risks (i.e., uncertainty) [15, 16]. **Figure 6** describes an acquisition strategy mapping framework and shows the mapping of requirement type to architecture solution type according to various market and technology risks. Section 4 describes a recommended MS&A approach for acquisition strategy development and optimization using this acquisition strategy mapping framework. Theoretically, for these games, the players can be the USG team and contractors to participate in the games playing action. In practice, during the pre-Milestone A, the USG team can also play the contractor role to determine the win-win acquisition strategy from both USG and contractor perspectives. Detailed description of the game engines can be found in [15, 16].

In practice, when the architecture solution does not converge to a single optimum solution, a brute force approach can be used to force the solution to converge to a single system architecture solution for acquisition strategy development and optimization. Since the brute force approach might not converge or lead to an

Requirements Classifications, Risk Assessment Classification, Acquisition Strategy Mapping, Architecture Solution Classification and Risk Assessment							
Requirement Type	Requirement Type Description	Market Uncertainty	Technology Uncertainty	Advanced Acquisition Strategy Mapping	Architecture Solution Type Classification	Architecture Risk Assessment	
						Technical & Performance Risks	Cost & Schedule Risks
Type 1	Firmed and fixed requirements with known Technology Enablers	Low	Low	FFP, FPEPA	Type 1 Solution: Conservative	Low	Low
Type 2	Well-defined requirements with some uncertainties on technology enabler and market	Low	Medium	FFIF, FPAF	Type 2 Solution: Innovative	Low	Medium
		Medium	Low			Medium	Low
		Medium	Medium			Medium	Medium
Type 3	Requirements are somewhat known with some market uncertainty but can not identify the exact technology enablers	Medium	High	CPIF	Type 3 Solution: More Innovative	High	Medium
Type 4	Requirements are somewhat known with some technology uncertainty but can not identify the exact company (or companies) to provide the technology enabler	High	Medium	CPAF, CPIF	Type 4 Solution: Less Conservative	Medium	High
Type 5	Unknown Requirements with unkhown technology enable and market	High	High	CPAF, CPFF	Type 5 Solution: Most Innovative	High	High

Figure 6.
Acquisition strategy mapping framework [15, 16].

optimum solution, [17] proposed a multiple-criteria decision model based on the Marquis de Condorcet principle found in the ELECTRE models for addressing the situations when the game models do not yield optimal outcome.

3.1.3 MS&A approach for program risk assessment in pre-milestone A

Based on existing US defense acquisition life cycle, the MS&A approach for pre-award phase at pre-Milestone A, the USG team often assesses program risk associated with the following nine pre-award events, including (i) Program Go-Ahead, (ii) Early Strategy and Issues Session (ESIS) (see AFI 63–138, is a key event), (iii) Acquisition Strategy Review Board (ASRB), (iv) Acquisition Strategy Panel (ASP) (see AFI 63–101, is a key event), (v) Acquisition Strategy Document (ASD) (is considered as a key event), (vi) Strategy Review Board (SRB), (vii) Source Selection Plan (SSP) (see 2011 DOD Source Selection Procedures), (viii) Request for Proposal (RFP) (is also considered as a key event), and (ix) Source Selection and Proposals Evaluation. The MS&A objective for the pre-award risk assessment is to provide MS&A models and tools for evaluating and assessment of the program and technical baseline (PTB) risks at each of the key events. As pointed out in [18, 19], there are nine PTB components, including five Program Baseline (TB) and four Technical Baseline (PB) components, as shown in **Figure 7**(**a**) and (**b**), respectively. Detailed description of these PB and TB components can be found in [18, 19].

Figure 8 proposes a MS&A approach for the program risk assessment of four TB and five PB components. The approach recommends a set of three MS&A models and tools, namely, (i) MS&A model and tool #1, (ii) MS&A model and tool #2 and (iii) MS&A model and tool #3 for supporting three MS&A tasks, including (i) Program risk quantification task on the roll-up program cost, schedule and performance risks, (ii) PB and TB risks Quantification task assessing impact on (key) pre-award acquisition event, and (iii) Task on prediction of PTB risk impact at each (key) pre-award acquisition event, respectively.

MS&A model and tool #1 is a set of mathematical models for evaluating the overall program risk based on individual TB and PB components' risks. The overall PTB risk is quantified in terms expected values of the likelihood and consequence that will be placed on the pre-award PTB risk management matrix. A notional PTB risk management matrix is depicted in **Figure 9**. MS&A model and tool #2 is a set of mathematical models and software tools for (i) Assessing the PB

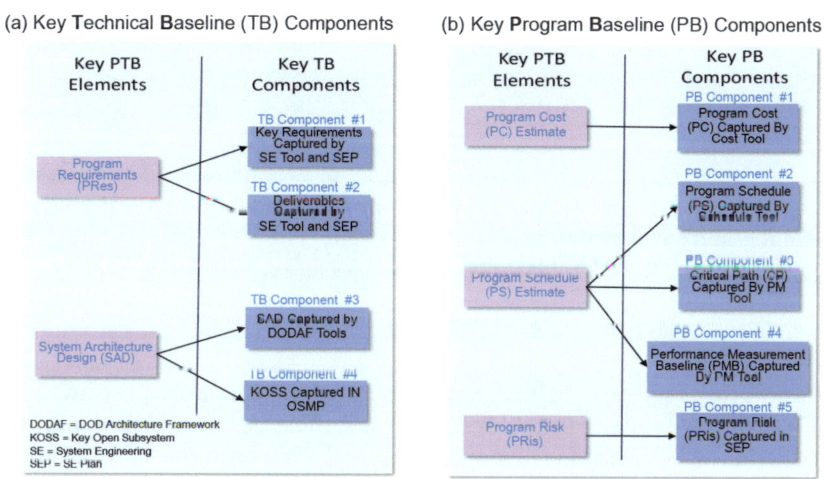

Figure 7.
Description of PTB elements [18, 19].

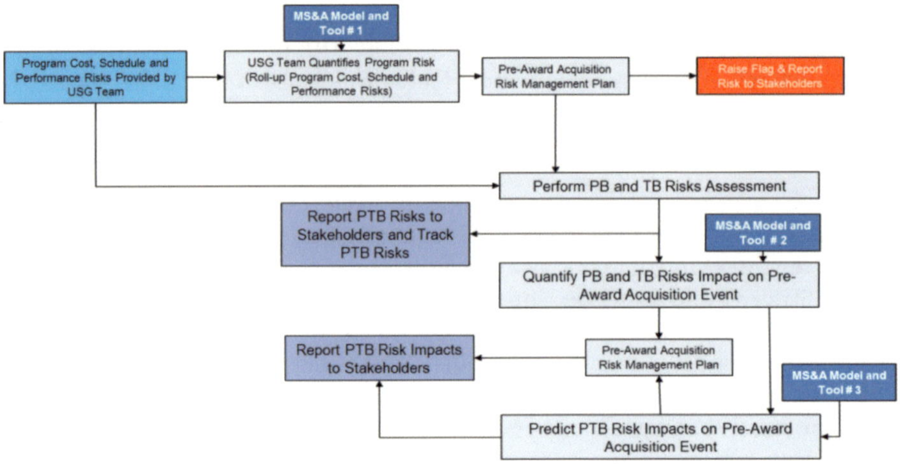

Figure 8.
MS&A approach for pre-award risk assessment.

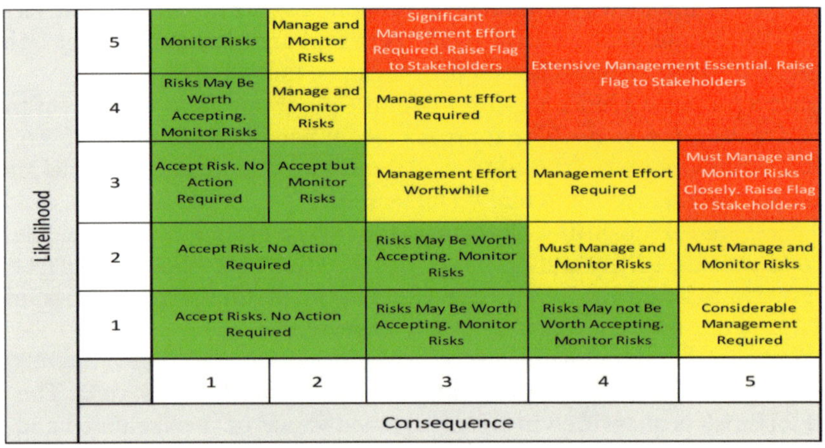

Figure 9.
A notional program risk management matrix.

and TB components' risks, (ii) Quantifying PB and TB risks impact on a specific pre-award event, and (iii) Evaluating PTB risk rolled up and quantification from individual PB and TB components' risks. The rolled up PTB expected likelihood and consequence values will also be placed on a program risk management matrix like **Figure 9**.

Finally, the MS&A model and tool #3 is a set of mathematical models and software tools for predicting the PTB risk at a future acquisition event given the risk assessment results at the current acquisition event. The PTB risk results are quantified in terms of expected likelihood and consequence values.

3.2 MS&A approach for system architecture analysis in post-milestone A

This section describes a MS&A approach for supporting the post-award phase of the DoD acquisition life cycle. At post-Milestone A, a contractor is already selected, and the USG team is responsible for working with the selected contractor to refine the USG reference system architecture and minimize associated technical and execution risks. **Figure 10** proposes a MS&A approach for post Milestone A.

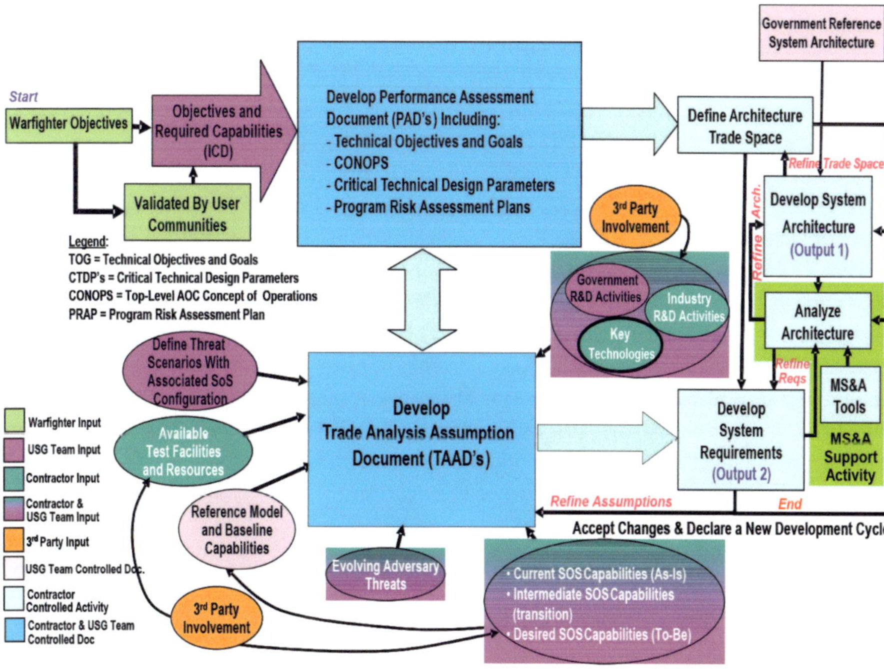

Figure 10.
Proposed MS&A approach for supporting post milestone A.

The approach shown in **Figure 10** shows the (i) required inputs including warfighter and stakeholder, Performance Assessment Documents (PADs) and Trade Analysis Assumption Documents (TAADs); (ii) desired MS&A activities and supporting MS&A models and tools; and (iii) essential outputs for supporting post Milestone A tasks. The figure is color coded to illustrate appropriate (i) warfighter input, USG Team input, activity and documents, (ii) contractor input, activity and documents, (iii) third party (i.e., related subcontractors) involvement, and (iv) joint USG and contractor teams' activities. USG team provides ICD, SOSE CONOPS and associated threat scenarios, government reference architecture, and warfighter needs. Using USG's inputs, including PADs, TAADs and SOSE perspective, the selected contractor team is responsible for developing desired trade space and performing the system architecture trade analysis and refine the government reference architecture (GRA) and providing the "best"(or optimum) system architecture solution and associated system requirements for the development of hardware prototype. The USG team serves as the final adjudicator of what is the "best," and define which Technical Performance Measures (TPMs) are more important than the others for meeting the warfighter and stakeholder needs, and which residual risks are of the most concern. Typical PADs include (i) Program Technical Objectives and Goals (TOG), (ii) Program TPMs, (iii) Top-Level CONOPS, and (iv) Program Risk Assessment Plan (PRAP). Trade Analysis Assumption Documents (TAAD's) with typical TAADs including (i) Adversary Capability Document (ACD), (ii) Scenarios Document (SD), (iii) Value Model Document (VMD), (iv) Master Test Plan (MTP), (v) Integrated Master Plan/ Integrated Master Schedule (IMP/IMS), (vi) System Capability Baseline (SCB), and (vii) Technology Maturity Baseline (TMB).

For post-Milestone A, the selected contractor is responsible for (i) the architecture analysis (what-if analyses) on the selected alternative SOSE architecture solutions, and (ii) providing all MS&A models and tools for activities supporting SOSE architecture analysis associated with GRA refinement. The contractor MS&A models and tools should be developed for supporting the following SOSE architecture analyses, including, at the minimum, (i) Technology Insertion Assessment: What available technologies could be inserted to gain a significant increase performance without unacceptable increased in risk, (ii) System Capabilities Evaluation: increases/ decreases in system capabilities vs. gains/losses in overall system performance, (iii) SOSE CONOPS Assessment: SOSE CONOPS Changes for increased performance vs. ease of integration, (iv) TPMs Evaluation: Benefits for not meeting threshold objective TPMs vs. not to exceed TPMs, (v) Threat/Scenarios Analysis: Benefits for not to address the full baseline operation under different threat/scenarios vs. Benefits to address scenarios beyond the baseline, (vi) Integrated Management Plan (IMP)/ Integrated Master Schedule (IMS) Assessment: Where would the USG derive benefit from changing quality standards, cost management system, award fee structure, the schedule for implementation, and (vii) Master Test Plan Analysis: Address changes in planned test facilities, test resources, or test restrictions that would provide overall benefit to fully testing the capabilities of the system.

4. MS&A approach for acquisition strategy development and optimization supporting pre-award phase

The proposed MS&A approach for the acquisition strategy development was derived from [15, 16], where an acquisition strategy is developed based on the selected SOSE architecture solutions and associated technical and program risks and cost and schedule risks. Based on the technology (TRL) and market (MRL) uncertainty risk assessment results associated with the selected SOSE architecture solutions, the proposed MS&A approach uses advanced acquisition strategy mapping framework, as shown in **Figure 6**, to select an appropriate contract type with associated optimum architecture solution. Depending on the TRL and MRL assessment results, an optimum contract type is chosen and appropriate game theoretical type is selected to optimize the contract parameters, including target price, sharing ratios (SR) and contract fees allowing maximum USG savings with "increased competition" or "increased number of bidders." Currently, References [15–17] only addressed the three contract types, including Firm Fixed Price (FFP), Fixed Price Incentive Firm (FPIF) and Cost Plus Incentive Firm (CPIF). **Figure 6** also describes requirements classifications, risk assessment classification, acquisition strategy mapping, and architecture solution classification and risk assessment for architecture risk assessment.

Figure 11 describes a recent advanced MS&A approach for supporting acquisition strategy development and optimization along with an example of an acquisition solution output [15, 16]. As shown in **Figure 11**, the approach requires input from warfighter and associated stakeholders, a set of optimum alternative system architectural solutions obtained from SOSE CONOPS assessment, and a predefined PCF[11] to evaluate USG saving and contractor profit along with the contract's parameters. Additionally, the required inputs to the proposed MS&A models and tools include USG architecture solution type, risk assessment results, cost distribution, corresponding contract type. The outputs include (i) Optimum acquisition strategy[12] and contract type and associated contract parameters, including

[11] PCF for USG is used to evaluate the USG saving/loss and associated payoffs; and for contractor bidding game, PCF is used to evaluate the contractor's profit/loss and associated payoffs.

[12] Optimum bidder strategy is based on Nash strategy. For non-optimum bidder strategy, contractors select their bid based on a fixed (or randomly assigned) percentage of cost and contract's parameters are optimized for maximum profit and minimum execution risk using assigned PCF.

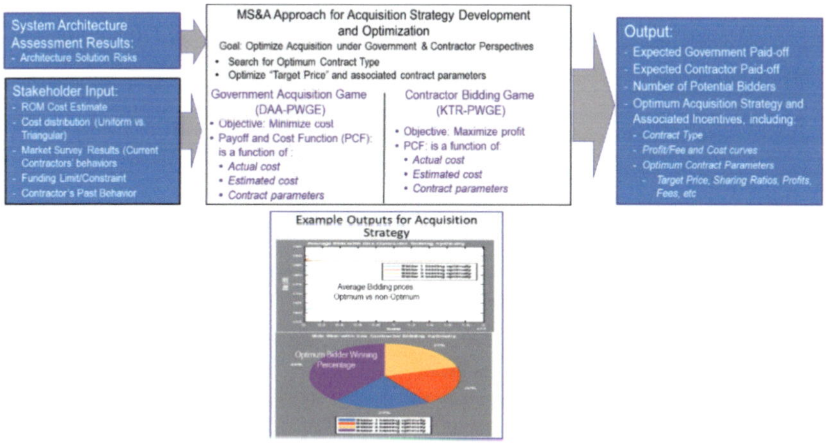

Figure 11.
MS&A approach for acquisition strategy development and optimization.

incentives, target price, SRs and fees, (ii) USG saving (payoff), (iii) contractor profit (payoff), (iv) Number of potential bidders (i.e., increase competition), and (v) Risk results in terms of technology (technical and performance) and program (cost and schedule) risks. The acquisition strategy depends on the program and technical risks assessment of the selected optimum architecture solution obtained from the SOSE-CONOPS-assessment model in Section 3.1.1 and system-architecture-assessment model in Section 3.1.2, hence these two MS&A models will be tightly coupled with the acquisition strategy development and optimization MS&A models discussed in this section.

5. Existing SOSE MS&A models and tools

This section provides a summary of existing MS&A models and tools for supporting pre-award phase of US DoD acquisition life cycle. Section 5.1 focuses on the models and tools for SOSE CONOPS assessment, Section 5.2 for system architecture assessment, and Section 5.3 on space systems acquisition strategy development and optimization using game theoretic and multi-criteria decision support system.

5.1 Models and tool supporting SOSE CONOPS Assessment

The MS&A models and tool implemented in Matlab[13] for SOSE CONOPS assessment presented in this section are derived from [10]. The models and Matlab tool focus on space SOSE [9, 20] in the presence of friendly RFI threats. The current Matlab models and tools include public open source databases for military, commercial and civilian satellites and ground systems. They can be used to evaluate key SOSE CONOPS performance metrics, including (i) Communication LM and communication link availability in terms of network score, and (ii) the three key resiliency metrics for measuring spectrum resiliency against RFI threats. The three key resiliency metrics are (a) Resilience Assessment Index against RFI (RAI-RFI), which is a measure of "Reconstitution" metric calculating the probability of a ground/satellite

[13] The Matlab tools were developed jointly by The Aerospace team and CSUF graduate student team. CSUF team includes Tom Free, Scott Digiambattista, Nicole Hemming-Schroeder, Catherine Osborne, Lauren Benson, Jordan Golemo, and Maria Heinze under the Industry Collaboration program between CSUF and The Aerospace Corporation. The CSUF program director is Prof. Charles Lee.

system being disrupted by RFI and its ability to reduce RFI by re-routing the desired signal to avoid RFI threats, (b) Spectrum Resiliency Assessment Index against RFI (SRAI-RFI), which is a measure of "Avoidance-Robustness-and-Reconstitution" metric for evaluation of the ability of a system that can access the spectrum and be able to response to a disruptive event - SRAI-RFI is a metric calculating the probability that a system can access to its allocated RF frequency band in the presence of RFI threats, and (c) Resilient Capacity against RFI (RC-RFI), which is a measure of "Avoidance, Robustness, Reconstitution, and Recovery." The RAI-RFI Model generates a "Heat-Map" to show areas impacted by RFI threats, SRAI-RFI Model generates a "Heat-Map" to show the likelihood that a communication system can access to the allocated frequency-band in the presence of RFI events, and RC-RFI Model generates SOSE communication LM and link availability for the "areas identified by RAI-RFI and SRAI-RFI" models. **Figure 12** shows the Matlab Graphic User Interface (GUI) of the Matlab tool describing 2-Dimension (2-D) and 3-D simulation of a notional SOSE CONOPS. The tool can generate a set of potential SOSE architecture solutions with minimum performance degradation in the presence of RFI threats.

5.2 Models and tools supporting system architecture assessment and multi-criteria decision support system

The available MS&A models and tool implemented in Matlab[14] for system architecture assessment presented in this section are collected from [18, 19, 21–24]. The Matlab models implemented static Bayesian games with (i) complete information games and associated pure and mixed games, and (ii) incomplete information games and associated mixed games. The Matlab models also incorporated both brute force and multi-criteria decision approaches. The brute force approach iterates the PCFs adjusting USG and contractors gains and losses until the contactors' architecture solutions converge to UGS solution. The brute-force's criterion is set to a minimum of two contractors' solutions are required to converge to USG solution [16, 24]. The multi-criteria decision approach implemented advanced ELECTRE II with five evaluation criteria, including market uncertainty, technological

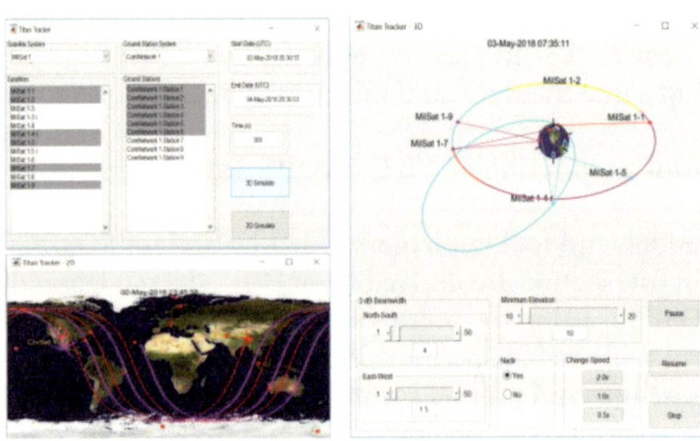

Figure 12.
Matlab GUI for SOSE CONOPS assessment.

[14] The Matlab tools were developed jointly by Aerospace team, University of Hawaii (UH) team, and North Carolina State University (NCSU) team. NCSU team includes Paul Vienhage, Heather Barcomb, Karel Marshall, William Black, and Amanda Coons under the Industry Collaboration (IC) program between NCSU and The Aerospace Corporation. NCSU IC program director was Prof. Hien Tran. UH team led by Prof. Tung Bui.

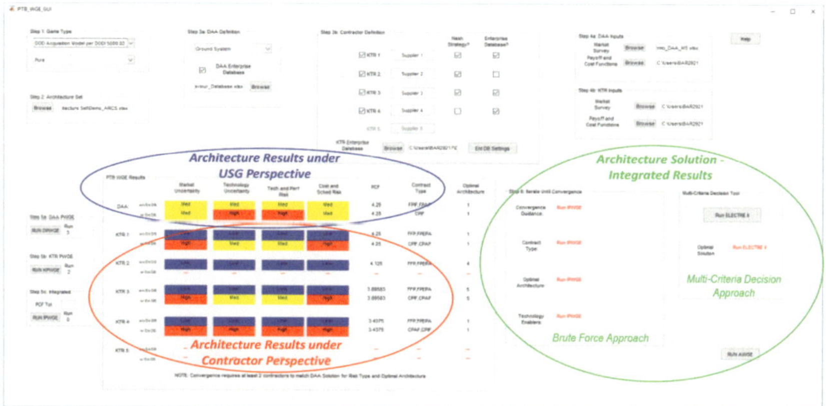

Figure 13.
Matlab GUI for system architecture assessment.

uncertainty, technical and performance risk, cost and schedule risks, and payoffs and costs [17]. **Figure 13** depicts the Matlab GUI for evaluating system architectures under both USG and contractor perspectives [16, 24]. The figure shows a notional case assuming Pure game, four potential contractors bidding, with and without enterprise databases. Practically, enterprise database includes program of records (i.e., past programs data from open sources). Non-enterprise open source database includes only survey data collected from the potential bidding contractors. Detailed description of the models and tool can be found in [15–17, 24]. It should be noted that the MS&A models and tool presented in this section are not intended to predict what contractor has the winning system architecture solution. The intention is to gain insight into the winning architecture solution based on contractors' supplied information for the development and selection of the USG reference architecture for Request for Proposal (RFP) preparation.

5.3 Models and tools supporting space systems acquisition strategy development and optimization using game theoretic

Like Section 5.2, the MS&A models and tools presented here were also implemented in Matlab [18, 19, 21–24]. The Matlab simulation models also implemented static Bayesian with complete and incomplete information games and associated pure and mixed games for the acquisition strategy development and optimization. The current Matlab models implemented three common contract types, including FFP, FPIF and CPIF. **Figure 14** illustrates the Matlab GUI of the tool for developing optimum acquisition strategy with associated contract parameters under both USG and contractor perspectives [16, 24]. The figure shows a notional case with Pure game, four potential contractors bidding, with and without enterprise databases. Note that enterprise database includes program of records with past programs data from public open sources. Non-enterprise database includes only notional survey data collected from potential bidding contractors.

Figure 14 also shows a notional use case for four contractors (a.k.a. Suppliers) bidding the contract assuming that (i) there are three contractors (Suppliers #1, #2 and #3) bidding using optimum Nash strategy and one contractor (Suppler #4) bidding using non-optimum strategy, (ii) the selected USG reference architecture is obtained from MS&A models and tool presented in Section 5.2, and (iii) CPIF is the selected contract type based on the risk assessment results for the selected system architecture solution. Bidding results show that Supplier #1 has

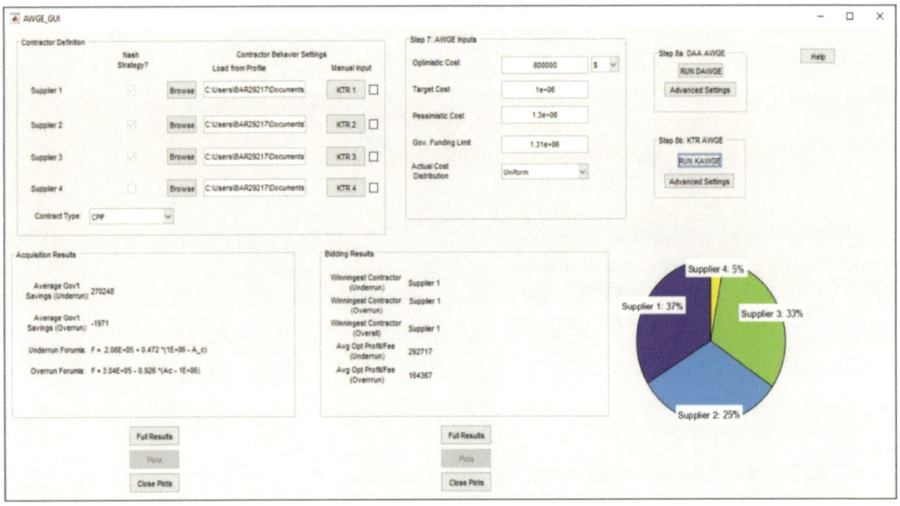

Figure 14.
Matlab GUI for system acquisition strategy development and optimization.

the winning bidding strategy with 37% wins, followed by Supplier #3 with 37% wins and Supplier # 2 with 25% wins. Supplier #4 has the lowest winning bid with 5% due to non-optimum bidding strategy, i.e. not using Nash strategy. Acquisition results captured the key features of the winning bidding strategy, including USG saving for overrun and underrun cases, underrun and overrun formulas.

Again, the MS&A models and tool presented in this section are not intended to predict what contractor has the winning bidding strategy. The intention is to gain insight into the winning bidding strategy based on the selected optimum architecture solution for the development and selection of the USG reference architecture for RFP preparation.

6. Conclusion and way-forward

The systems-of-systems MS&A approaches presented in this chapter focused on recent advanced framework, processes and available models and tools for supporting pre- and post-Milestone A of the US defense acquisition life cycle with capability-based acquisition approach. Proposed MS&A approaches were derived from the USG point of view using systems-of-systems perspective to address optimum reference system architecture solu¬tion and associated acquisition strategy for acquiring the selected solution meeting desired cost, schedule and technical performance. The proposed MS&A approaches and associated Matlab models and tools were primarily focused on pre-Milestone A and developed based on SOSE CONOPS modeling and simulation of resilient space SOSE operations, Bayesian games combined with war-gaming concept and multi-criteria decision support system for optimizing system architecture solutions and associated acquisition strategy. Available Matlab tools were presented for assessing space SOSE CONOPS, evaluating alternative system architecture solutions and optimizing acquisition strategy of common contract types, such as FFP, FPIF and CPIF [15–24]. In general, the systems-of-systems MS&A approaches presented here can be extended to support other non-defense system and acquisition life cycle from non-government perspectives. Existing Matlab models and tools presented in Section 5 can also be extended to non-space SOSE CONOPS, Bayesian dynamic games with other contract types (such as FPEPA, FPAF, CPAF and CPFF).

The author hopes that this chapter provides MS&A concepts and source of ideas for the readers to develop commercial systems-of-systems frameworks, processes, models and tools for supporting of their own MS&A works.

Acknowledgements

The author would like to thank his esteemed colleagues, Professors Charles Lee and Sam Behseta at California State University in Fullerton, Ms. Navneet Mezcciani and Mr. Garick Lue-chung at The Aerospace Corporation for their continuous support. He also wants to express his deep appreciation to his wife, Thu-Hang Nguyen, for her constant moral support during the process of writing this chapter.

Conflict of interest

The author declares no conflict of interest. The MS&A approaches and system acquisition views presented in this chapter are those of the author and do not reflect endorsement of California State University in Fullerton or The Aerospace Corporation or the US DoD.

Author details

Tien M. Nguyen[1,2]

1 Center for Computational and Applied Mathematics, California State University, Fullerton, USA

2 The Aerospace Corporation, El Segundo, California, USA

*Address all correspondence to: tmnguyen57@fullerton.edu

References

[1] CJCSI 5123.01H "Charter of the Joint Requirements Oversight Council (JROC) and Implementation of the JCIDS," 31 Aug 2018.

[2] John Rausch, "Joint Capability Integration & Development System (JCIDS) Overview," Capabilities and Analysis Division, The Joint Staff, J-4.

[3] Capabilities-Based Assessment Handbook, Air Force Materiel Command (AFMC) OAS/A5, Office of Aerospace Studies, 10 March 2014.

[4] DoD Instruction 5000.02 "Operation of the Adaptive Acquisition Framework" dated 23 January 2020.

[5] Capabilities Based Requirements Development, Air Force Instruction 10-601, 27 April 2021.

[6] US DOD, Systems Engineering Guide for Systems of Systems, Version 1.0, August 2008.

[7] Tien M. Nguyen, Tung X. Bui, "Recent Trends in Systems-of-Systems Design, Modeling, Simulation and Analysis for Complex Systems, Gaming and Decision Support," Chapter 1, SOS Perspectives and Applications - Design, MS&A, Gaming and Decision Support, Book Edited by Tien M.Nguyen, ISBN 978-1-83968-328-2, to be published in July 2021.

[8] Air Force Modeling and Simulation (M&S) Strategic Plan 2006-2023.

[9] Tien M. Nguyen, "SOSE, SOSE CONOPS, SOSE Architecture Design Approach: A Perspective on Space and Airborne Systems," Chapter 4, SOS Perspectives and Applications - Design, MS&A, Gaming and Decision Support, Book Edited by Tien M.Nguyen, ISBN 978-1-83968-328-2, to be published in July 2021.

[10] Tien M. Nguyen, Charles Lee, Tom Freeze, Andy T. Guillen, "Systems-of-systems enterprise architecture CONOPS assessment approach and preliminary results," Proceedings Volume 11422, Sensors and Systems for Space Applications XIII; 114220M (2020).

[11] USAF Space Command, "Resiliency and Disaggregated Space Architectures,", Distribution A: Approved for public release; distribution unlimited.

[12] Brian Connett, "RESILIENT AND FRACTIONATED CYBER PHYSICAL SYSTEM," Thesis, Naval Post Graduate School, September 2014.

[13] Gary McLeod, George Nacouzi, et. al., "Enhancing Space Resilience Through Non-Materiel Means," Research Report, RAND Corporation, 2016, www.rand.org/t/RR1067.

[14] Ron Burch, "A Method for Calculation of the Resilience of a Space System," IEEE Military Communications Conference Proceedings, 2013.

[15] Tien M. Nguyen, Andy Guillen, Sumner Matsunaga, Hien T. Tran, Tung X. Bui, "War-Gaming Applications for Achieving Optimum Acquisition of Future Space Systems," a Book Chapter in the book titled "Simulation and Gaming," published by INTECH-Open Science-Open Minds, ISBN: 978-953-51-3800-6, DOI: 10.5772/intechopen.69391, 2018.

[16] Tien M. Nguyen, Hien Tran, Andy Guillen, Tung X. Bui, Sumner Matsunaga, "Acquisition War-Gaming Technique for Acquiring Future Complex Systems: Modeling and Simulation Results for Cost Plus Incentive Fee Contract," Mathematics 2018, 6, 43. URL: https://doi.org/10.3390/math6030043.

[17] Tien M. Nguyen, Tom Freeze, Tung X. Bui, Andy Guillen,

"Multi-criteria decision theory for enterprise architecture risk assessment: theory, modeling and results," Proc. SPIE 11422, Sensors and Systems for Space Applications XIII, 114220P (29 April 2020); doi: 10.1117/12.2559317.

[18] Tien M. Nguyen, Andy T. Guillen, Sumner S. Matsunaga, "A Resilient Program Technical Baseline Framework for Future Space Systems," Invited paper, Proceedings of SPIE Vol. 9469 946907-946901, 2015.

[19] Tien M. Nguyen, Andy T. Guillen, James J. Hant, Justin R. Kizer, Inki A. Min, Dennis J. L. Siedlak and James Yoh, "Owning the Program Technical Baseline for Future Space Systems Acquisition: Program Technical Baseline Tracking Tool," Proceedings of SPIE Vol. 10196-7, 2017.

[20] John Nguyen, "SOS Taxonomy: Space and Airborne Systems Perspective," Chapter 2, SOS Perspectives and Applications - Design, MS&A, Gaming and Decision Support, Book Edited by Tien M.Nguyen, ISBN 978-1-83968-328-2, to be published in July 2021.

[21] Tien M. Nguyen, and Andy T. Guillen, "War-Gaming Application for Future Space Systems Acquisition," Proceedings of SPIE, 2016 SPIE Conference Baltimore Convention Center, Baltimore, Maryland, 17 - 21 April 2016.

[22] Tien M. Nguyen and Andy Guillen, "War-Gaming Application for Future Space Systems Acquisition: Part 1 - Program and Technical Baseline War-Gaming Modelling Approach," 2017 SPIE Proceedings, Vol. 10196-7, Anaheim, California, United States, 9 - 13 April 2017.

[23] Tien M. Nguyen and Andy Guillen, "War-Gaming Application for Future Space Systems Acquisition: Part 2 - Acquisition and Bidding War-Gaming Modelling Approaches," 2017 SPIE Proceedings, Vol. 10196-7, Anaheim, California, United States, 9 - 13 April 2017.

[24] Paul Vienhage, Heather Barcomb, Karel Marshall, William A. Black, Amanda Coons, Hien T. Tran, Tien M. Nguyen, Andy T. Guillen, James Yoh, Justin Kizer, Blake A. Rogers, "War-Gaming Application for Future Space Systems Acquisition: MATLAB Implementation of War-Gaming Acquisition Models," Sensors and Systems for Space Applications IX, Proceedings of SPIE Vol. 10196-7, 2017.

Section 4

Materials Science and Engineering

Chapter 7

Thermomechanical Analysis of Ceramic Composites Using Object Oriented Finite Element Analysis

Satyanarayan Patel

Abstract

This chapter discussed the object oriented finite element (OOF2)-based studies for ceramic composites. OOF2 is an effective method that uses an actual microstructure image of the material/composites for simulation. The effect of filler inclusions on the thermomechanical properties (coefficient of thermal expansion, thermal conductivity, Young's modulus, stress and strain) is discussed. For this purpose, various ceramics composites (thermal barrier coating and ferroelectric based) are considered at homogeneous and heterogeneous temperature/stress conditions. The maximum stress is found at the interface of the filler/matrix due to their mismatch of thermal expansion coefficient. Further, residual and localized interface stress distributions are evaluated to analyze the composite's failure behavior. The possible integration of OOF2 with other simulation techniques is also explored.

Keywords: OOF2, Ceramics composite, Object Oriented Finite Element, Thermal stress/strain, Thermomechanical analysis

1. Introduction

Ceramics composites are a significant field of research for the industrialist and researcher [1–6]. This is because of their wide variety of applications and better mechanical properties such as higher strength, toughness or fracture, etc. Apart from the superior mechanical properties, ceramic composites are extensively used for the thermal barrier coating (TBC) [7–9] and nuclear fuel cell [6] applications due to high thermal expansion coefficient (α) [10–12], thermal shock resistance [3, 6] and thermal conductivity (λ) [1, 3, 5, 6, 11]. Many researchers have studied ceramics composites' effective thermal and mechanical properties with changes in compositions using different modeling approaches [4, 5, 13, 14]. However, the prediction of precise thermal and mechanical properties by various numerical or simulation techniques has limited success so far [4]. Several modeling methods are used to obtain ceramics composites' thermal and mechanical properties [4, 5, 13, 14]. These methods are categorized in two ways (i) macroscopic and (ii) microscopic. The macroscopic approach is easy to implement and predicts the average or global response of composites. Thus, it considers the volumetric effect of individual phases in the composites. However, the effect of size, shape, orientation, and arrangement of individual phases related to microstructure is wholly ignored.

This parameter plays a vital role in the effective thermal and mechanical properties of composites. Therefore, many finite element methods (FEM) are widely applied to predict/analyze and improve the thermomechanical properties of composite materials [4, 5, 13, 14]. FEM can easily integrate the microstructural details of composites and are also computationally intensive [4].

In this direction, the Object-oriented finite element method (OOFEM) can be an effective method in the FEM analysis. This method takes care of complexities of microstructure such as the effect of size, shape, orientation, and arrangement of individual phase, particularly in multiple component microstructure [2]. OOFEM is considered the actual microstructure of the materials compared to conventional FEM, where a "unit cell" model is used to predict material properties [2]. Moreover, the microstructure boundary conditions can be easy to implement. In this context, various researchers have used open-source software OOF2 2D version [1, 3, 4, 7, 15, 16]. The OOF2 software is developed by the National Institute of Standard and Technology (NIST) USA, an open-source tool [http://www.ctems.nist.govt./oof/oof2/index:html#download]. The OOF2 modeling uses microstructural images as input and considered individual phase grain size, shape, local orientation and distribution, etc., with their mechanical and physical properties for analysis [4, 5, 10, 15, 16]. It is used to predict composite thermal and mechanical properties such as λ, α, Young's modulus (Y) and thermal/mechanical stress–strain contour on microstructure images. Similarly, the residual thermal stresses (σ_r), the effect of thickness, λ and cracking in TBC due to stress relaxation are also investigated [3, 8, 9, 11, 17]. Additionally, these (σ_r, Y, λ and α) properties of composite with respect to filler content and operating conditions are analyzed. In this direction, numerous composites, i.e., $0.94Na_{1/2}Bi_{1/2}TiO_3$–$0.06BaTiO_3/ZnO$ (NBT-6BT-ZnO), Ni-Al_2O_3, Al/B_4C, Al-TiB_2 Al-MgO, WC-Al_2O_3, AlN-TiN, and Al-TiO_2 are investigated and results are supported by various other techniques [2, 4, 10, 12, 18–21].

S. Patel *et al.* [2, 19–21] predicted the mechanical and thermal properties of NBT-6BT-ZnO, Al-MgO, WC-Al_2O_3 and AlN-TiN composites by the OOF2 method. The simulation results are validated with different analytical models. They show that an increase in the filler content increases the local stress concentration, which can be the start point of failure in the material. S. Patel *et al.* [2, 19–21] also studied filler orientation, gradient and uniform temperature environment effect on the thermomechanical properties. Neeraj *et al.* [4, 10] obtained Y of Ni-Al_2O_3 composite by OOF2 and compared it with the ultrasonic measurement with the possibility of crack initiation due to stress distribution. Andrew *et al.* [16, 22] studied the complex microstructure using the OOF2 FE tool and reported that the quality of the results depends on the quality of the generated mesh. The set of generics is modified, which results in improve the quality of the 2D mesh. Elomari *et al.* [23] have analyzed thermal expansion behavior between matrix and reinforcement in terms of silica layer formed during oxidation, size, and thermal stress. Chawla *et al.* [13, 14] performed a microstructure-based simulation to predict the thermomechanical behavior of composite. Plasma-sprayed Y_2O_3-stabilized ZrO_2 analysis was carried out to measure Y of the actual coating and compared with experimentally observed values [24]. The magnitudes of σ_r in textured and untextured Al_2O_3 are obtained concerning grain orientations with the help of OOF2 [17]. Thermal shock resistance and λ of yttria-stabilized zirconia (YSZ)-Al_2O_3 and $3Al_2O_3{\cdot}2SiO_2$ are computed and compared by analytical methods [3]. Further, λ is analyzed for 4-phase and 3-phase composite of Y_2O_3 stabilized ZrO_2-Al_2O_3-$MgAl_2O_4$-$LaPO_4$ and CeO_2-$MgAl_2O_4$- $CeMgAl_{11}O_{19}$, respectively to use them with nuclear fuel [6]. Moreover, porous W/CuCrZr composites micrographs with OOF2 are simulated for tensile deformation and thermal conduction behavior [7].

Recently, OOF2 was used to obtain the stress distribution and Young's modulus in the thermally treated Mg-9 wt.%Li-7 wt.%Al-1 wt.%Sn alloy to enhance the wear resistance [24–26]. Further, continuously reinforced concrete pavement α correlation with spalling, performance and mechanical behavior is also studied with OOF2 and results are compared with commercial FE software [27]. Furthermore, the λ and thermal effusivity are simulated for the different regions of TBCs as synthesized by thermal spray processes and suspension plasma spray for gas turbine applications [28]. It was found that modeling provides good results [28]. Moreover, Si_3N_4 welding with 316L stainless steel is designed using Mo/Ag composite as interlayer and the residual tensile stress in the interlayer is performed [29]. It can be said that the OOF2 provides a better transition between the mechanical/thermal behavior of a heterogeneous material at the macro-scale and the mechanical/thermal response of its constituent phases. It provides adequate information about microstructures/phases (volume fraction, distribution, orientation) to ensure a realistic assessment of thermo-mechanical properties [30–32]. More recently, a study was performed on ferritic-pearlitic based steel and found that the predicted results was in good agreement with the experimental results [31]. They found that the microstructural morphology plays a vital role in strain partitioning, strain localization and formability of the ferritic-pearlitic steels [31].

Recently, OOF2 is also used for meshing and FE solutions are obtained by integrating meshing with ABAQUS or MATLAB. In this direction, many researchers have used OOF2-ABAQUS to get the shear stress distribution [26, 33], elastic modulus, crack distribution in the materials/composites [32, 34], residual stresses [29], and thermal properties [28]. Devi lal *et al.* [24, 26] fabricated 7 wt.% Y_2O_3 stabilized ZrO_2 beam to study the presence of microcracks/pores during bending by OOF2 with ABAQUS. Similarly, crack analysis of duplex stainless steel (ferrite + austenite phases) strength and hydrogen diffusion characteristics are investigated at the microstructure scale [30, 32]. Most recently, the space charge distribution between two-grain boundaries or interface in bulk is also studied [35]. Moreover, to generate input data sets for machine learning, an OOF2 based simulation is performed with a variety and radii of pores for brittle porous materials failure analysis [36]. As discussed above, OOF2 is used by various researchers for numerous applications and integration with other software. Hence, this work focused on the ceramic composites' thermal/mechanical properties prediction.

This chapter discussed the detailed OOF2 analysis procedure with boundary conditions and assumptions for ceramics composites. The thermal (λ and α) and mechanical (Y) properties are predicted and compared with other analytical methods. In FE analysis, thermal stress–strain contour and heat flux are used to indicate the different temperature conditions on microstructure images. The local stress and σ_r distribution interpenetrating phase and particle-reinforced structure are also studied.

2. Method

2.1 Microstructure image and mapping

Figure 1(a) shows the workflow procedure of OOF2 analysis. The complete OOF2 simulation and analysis process consists of six steps. In the first step, a microstructure image is needed as input; for this purpose, materials scanning electron microscopy (SEM) or back-scattered electron (BSE) images can be used. A BSE microstructure image of NBT-6BT/ZnO composites contains 4% ZnO by volume is used for analysis, as shown in **Figure 1(b)**. The image consists of a length of 24 μm

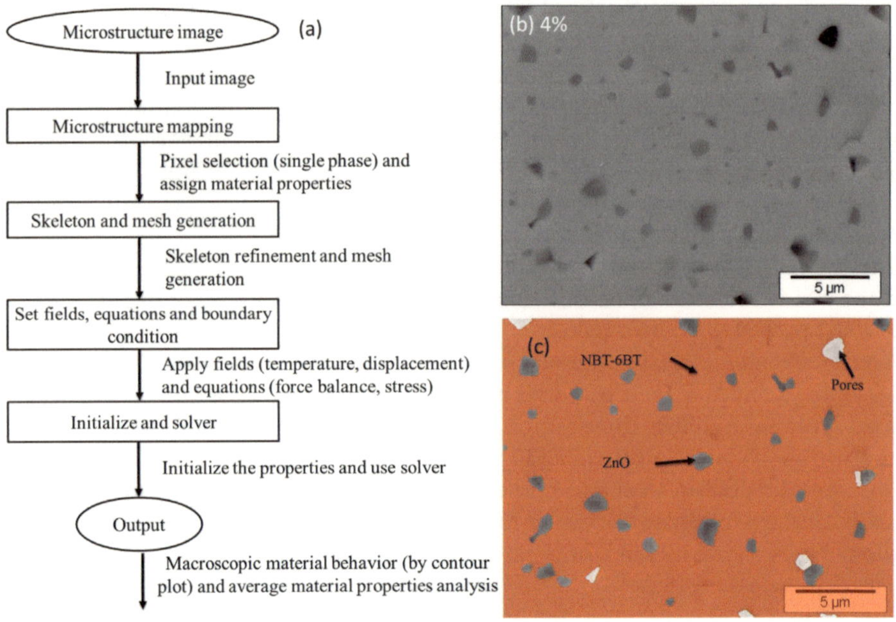

Figure 1.
(a) Outline of OOF2 analysis procedure (b) microstructural image of composites having NBT-6BT-4% ZnO, (c) single-phase material pixel selection, assign group and properties. (Reprint with permission from Ref. [2]. Copyright© 2020 Elsevier BV).

and size of 645 × 484 pixels. In the BSE microstructure, contrast is used to differentiate NBT-6BT (matrix) and ZnO (inclusion or filler phase). The single-phase material pixels are selected by pixel selection and grouped to give each phase thermal and mechanical properties. **Figure 1(c)** illustrates that the red color pixel belongs to NBT-6BT and the gray color pixel represents the ZnO. It is to be noted that the composite is not fully dense; thus, pores (~4%) are also obtained, as shown in **Figure 1(c)** in white color. Then selected single phase pixel materials are assigned with their thermal and mechanical properties, i.e., Poisson ratio, Y (GPa), λ (W/m-K) and α (K^{-1}).

Similar kinds of pixel selection and properties assignment are used by many researchers where two-phase materials or composites such as Ni-Al [37], Al-MgO [19], WC-Al_2O_3 [20] and AlN-TiN [21] for various applications. Moreover, a varying ZnO composition from 4–10% is also studied by Manish *et al.* [2] for NBT-6BT-ZnO composites. In the next step, microstructural finite element meshing is created by a skeleton, as shown in **Figure 2**. The geometry of the mesh explained the geometry of the skeleton. It does not cover the information about the finite element shape functions, equations, fields, etc. The skeleton represents the FE discretization of the microstructural. It is composed of quadrilateral and triangular elements, nodes, and segments. In the OOF2, the skeleton is refined by various tools, as discussed in Section 2.2.

2.2 Skeleton modification and refinement

Figure 2(a) and **(b)** show the nodes of a skeleton element where nodes are on the element's corner and edges. However, the other FEM tools generally do not consider the nodes along their edges or interiors. In the refinement process, nodes of the edge move along the edge only. Moreover, the interior nodes are free to move

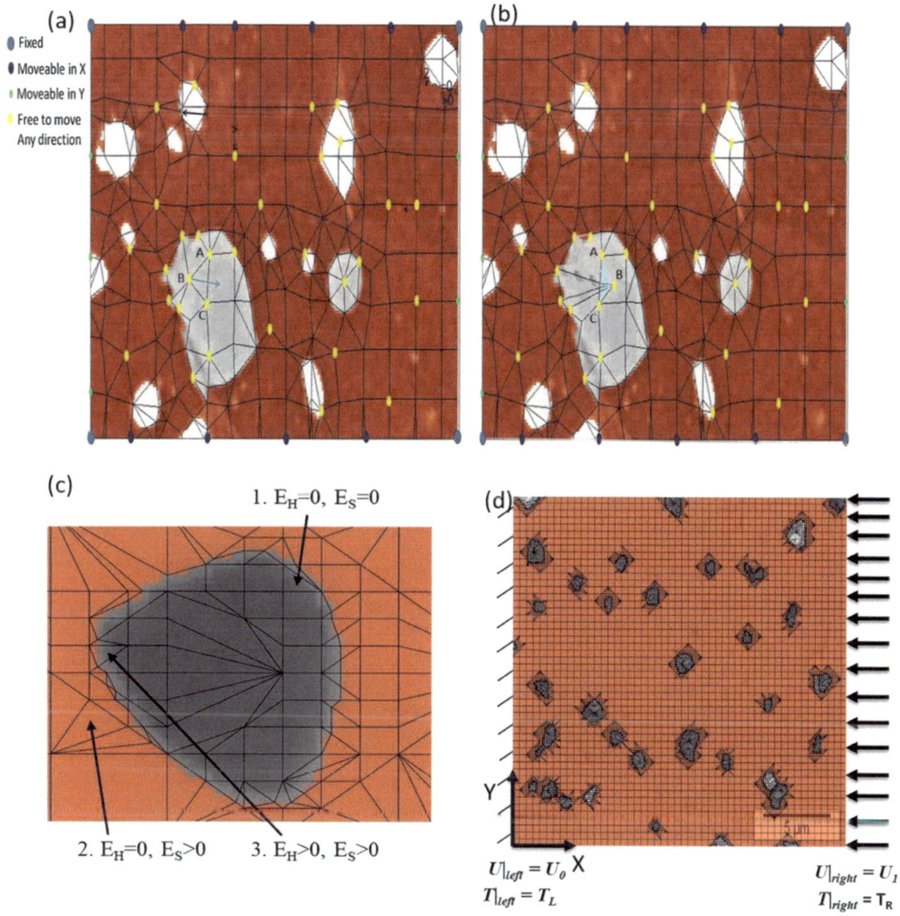

Figure 2.
(a) and (b) Microstructural meshing with node movement, (Reprint with permission from Ref. [2] Copyright © 2014 Wiley Publishing Ltd) (c) energy functional approach used during skeleton mesh refinement process [38], (d) final mesh with boundary conditions of NBT-6BT-4% ZnO (Reprint with permission from Ref. [21] Copyright© 2020 Elsevier BV).

anywhere. These node's movement can also be restricted by the pin node tool. In the **Figure 2(a)**, an element (A, B, C) is shown, which can create an illegal element via movement, as shown in **Figure 2(b)**. The node 'B' motion (see **Figure 2(a)**) generates two illegal elements (**Figure 2b**). The illegal element refers to an element that breaks the ordering; three collinear nodes also create illegal elements (nonconvex quadrilaterals). The highlighted quadrilateral element represents the illegal element because its nodes (A, B, C) are not convex (**Figure 2b**).

This type of node movement and element creation depends on the mesh energy parameters. Further, the FE skeleton elements refinement process increases the mesh quality and reduces functional energy parameter ($\ddot{E}$) [36]. Two types of effective energy function maintain the mesh quality as shape energy ($\ddot{E}_S$) and homogeneity energy ($\ddot{E}_H$). The $\ddot{E}_H$ is given as [15]:

$$\ddot{E}_H = 1 - \ddot{E}_{HM} \tag{1}$$

where $\ddot{E}_{HM}$ is microstructural components homogeneity energy. $\ddot{E}_H$ can be reduced when all components of the microstructure are encompassing a single-phase material. In the case of triangular element $\ddot{E}_S$ is defined as [13]:

$$\ddot{E}_S = 1 - 4\sqrt{3}\,A_T/l^2 \tag{2}$$

where A_T and l are triangle element area and side length, respectively. For the equilateral triangular component, $\ddot{E}_S$ is zero and equal to one when all the vertices are collinear (zero aspect ratio). It can also be estimated for quadrilateral elements using a quality factor γ_i for each corner i as [16]:

$$\gamma_i = \frac{A_{par}}{l_0^2 + l_1^2} \tag{3}$$

where A_{par} is, the area of the parallelogram formed by l_0 and l_1 edge length element converges at node i. $\gamma_i < 1$ at a corner where the two converging edges have different lengths or meet at an acute/obtuse angle. $\gamma_i = 0$ when edges are collinear or the length of one edge is zero (degenerate cases). Shape energy for a quad element is estimated by weighted average from the γ_S at the corner with the minimum γ (γ_M) and the corner opposite to it (γ_0). The weighted average method is utilized to make sure that each node movement affects the total energy and quadrilateral element $\ddot{E}_S$ can be defined as [13, 15]:

$$\ddot{E}_S = 1 - [(1-\eta)\bullet\gamma_m + \eta\gamma_0], \tag{4}$$

where $\eta = 10^{-5}$ is an arbitrary small parameter. It is needed to avoid pathologies that arise due to the shape energy has no dependence on the position of one of the nodes. The energy functional parameter ($\ddot{E}$) is expressed by [13, 16]:

$$\ddot{E} = \delta\ddot{E}_H + (1-\delta)\ddot{E}_S, \tag{5}$$

For $\delta = 0$, $\ddot{E}_S$ will note benefaction in $\ddot{E}$ of the elements. There will be a trade-off between $\ddot{E}_S$ and $\ddot{E}_H$ when $0 < \delta > 1$. Some skeleton refinement processes are used to achieve the optimum homogeneity and shape of the element. These energy functions elements are depicted in **Figure 2(c)**.

The homogeneity energy ($\ddot{E}_H$) of the elements is contributing the total $\ddot{E}$ when $\delta = 1$ and skeleton refining δ is several according to homogeneity index. For the mesh refinement, firstly, we focus on reduce $\ddot{E}_H$ of an element by assigning a higher value of δ. **Figure 2(c)** displays different types of $\ddot{E}_S$ and $\ddot{E}_H$ components. In the equilateral triangular element 1, as shown in **Figure 2(c)**, the contribution of $\ddot{E}_S$ is zero. Further, element 1 also contain single material; hence the $\ddot{E}_H$ value of the element is also zero. $\ddot{E}_S$ rise for elements 2 and 3 as the triangular shape deviates from that of an equilateral triangle. Similarly, element 2 also has $\ddot{E}_H$ as zero because of homogeneous phase content. Therefore, $\ddot{E}_S$ and $\ddot{E}_H$ of elements 2 and 3 contribute to the total energy function parameter ($\ddot{E}$).

Subsequently, the skeleton is generated and modified by various refinement tools, i.e., snap refine, anneal, snap node, merge triangle, refine, smooth, etc. In the snap anneal process, each node tries to move and turns to a pixel boundary. Snap refine skelton modifier tool combines the most likable refine and snap nodes features in a single method. It is subdivided into components along pixel boundaries. Snap refine introduces additional edges and new nodes that can be made to follow pixel category boundaries. So, it can also be more flexible than snap nodes in fitting a skeleton, as shown in **Figure 2(c)**. In a smooth tool, nodes move to the average positions of their nearby element. The node satisfied the acceptance criteria and the move is accepted. Smoothing of the skeleton requires several iterations. The merge triangle modifier tool is used to merge neighboring homogeneous triangles to form a quadrilateral element. Rationalize is used to fix an irregular-shaped component of a microstructure skeleton by removing their immediate neighbors. Relax tool moves nodes and improves element shape and homogeneity. Thus, each refinement tool has its refinement process. The final skeleton explains only the mesh's geometry; it does not cover any information about the finite element shape functions, field, equations, etc. The skeleton creates FE mesh, and the final mesh generation is shown in **Figure 2(d)**. The accuracy of FE solutions be dependent on the homogeneity index and mesh quality. The FE mesh has a 0.98 homogeneity index with 6039 nodes and 8373 elements [38].

2.3 Micro-scale fields and equations

In the case of temperature and displacement applied on the image, the force balance equation and temperature field should be defined in-plane. Fields are defined on two and more graphically overlapping subproblems on a FE mesh. For each subproblem, the field's value is the same. These values are only stored on a FE mesh. OOF2 used 3D material but solved 2D problems. OOF2 has a generalized plane strain and plane stress approach, implemented by an in-plane elastic modulus. In this software, two types of equations are used plane-flux and divergence. It is activated and deactivated on subproblems. The static divergence equation is expressed as [39]:

$$\nabla \cdot flux + force = 0 \tag{6}$$

where force is generalized force, i.e., heat equation ($\nabla \cdot J = -\partial U/\partial T$, where $-\partial U/\partial T$ is a change of energy density) and force balance equation ($\nabla \cdot \sigma = f$, where f is actual force). The nonstatic equation is a general form of divergence equation (-time-dependent version) and given by [39]:

$$M\frac{\partial^2}{\partial t^2}field + C\frac{\partial}{\partial t}field + \nabla \cdot flux + force = 0, \tag{7}$$

Heat equation for the first-order problem, M term will be zero. C is the second problem that may be zero. Plane flux equation is a generalization of plane stress and the out of plane component flux is zero and can be written as [39]:

$$(Flux)_z = 0 \tag{8}$$

It can solve nonlinear Eqs. (6)-(8) if the nonlinearity is limited to force and flux terms.

2.4 Boundary conditions

OOF2 FE tool provides five boundary conditions: Dirichlet, floating, periodic, Navman, and generalized force. This work uses Dirichlet boundary conditions on the microstructure image, as shown in **Figure 2(d)**. The Dirichlet boundary condition has identified the value of one component of a field along with a boundary. In both cases, the temperature and displacement application force balance equation and temperature field are defined in-plane. However, when plane strain is used, then the displacement is defined, active, and in-plane. Similarly, for thermal stress force balance equation is used when the temperature is defined, active, and in-plane. The periodic boundary conditions are applied on the left and right vertical edges. The microstructure edge is considered at a constant temperature and/or displacement. In the present work, boundary conditions are depicted in the

Figure 2(d), where the left side edge is fixed or at zero displacements (i.e., $U|_{left} = U_0 = 0$). When a temperature gradient is used, the same edge is considered under a low temperature ($T|_{left} = T_L$). However, the right edge is kept as defined displacement ($U|_{right} = U_1$). When a temperature gradient is used, the same edge is considered for higher temperature ($T|_{right} = T_R$). Moreover, the governing equations are solved iteratively and the convergence criterion is fixed as 10^{-13}. Finally, thermal or elastic stress/strain contours are plotted for the analysis (discussed in Section 4). It is to be mentioned that these steps are previously discussed in detail by several researchers [2, 13, 16, 19, 20, 40].

2.5 Assumption

The matrix and individual filler properties are considered isotropic, linear elastic and homogeneous. It is considered that matrix and filler interfaces have good bonding. The homogeneity index of mesh is obtained as 98% and the same is used for simulation. In the selected temperature range, material properties are assumed to be constant. In the simulation 2D microstructure is used; thus, the effect of thickness is not considered. The OOF2 software maintained vector (displacement), scalar (temperature) and tensor fields on two and more graphically overlapping sub-problems of a FE mesh. In the case of uniform temperature application, all the edges temperature fixed at the same temperature. The basic iterative matrix equation solver is applied to obtain the solution. The symmetric and asymmetric matrixes are solved using Conjugate Gradient and generalized minimum residual method, respectively. In both cases, a Bi-conjugate gradient with the incomplete lower upper (ILU) preconditioner is used for FE simulation. Once field equations achieve the solution, thermal and mechanical properties are obtained with the help of averaging the applied field.

3. Comparison of analytical and simulation analysis

3.1 Young's modulus

Young's modulus (Y) is predicated using OOF2 analysis and results are compared with the analytical method. It is well known that the NBT-6BT ceramics is brittle in nature and have lower fracture toughness. Therefore, ZnO addition can improve the *Y* and elevate other properties of the composite. The left edge ($d|_{left}$) of microstructure is considered at zero displacement and right edge ($d|_{right}$) is given a uniform displacement as:

$$d|_{left} = d_0(d_x = 0) \quad (9)$$

$$d|_{right} = d_1(d_x = 0.001) \quad (10)$$

A similar boundary condition is depicted in the **Figure 2(d)**. Y of composites is calculated by:

$$Y_{avg} = \frac{stress}{strain} = \frac{\sigma_{avg}}{\varepsilon_{avg}} \quad (11)$$

Where ε_{avg} is average strain and σ_{avg} is average stress. The small displacement is applied to the microstructure; hence, it can be assumed that only elastic deformation occurs. The calculated Y is given in the **Table 1**.

The predicated Y is also compared with various theoretical models, i.e., Reuss rule of mixture [41], Voigt rule mixture [42] and Hashin and Shtrikman [43] are also given in **Table 1**. Many researchers considered these methods to estimate *Y* of composites and compared them with OOF2 results [2, 13, 19, 44, 45]. The estimated value by OOF2 is in close agreement with that of theoretical methods values. Similar studies are also considered with various compositions by variation of filler content [2, 13, 19, 44, 45]. As the filler volume increases, the effective Y of composites also increases in most cases. However, the composites Y variations depend on the difference between matrix and filler Y.

3.2 Thermal expansion coefficient and thermal conductivity analysis

In order to predict the thermal properties of NBT-6BT- ZnO composites, the α is estimated by OOF2 analysis. The α is predicted when the microstructure edges are at a 10°C temperature difference. The thermal boundary conditions are depicted in the **Figure 2(d)**, where the left side edge is kept at low temperature ($T|_{left}$ = T_L = 30°C). However, the right edge is kept at a higher temperature ($T|_{right}$ = T_R = 40°C). Moreover, the bottom and top edges are kept as adiabatic. The α is obtained by:

$$\alpha = \frac{\varepsilon_{avg}}{\Delta t} \tag{12}$$

where ΔT is the temperature difference. The α predicted with the help of OOF2 analysis is given in **Table 1**. The α is also estimated with the help of different theoretical methods, i.e., Kerner's model [46], Rule of a mixture, Turner's model [47], and Schapey's model [48] and given in **Table 1**. It can be said that the OOF2 predicted value of α is in good agreement with different theoretical methods. A similar analysis is performed with different filler (ZnO) compositions based composited and found that as the filler content increases, α decreases [2]. This is because the filler's α is lower than the matrix α. A similar study is also performed to analyze the behavior of α in composites with different filler content by many researchers [1, 2, 12, 13, 49].

Figure 3(a) and **(b)** show the boundary conditions for WC-Al_2O_3 and AlN-TiN based composites to obtain the λ. In this case, the top and bottom edges are considered insulating to ensure heat transfer in one direction only.

The heat flux equation is solved by the conjugate gradient method with the help of a linear solver. The equivalent effective λ of composites is expressed by Fourier's law:

Methods	Young's modulus (GPa)	Methods	The effective α ($\times 10^{-6}$°C^{-1})
OOF2	132.80	OOF2	6.41
Voigt	136.08	Rule of Mixture	6.89
Reuss	134.98	Kerner's Models	6.86
Hashin and Shtrikman lower limit	135.53	Turner' Models	6.81
Hashin and Shtrikman upper limit	135.41	Scharpery' Models	6.86

Table 1.
Comparison of Young's modulus (GPa) and the effective α ($\times 10^{-6}$°C^{-1}) obtained using OOF2 analysis and various theoretical methods [2].

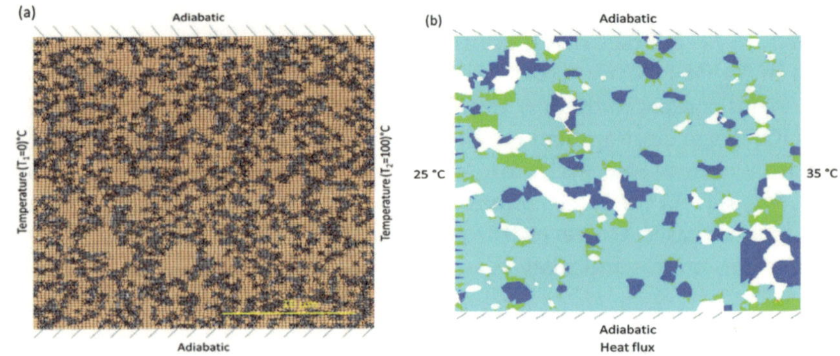

Figure 3.
(a) Boundary condition for WC-Al_2O_3 (Reprint with permission from Ref. [20] Copyright© 2014 Imperial College Press) and (b) heat flux profile of AlN-5%TiN composites for thermal conductivity analysis. (Reprint with permission from Ref. [21] Copyright© 2014 Wiley Publishing Ltd).

$$\lambda = A\frac{Q_{avgdx}}{dT} \tag{13}$$

where Q_{avg} is average heat flux, A is the cross-sectional area, dT is temperature difference and x is width of the microstructure.

The λ of both composites WC-Al_2O_3 and AlN-5%TiN is predicated by OOF2 and found that results agree with various theoretical methods/models (effective-medium theory arithmetic and harmonic mean model, geometric mean model Lewis and Nielsen method, and Maxwell method) [20, 21]. A similar investigation is performed with various Al_2O_3 and TiN content and found that as the Al_2O_3 and TiN increase, λ of composites decreases and increases, respectively [20, 21].

Thus, the increase or decrease of composites λ is dependent on the filler's λ. The Y, α and λ analysis shows that OOF2 can effectively predict composites' thermal and mechanical properties. In order to look in-depth at the thermal and elastic stress/strain behavior under uniform and gradient temperature conditions, various contour plots are discussed in Section 4.

4. Microstructural stress-strain analysis

4.1 Stress-strain analysis under uniform temperature

A homogenous temperature is applied to the microstructure and thermal stress distribution is analyzed in the composites. **Figure 4(a)** shows typical thermal stress (σ_{xx}) contour plots when a homogenous temperature of 500°C is applied on NBT-6BT-4%ZnO composite. It is found that the stresses are highest at the sharp edges/interfaces of matrix and filler. At the interface, the stress distribution region is varied and nonuniformly distributed in the filler region. This is because the stress variation depends on the size of the filler particles/grains [50]. If the applied temperature is sufficiently large, local stress can reach above the yield strength, resulting in thermal cracking of the composite [17]. It is important to note that a critical radius is needed for the thermal or residual stress-based crack initiation in the composites. However, most of the case length of maximum stress is small compared to the critical length of crack initiation. Thus, the residual or thermal stress does not form a crack in the composites. Therefore, composites will not fail under such circumstances and can be utilized at higher temperatures. Moreover, when the composite is cool down to room temperature, the residual stresses play a

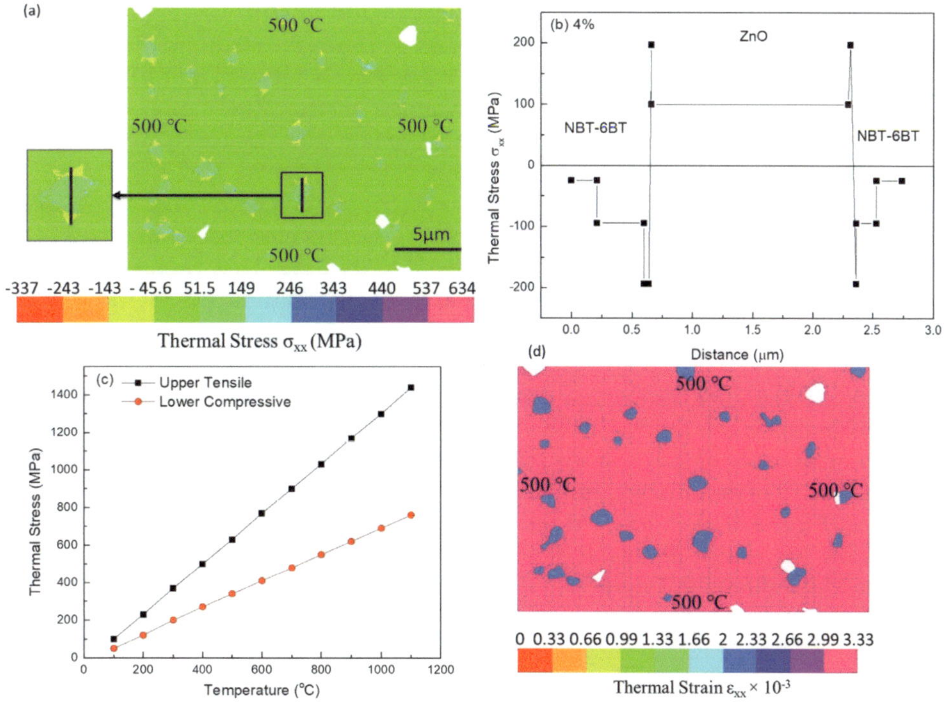

Figure 4.
Thermal stress (a) contour plots when a homogenous temperature of 500°C is applied on microstructure/image and (b) variation across the line scan as shown in (a), (c) thermal stress variation under different uniform temperature conditions, (d) thermal strain contours when a homogenous temperature of 500°C is applied. All plots are shown for the NBT-6BT-4%ZnO composite. (Reprint with permission from Ref. [2]. Copyright© 2020 Elsevier BV).

vital role. The magnitude of the residual stress varies according to the applied temperature; however, the exact value of residual stress is hard to predict by OOF2. Further, the filler has more stress distribution compared to the matrix because filler/ZnO have lower α (4.3×10^{-6}°C^{-1}) compared to matrix/NBT-6BT (7×10^{-6}° C^{-1}) [2]. However, if the matrix has lower α compared to filler, then vice versa stress distribution can be observed. This type of stress behavior is reported for AlN-TiN composites where filler (TiN) and matrix (AlN) has an α of 9.1×10^{-6}°C^{-1} and 4.6×10^{-6}°C^{-1}, respectively [21].

In order to look in-depth at stress distribution around the filler/matrix interface, a line scan is used (zoom part in the **Figure 4(a)**). The corresponding stress variation is depicted in **Figure 4(b)**. At the interface, stresses increase to ~10 and 3 times of average stress in matrix and filler regions. Moreover, the filler region has ~3 times higher stress distribution compared to the matrix region. This is because of the difference between Y and α of filler and matrix of composites. It is found that filler consists of tensile stress of ~60 MPa, whereas matrix has a compressive stress of ~20 MPa. The compressive and tensile stress distributions are observed in different regions of composites. When the temperature is applied, one material expands more than others because of a mismatch of α, which compresses the other materials in some regions. However, in NBT-6BT-ZnO composites, the compressive stress value is lower than tensile stress.

In the same way, the homogenous temperature is varied from 100 to 1100°C and contours are plotted at each temperature. It is found that stress distribution is similar to **Figure 4(a)**, whereas the magnitude of stress is varied according to applied temperature. **Figure 4(c)** shows the maximum stresses (tensile and compressive) when a homogenous temperature of 100 to 1100°C is applied. As the applied temperature increases, both tensile and compressive stress increases. This is because of that at higher temperature large mismatched between α of filler and matrix is obtained. A similar trend is also observed in other ZnO content compositions. Moreover, it was found that as the filler content increases magnitude of the stress (tensile and compressive) also increases. It is because the higher volume content of filler increases the larger surface area for the mismatch of α. A similar stress–strain simulation is performed on various ceramics composites such as Al_2O_3-SiC [51], Ni-Al_2O_3 [10], WC-Al_2O_3 [20], AlN-TiN [21], etc. They found that with an increase in temperature and filler volume, stress distribution also increases. The detailed discussion on the interface stress variation due to mismatch of the α and Y in composites is also discussed by Chawla *et al.* [13, 14] and Wang *et al.* [45].

Figure 4(d) shows the thermal strain contour when a homogenous temperature of 500°C is applied on the NBT-6BT-4%ZnO composite. The matrix has a high strain compared to filler due to the mismatch of α. Further, strain looks homogenous in both filler and matrix, resulting in constant stress in NBT-6BT and ZnO. However, **Figure 4(a)** shows that the stresses are nonuniformly distributed in the composites. Moreover, **Figure 4(d)** shows positive strain only, whereas negative stress is also observed in **Figure 4(a)**. When a temperature is applied on the composites, each grain/interface regions of matrix/filler consist of different thermal strain distribution, which results in stress variation/distribution at the interface. At some point, the interface experiences large compressive stress that can be negligible or act on a small length compared to tensile stress. Further, it was found that as the applied temperature increases, the strain also increases. It is a result of an increase in α with temperature for both filler and matrix. The thermal strain variations also alter with an increase in ZnO/filler content in compositions. This is because the higher volume content of filler increases the larger surface area for the mismatch of α. In the same way, Al-MgO [19], WC-Al_2O_3 [20], Ni-Al_2O_3 [4, 10], AlN-TiN [21], etc., are also studied to look at the effect of uniform temperature and filler content. Apart from these, numerous other composites are also studied for better strain control via tailoring α of the filler/matrix [5, 10, 18, 19, 23, 41, 51]. Moreover, the effective α in composites analysis is potentially used in TBC and thermal resistance-based applications [7–9].

4.2 Stress-strain behaviors under gradient temperature

The OOF2 analysis is performed under a gradient temperature condition of 0–800°C for composites (AlN-5%TiN). For this purpose left and right edge is applied with a temperature of 0°C and 800°C. However, the bottom and top edges are considered adiabatic, as depicted in **Figure 5(a)**. The resultant thermal stress contour under gradient temperature is shown in the **Figure 5(a)**. It is observed from **Figure 5(a)** as the temperature increases from left to right, corresponding stresses also increase. The stresses are found to be maximum at the interface of the filler/matrix and nonuniformly distributed. Further, the stress distribution is localized at the filler region and increases composite failure/crack risk. As the filler content increases, the risk of failure is considerably higher due to clustering. The inclusion of the filler can decrease or increase the stress distribution in composites which again depends on the difference of Y and α of matrix/filler. Moreover, the possibility of crack generation depends on the length maximum stress distribution at the interface or critical length of the crack. The reason for the same is already discussed in Section 4.1 and the same is also valid here. Interestingly, it is found that the stress distribution largely varies according to the orientation of the filler grain. The inset of **Figure 5(a)** shows stress variation around two grains (i) and (ii). It can be seen

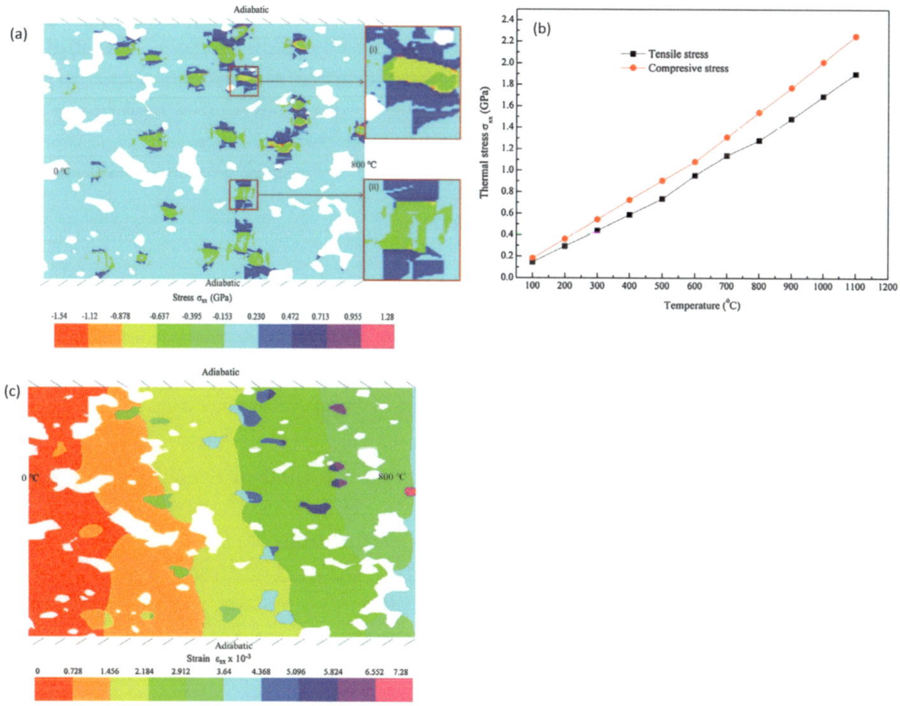

Figure 5.
(a) Thermal stress contour when a gradient temperature of 0–800°C is applied, (b) maximum upper (tensile) and lower (compressive) thermal stress and (c) thermal strain contours when a gradient temperature of 0–800° C is applied for AlN-5%TiN composites. (Reprint with permission from Ref. [21]. Copyright © 2014 Wiley Publishing Ltd).

that in the region (i) (**Figure 5(a)**), filler/grain is aligned with the applied temperature consists of higher stress compared to grain (ii) which is almost perpendicular to applied temperature. However, both the grain (i) and (ii) regions have similar temperature zones. Alike behavior is also observed in entire composites. Similar behavior is also observed in numerous compositions of Al-MgO [19], WC-Al_2O_3 [20], NBT-6BT-ZnO [2], Ni-Al_2O_3 [4, 10], etc. Further, with an increase in filler compositions, the magnitude of the stress increases, whereas the grain orientation behavior remains the same. It can be said that the various strength of composites can be fabricated via preferred grain orientation. Moreover, a large amount of thermal/residual stress can be increased/decreased in composites according to the direction of the grain (transverse and longitudinal). Hence, if composite is synthesized by oriented grain, i.e., filler as platelets/nanorods instead of power can show better performance. Recently, $BaTiO_3$:100xZnO composite ceramics are made with a micrometer and nanometer scaled power, supporting these observations [52].

Similar to Section 4.1, tensile and compressive stress is observed in the gradient temperature conditions, whereas the magnitude is lower than that of uniform temperature conditions. The tensile and compressive stress increases with applied temperature, as shown in the **Figure 5(b)**. The left side edge temperature is fixed as 0°C, whereas the right side temperature varies from 100 to 1100°C. The stresses linearly increase with the applied temperatures. Several researchers have reported alike trends in various composites. Thermal strain contour under a gradient temperature condition of 0–800°C for composites (AlN-5%TiN) is presented in **Figure 5(c)**. It is observed that as the temperature increases from left to right, the corresponding strain also increases. The filler has a higher strain than the matrix because of filler has higher α compared to the matrix. Further, a large number of pores are observed in the AlN-TiN composites. When a temperature is applied, filler/matrix can easily expand at pores; hence, the variation in stress distribution is not observed. These types of strain contour and analysis are also discussed in the literature [2, 5, 13, 20]. It can be said the OOF2 can be an effective tool to see the effect of grain orientation or stress concentration at the interface.

4.3 Elastic stress-strain analysis

The elastic stress-strain is studied in the composite by deforming microstructure, as presented in **Figure 6**. Thus, the left edge is kept as fixed $U|_{left} = 0$ and the right edge $U|_{\text{right}} = 0.3$ μm is deformed and resultant stress-strain contours are depicted in **Figure 6** for Al-10% MgO composites. **Figure 6(a)** shows that the matrix (Al) has lower stress as compared to filler (MgO). This is due to the fact that the matrix has lower Y than filler as 70 GPa and 294 GPa, respectively [19]. Further, maximum stresses are obtained at the interface of the filler and matrix in an entire composite. It gives crucial information about the composite that interfaces play a vital role in effective load transfer between filler and matrix. The sharp edges of the grain have enormous stress (stress intensification occurs) compared to smooth edges. These edges of filler act as stress concentration sites and are more pronounced when filler/grain is oriented in the direction of the applied load. Hence, if the uniform size of grain/filler is added to the matrix, then the strength of the composites can be improved. Similar behavior is also observed when the filler content is increased from 5–15%. However, the magnitude of stress increases with the inclusion of filler due to more frequent particle-to-particle contact. It is also observed that grain orientation also plays a vital role in stress distribution. The effect of filler orientation on elastic stress distribution is the same as discussed in Section 4.2.

The stress distribution can help to obtain the composite failure analysis. If the length of stress distribution (higher than yield stress) is more than the critical radius, crack generation will occur. Another software integration or analytical calculation is needed to find the exact failure stress analysis because OOF2 cannot provide direct failure analysis. The elastic behavior under various deformed conditions in different composites is previously studied [2, 5, 13, 20, 21]. They also found a similar type of stress–strain contour. However, the magnitude of the stress varies according to composite compositions. Effect of Y in matrix and filler interface, load transfer and orientation effect in composites are discussed by Chawla *et al.* [13] and

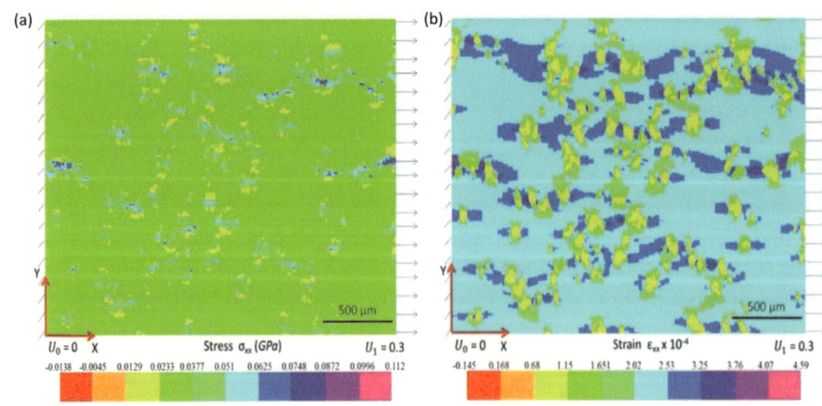

Figure 6.
(a) Elastic stress (b) elastic strain contours for composites under μm displacement. Al-10%MgO composites. (Reprint with permission from Ref. [19]. Copyright © World Scientific Publishing Company).

Wang *et al.* [33], respectively [45]. They found that closely spaced filler encourages more deformation within the matrix. Moreover, the neighbor of weak filler zones evolves and, in the end, leads to mechanical failure. These low and high stresses can be calculated and used to forecast the composite's failure displacement and temperature. Additionally, **Figure 6(b)** depicts the contour of strain variation under the elastic displacement condition. The matrix has a higher strain compare to filler and is concentrated at interfaces. A higher strain is observed at closely placed filler because of a large degree of constraint region. Other researchers in different ceramic composites previously observe a similar effect. This region can lead to failure of the composites; in this direction, matrix, filler and filler-matrix interface failure criteria are provided by Ha *et al.* [53]. It can be said that the geometric inhomogeneity in composites produces the nonuniform strain/stresses due to mechanical or thermal loadings. Thus, failure can initiate at any point within the composite, filler, matrix, or filler-matrix interface and consists of different failure processes according to the initiation of critical points. Recently, a similar boundary condition (shown in **Figure 6**) is also applied for NBT-6BT/ZnO composite to obtain the elastic stress–strain response [2, 38]. It was found that the stresses are concentrated at the sharp edges or interface of the composites and increase with the ZnO concentration. However, strain is uniform throughout the composites for particular compositions and varied with ZnO content [2, 38]. Further, the predicated stress–strain are compared with experimental data, which shows good agreement and validates findings by OOF2 [2].

In summary, ceramics composites have a wide variety of applications in aerospace, defense, energy, medical and transport sectors due to higher strength, toughness or fracture, etc. [54]. They are also extensively used for the TBC [7–9] and nuclear fuel cell [6] applications due to high α [10–12], thermal shock resistance [3, 6] and better λ [1, 3, 5, 6, 11]. However, predicting precise ceramics composites' thermal and mechanical properties by numerical or simulation techniques has limited success so far. In this direction OOFEM based analysis only take care of the effect of size, shape, orientation, distribution and arrangement of individual matrix/filler phases during the analysis. Hence, considering these individual phases properties, it is used to predict ceramics composite λ, α, Y, σ_r, wear, cracking analysis and thermal/mechanical stress–strain contour on microstructure images. It can be said that OOF2 is an effective tool to obtain real microstructural analysis compared to other existing FE analysis techniques. Moreover, it provides the input meshing/parameter for various commercial software to achieve real-time thermal/mechanical/electric analysis. Thus, ceramic composites' effective thermal/mechanical properties (λ, α, Y) should be analyzed with OOF2 to obtain the real interaction between filler/matrix or distribution of σ_r, wear and crack in individual phases.

4.4 Integration of OOF2 with other software

The OOF2 software is also integrated with other software where composite failure analysis is performed. In this direction, OOF2 is utilized for mesh generation because it uses the actual microstructure of composites. Then the mesh is imported into the other software such as ABAQUS for further analysis. In this direction, OOF2 is used for Al-Al_2O_3 composites, matrix, pores and particles selection in microstructure; then mesh is generated and converted to two dimensional-FE to analyze in ABAQUS [55]. Similarly, W/Cu composite is considered for preprocessing by OOF2; however, the commercial FEM code ABAQUS is used as a solver [11]. Moreover, SiC/SiC composite microstructure generates FE by OOF2 from simplified images of the composite cross-sections. The final FE mesh is exported directly into the ABAQUS to analyze effective material properties and stress distributions [56]. OOF2 and ANSYS are used to obtain the λ based on 2D and 3D calculations, respectively [57]. They found that a power-law function can accurately define the relationship between 2D and 3D results [57]. Further, they concluded that various microstructure images with pores are also valid for λ and 2D image-based model is easy to implement [57]. The YSZ based TBC image is imported in OOF2 and a mesh is constructed; then, a nonlinear elastic–plastic simulation model of micro-indentation is used for further simulation performed by ABAQUS software [58]. Finally, a three-dimensional (3D) OOFEM is also developed by NIST USA to simulate materials' overall mechanical, dielectric, or thermal properties using actual or simulated micrographs [59]. In summary, it can be said that OOF2 can be used in varieties of microstructure mesh generation, which is further integrated into other software for detailed analysis.

5. Conclusions

OOF2 based FE analysis is discussed for various ceramics composites. The thermal (thermal expansion coefficient, thermal conductivity) and mechanical (Young's modulus, residual stress) properties of the composites are predicated by OOF2 simulation and results are comparable with the analytical methods/models. A uniform temperature condition is applied to the composites and stress–strain distribution is analyzed. The maximum stress is found at the interface of the filler/matrix due to their mismatch of thermal expansion coefficient. The compressive and tensile stress distribution are observed in different composites with variation in their magnitude and localized at the interface. Similarly, a temperature gradient is also applied on the composites and found that the stress distribution depends on the orientation of the filler/grain. The thermal stress behavior is altered by fabricating composites via nano/micro-grain or making all grains parallel/perpendicular to the applied temperature field. The residual stresses are also varied according to the size, orientation and shape of the filler. Elastic stress also consists of a similar effect with filler size and orientation. A number of ceramics composites show analogous behavior at uniform/gradient temperature and displacement conditions. OOF2 mesh generation can be integrated with other software for failure analysis.

Acknowledgements

The authors want to thanks Mr. Manish Kumar Meena for providing one figure and write-up.

Conflict of interest

The authors declare no conflict of interest.

Author details

Satyanarayan Patel
Department of Mechanical Engineering, Indian Institute of Technology Indore, Indore, Madhya Pradesh, India

*Address all correspondence to: spatel@iiti.ac.in

References

[1] Ray N, Kempf B, Mützel T, Froyen L, Vanmeensel K, Vleugels J. Effect of WC particle size and Ag volume fraction on electrical contact resistance and thermal conductivity of Ag-WC contact materials. Materials & Design. 2015;85: 412-22.

[2] Meena MK, Kumar M, Lalitha K, Patel S. Thermomechanical analysis of $0.94Na_{1/2}Bi_{1/2}TiO_3$-$0.06BaTiO_3/ZnO$ composites using finite element method. Journal of Alloys and Compounds. 2021; 854:157161.

[3] Angle JP, Wang Z, Dames C, Mecartney ML. Comparison of two-phase thermal conductivity models with experiments on dilute ceramic composites. Journal of the American Ceramic Society. 2013;96(9):2935-42.

[4] Sharma NK, Pandit SN, Vaish R, Srivastava V. Effective Young's Modulus of Ni-Al_2O_3 composites with particulate and interpenetrating phase structures: A multiscale analysis using object oriented finite element method. Computational Materials Science. 2014;82:320-4.

[5] Bakshi SR, Patel RR, Agarwal A. Thermal conductivity of carbon nanotube reinforced aluminum composites: A multi-scale study using object oriented finite element method. Computational Materials Science. 2010; 50(2):419-28.

[6] Angle JP, Nelson AT, Men D, Mecartney ML. Thermal measurements and computational simulations of three-phase (CeO_2-$MgAl_2O_4$-$CeMgAl_{11}O_{19}$) and four-phase (3Y-TZP-Al_2O_3-$MgAl_2O_4$-$LaPO_4$) composites as surrogate inert matrix nuclear fuel. Journal of Nuclear Materials. 2014;454(1):69-76.

[7] Zivelonghi A, Brendel A, Lindig S, Nawka S, Kieback B, You JH. Microstructure-based analysis of thermal- and mechanical behaviors of W/CuCrZr composites and porous W coating. Journal of Nuclear Materials. 2011;417(1):536-9.

[8] Torkashvand K, Poursaeidi E, Ghazanfarian J. Experimental and numerical study of thermal conductivity of plasma-sprayed thermal barrier coatings with random distributions of pores. Applied Thermal Engineering. 2018;137:494-503.

[9] Kyaw S, Jones A, Jepson MAE, Hyde T, Thomson RC. Effects of three-dimensional coating interfaces on thermo-mechanical stresses within plasma spray thermal barrier coatings. Materials & Design. 2017;125:189-204.

[10] Sharma NK, Misra RK, Sharma S. Thermal expansion behavior of Ni-Al_2O_3 composites with particulate and interpenetrating phase structures: An analysis using finite element method. Computational Materials Science. 2014; 90:130-6.

[11] Zivelonghi A, You J-H. Mechanism of plastic damage and fracture of a particulate tungsten-reinforced copper composite: A microstructure-based finite element study. Computational Materials Science. 2014;84:318-26.

[12] Sharma NK, Misra RK, Sharma S. Experimental characterization and numerical modeling of thermo-mechanical properties of Al-B_4C composites. Ceramics International. 2017;43(1, Part A):513-22.

[13] Chawla N, Patel B, Koopman M, Chawla K, Saha R, Patterson B, et al. Microstructure-based simulation of thermomechanical behavior of composite materials by object-oriented finite element analysis. Materials Characterization. 2002;49(5):395-407.

[14] Chawla N, Shen YL. Mechanical behavior of particle reinforced metal

matrix composites. Advanced engineering materials. 2001;3(6):357-70.

[15] Langer SA, Fuller ER, Carter WC. OOF: an image-based finite-element analysis of material microstructures. Computing in Science & Engineering. 2001;3(3):15-23.

[16] Reid AC, Lua RC, García RE, Coffman VR, Langer SA. Modelling microstructures with OOF2. International Journal of Materials and Product Technology. 2009;35(3-4): 361-73.

[17] Vedula VR, Glass SJ, Saylor DM, Rohrer GS, Carter WC, Langer SA, et al. Residual-stress predictions in polycrystalline alumina. Journal of the American Ceramic Society. 2001;84(12): 2947-54.

[18] Akbari MK, Shirvanimoghaddam K, Hai Z, Zhuiykov S, Khayyam H. Nano TiB_2 and TiO_2 reinforced composites: a comparative investigation on strengthening mechanisms and predicting mechanical properties via neural network modeling. Ceramics International. 2017;43(18):16799-810.

[19] Patel S, Vaish R, Chauhan VS, Bowen C. Microstructural finite element modeling and simulation on Al-MgO composites. International Journal of Computational Methods. 2015;12(05): 1550030.

[20] Patel S, Vaish R. Finite element analysis of WC-Al_2O_3 composites. International Journal of Computational Materials Science and Engineering. 2014;3(01):1450002.

[21] Patel S, Vaish R, Sinha N, Bowen C. Finite element analysis of the microstructure of AlN-TiN composites. Strain. 2014;50(3):250-61.

[22] Reid ACE, Langer SA, Lua RC, Coffman VR, Haan S-I, García RE. Image-based finite element mesh construction for material microstructures. Computational Materials Science. 2008;43(4):989-99.

[23] Elomari S, Skibo M, Sundarrajan A, Richards H. Thermal expansion behavior of particulate metal-matrix composites. Composites Science and Technology. 1998;58(3-4):369-76.

[24] Lal D, Kumar P, Sampath S, Jayaram V. Low-temperature stiffening of air plasma-sprayed 7 wt% Y_2O_3-stabilized ZrO_2. Journal of the American Ceramic Society. 2020;103(3):2076-89.

[25] Maurya R, Mittal D, Balani K. Effect of heat-treatment on microstructure, mechanical and tribological properties of Mg-Li-Al based alloy. Journal of Materials Research and Technology. 2020;9(3):4749-62.

[26] Lal D, Kumar P, Sampath S, Jayaram V. Hysteretic and time dependent deformation of plasma sprayed zirconia ceramics. Acta Materialia. 2020;194:394-402.

[27] Choi P, Poudyal L, Rouzmehr F, Won M. Spalling in Continuously Reinforced Concrete Pavement in Texas. Transportation Research Record. 2020;2674(11):731-40.

[28] Uczak de Goes W, Markocsan N, Gupta M. Microstructural Changes in Suspension Plasma-Sprayed TBCs Deposited on Complex Geometry Substrates. Coatings. 2020;10(7):699.

[29] Guo S, Sun L, Zheng Z, Fang J, Wen Y, Liu C, et al. Microstructure and corrosion behavior of Si3N4/316L joints brazed with Ag-Cu/Ag/Mo/Ag/Ag-Cu-Ti multilayer filler. Electrochimica Acta. 2021;379:138193.

[30] Costa APO, Sousa RO, Ribeiro LMM, Santos AD, de Sá JMAC. Multiscale Modeling for Residual Stresses Analysis of a Cast Super Duplex Stainless Steel. In: da Silva LFM, editor.

Materials Design and Applications III. Cham: Springer International Publishing; 2021. p. 47-63.

[31] Isavand S, Assempour A. Effects of Microstructural Morphology on Formability, Strain Localization, and Damage of Ferrite-Pearlite Steels: Experimental and Micromechanical Approaches. Metallurgical and Materials Transactions A. 2021;52(2):711-25.

[32] Ogita G, Matsumoto K, Mochizuki M, Mikami Y, Ito K. Evaluation of Hydrogen-induced Cracking Behavior in Duplex Stainless Steel by Numerical Simulation of Stress and Diffusible Hydrogen Distribution at the Microstructural Scale. ISIJ International. 2021;61(4):1135-42.

[33] Gwalani B, Olszta M, Varma S, Li L, Soulami A, Kautz E, et al. Extreme shear-deformation-induced modification of defect structures and hierarchical microstructure in an Al–Si alloy. Communications Materials. 2020; 1(1):85.

[34] Nguyen BN, Henager CH, Wang J, Setyawan W. Tailoring ductile-phase toughened tungsten hierarchical microstructures for plasma-facing materials. Journal of Nuclear Materials. 2020;540:152382.

[35] Lund J, Vikrant KSN, Bishop CM, Rheinheimer W, García RE. Thermodynamically consistent variational principles for charged interfaces. Acta Materialia. 2021;205: 116525.

[36] Karakoç A, Keleş Ö. A predictive failure framework for brittle porous materials via machine learning and geometric matching methods. Journal of Materials Science. 2020;55(11): 4734-47.

[37] Sharma NK, Pandit S, Vaish R. Microstructural modeling of Ni-composites using object-oriented finite-element method. International Scholarly Research Notices. 2012;2012.

[38] Meena MK. Microstructural finite element modeling and simulation on NBT-6BT-ZnO composites. Jaipur, India: Malaviya National Institute of Technology Jaipur; 2020.

[39] https://www.ctcms.nist.gov/~langer/oof2man/Section-Concepts-Mesh.html. Accessed: 2021-07-27

[40] Molaro JL, Byrne S, Langer SA. Grain-scale thermoelastic stresses and spatiotemporal temperature gradients on airless bodies, implications for rock breakdown. Journal of Geophysical Research: Planets. 2015;120(2):255-77.

[41] Wong C, Bollampally RS. Thermal conductivity, elastic modulus, and coefficient of thermal expansion of polymer composites filled with ceramic particles for electronic packaging. Journal of applied polymer science. 1999;74(14):3396-403.

[42] Buryachenko VA, Kreher WS. Internal residual stresses in heterogeneous solids-A statistical theory for particulate composites. Journal of the Mechanics and Physics of Solids. 1995;43(7):1105-25.

[43] Hashin Z, Shtrikman S. A variational approach to the theory of the elastic behaviour of multiphase materials. Journal of the Mechanics and Physics of Solids. 1963;11(2):127-40.

[44] Ziegler T, Neubrand A, Piat R. Multiscale homogenization models for the elastic behaviour of metal/ceramic composites with lamellar domains. Composites Science and Technology. 2010;70(4):664-70.

[45] Wang M, Pan N. Elastic property of multiphase composites with random microstructures. Journal of Computational Physics. 2009;228(16): 5978-88.

[46] Kerner EH. The elastic and thermo-elastic properties of composite media. Proceedings of the Physical Society Section B. 1956;69(8):808-13.

[47] Turner PS. The problem of thermal-expansion stresses in reinforced plastics. 1942.

[48] Rosen BW, Hashin Z. Effective thermal expansion coefficients and specific heats of composite materials. International Journal of Engineering Science. 1970;8(2):157-73.

[49] Shen YL. Thermal expansion of metal-ceramic composites: a three-dimensional analysis. Materials Science and Engineering: A. 1998;252(2):269-75.

[50] Cannillo V, Leonelli C, Boccaccini AR. Numerical models for thermal residual stresses in Al_2O_3 platelets/borosilicate glass matrix composites. Materials Science and Engineering: A. 2002;323(1-2):246-50.

[51] Hoskins PB. Thermoelastic behavior of Al_2O_3-SiC nanocomposite via microstructure-based finite element analysis 2017.

[52] Wang X, Ren P, Wang J, Xu J, Xi Y. Multi-phase coexistence and temperature-stable dielectric properties in $BaTiO_3$/ZnO composite ceramics. Journal of the European Ceramic Society. 2020;40(5):1896-901.

[53] Ha SK, Jin KK, Huang Y. Micro-mechanics of failure (MMF) for continuous fiber reinforced composites. Journal of Composite materials. 2008;42 (18):1873-95.

[54] Binner J, Porter M, Baker B, Zou J, Venkatachalam V, Diaz VR, et al. Selection, processing, properties and applications of ultra high temperature ceramic matrix composites, UHTCMCs – a review. International Materials Reviews. 2020; 65(7):389-444.

[55] Gudlur P, Muliana A, Radovic M. The effect of microstructural morphology on the elastic, inelastic, and degradation behaviors of aluminum-alumina composites. Mechanics Research Communications. 2014;57: 49-56.

[56] Mital SK, Goldberg RK, Bonacuse PJ. Two-dimensional non-linear finite element analysis of CMC microstructures. Composites Part B: Engineering. 2014;57:144-54.

[57] Qiao J-H, Bolot R, Liao H, Bertrand P, Coddet C. A 3D finite-difference model for the effective thermal conductivity of thermal barrier coatings produced by plasma spraying. International Journal of Thermal Sciences. 2013;65:120-6.

[58] Fizi Y, Mebdoua Y, Lahmar H. Simulation analysis of mechanical properties and adhesion behavior of thermal barrier coatings. The Third International Conference on FRACTURE MECHANICS: FRACT'3, November 27-30, 2016; FRACT'3, November 27-30, 2016.

[59] Coffman VR, Reid ACE, Langer SA, Dogan G. OOF3D: An image-based finite element solver for materials science. Mathematics and Computers in Simulation. 2012;82(12):2951-61.

Chapter 8

Investigation of Strain Effect on Cleavage Fracture for Reactor Pressure Vessel Material

Kushal Bhattacharyya

Abstract

Failure mechanism of 20MnMoNi55 steel in the lower self of ductile to brittle transition (DBT) region is considered as brittle fracture but it has been observed from the experimental analysis of stress-strain diagram that clear plastic deformation is shown by the material before failure. Therefore, strain correction is implemented in the cleavage fracture model proposed by different researchers in the lower self of the DBT region with the help of finite element analysis. To avoid a huge number of experiments being performed, Monte Carlo simulation is used to generate a huge number of random data at different temperatures in the lower self of the DBT region for calibration of the cleavage parameters with the help of the master curve methodology. Fracture toughness calculated after strain correction through different models are validated with experimental results for the different probability of failures.

Keywords: fracture toughness, plastic strain, reactor pressure vessel, master curve, finite element analysis

1. Introduction

Regular maintenance of the reactor pressure vessel (RPV) is an important criterion that has to be considered where safety is the prime requirement for any country. In that respect embrittlement of the RPV material has to be quantified concerning reference temperature T_0. In the last few decades, several researchers tried to quantify this embrittlement nature of ferritic steel in RPV materials. Among them, the work done by Kim Wallin with the development of ASTM E1921 and master curve [1–3] proved to be quite impressive and acceptable in quantifying embrittlement for different ferritic steels used in reactor pressure vessels with the help of reference temperature T_0.

It has been observed in our previous work that references temperature (T_0) predicted for 20MnMoNi55 steel is constraint dependent [4] and it varies with different crack length, thickness, and geometry of the specimen. It also shows the variation with test temperature and censor parameter "m".

This observation is also predicted by different researchers [5–10] working in this field of ductile to brittle transition (DBT) region for different RPV materials.

Therefore, in the last few years, the main aim of the researchers was focused to study the constraint effects of reference temperature T_0 for RPV materials. Finite element analysis is considered a useful tool to study the stress distribution near the crack tip of the specimen at different temperatures in the DBT region. With this aim in mind some researchers try to capture the constraint effect with the help of T-stress [11–15], Q-stress [16–18] and triaxiality [19]. The author also tried to capture the loss of constraint effect on the reference temperature T_0 for these RPV materials with the help of these three stress-based parameters in his earlier work [20]. A satisfactory correlation of the constraint effect in the upper domain of the ductile failure-dominated region of the DBT region is observed. But the brittle failure-dominated region or the lower transition region remains untouched. Therefore, many researchers tried to address the lower transition region with different cleavage failure models. Among them, the work is done by Beremin [21] and his co-authors prove to be challenging for the RPV materials at the brittle failure-dominated portion of the DBT region. The author is motivated to explore a brittle failure model to capture the constraint effect satisfactorily in the lower transition region. With that framework in mind, the widely accepted Beremin model [21] is used to study the constraint effect on master curve and reference temperature (T_0) by different researchers for 20 years or more. Beremin cleavage fracture model provides a good correlation between localized stress pattern prevailing near the crack tip with that of global parameters of fracture like J-integral, applied to load, and fracture toughness with the help of Weibull stress a parameter of stress as coined by Beremin. But the calibration of the constant parameters which remain unique for the material is a challenging job because it demands to test more than 30 specimens to predict a reliable result as predicted by Khalili and Kromp [22]. But testing such a huge number of tests for a given material and at a given test temperature is much expensive. Therefore, many authors utilized the Monta Carlo simulation to generate a huge random number of data from the 6 experimental fracture toughness results to determine the parameters at different temperatures. The entire process is explained in the author's previous work [23].

Beremin model is entirely focused on brittle fracture where no strained effect is considered but it has been observed in the experimental stres-strain diagram for the material 20MnMoNi55 steel a huge plastic deformation is observed in the material before failure even at −110°C and the plastic region diminishes as it moves towards −150°C. Therefore, strain correction is required in the model for proper calibration of the Beremin parameters for the material. Beremin himself in his work felt the requirement for strain correction and he simultaneously developed a model considering the effect of strain in calculating Weibull Stress [21]. Recently, Ruggieri in his work [24, 25] utilized the strain effect by different models for a similar type of RPV material A515 Gr 65 pressure vessel steel. But he uses the toughness scaling model to calibrate the Beremin model parameters.

In December 2017, Claudio Ruggieri, Robert H. Dodds Jr. [26] through their work focuses on the importance of plastic strain effects into the probabilistic framework in brittle fracture.

In November 2019, Claudio Ruggieri [27] through his work proposed a probabilistic, micromechanics-based model which incorporates plastic strain effects on cleavage fracture and its dependence on the microcrack distribution. The model utilized a plastic-strain based form of the Weibull stress to capture the differences in brittle fracture toughness for a reactor pressure vessel (RPV) steel due to constraint loss.

In the present work fracture toughness of 20MnMoNi55 steel is determined with the help of three-point bending (TPB) at −100°C, −110°C, −120°C, −130°C,

−140°C, at reference temperature T_0 and master curve is calculated for all of this results. Then elastoplastic finite element analysis of each fracture specimen is performed taking the boundary condition from the experimental results. The stress-strain diagram obtained from the test performed by the previous researchers [27] for this material, provides the material properties required for the FEA. Beremin model parameters are calculated using strain correction as proposed by the modified Beremin model, local criterion using the distribution of particle fracture stress, following exponential dependence of eligible microcracks on ϵ_p (plastic strain) and under influence of plastic strain on microcrack density. The Weibull modulus (m) and Webull scale parameter (σ_u) are calibrated by Monte Carlo simulation for temperatures −100°C, −110°C, −120°C, −130°C, −140°C from the experimental results as explained in the previous work [24]. Then $C_{m,n}$ another Beremin parameter is also calibrated for the above-mentioned temperatures as described in the previous work of the author [28]. Once the $C_{m,n}$ is calculated fracture toughness can be predicted for different temperatures from the above-mentioned models for different probability of failures. The predicted fracture toughness is compared with the experimental values.

In this work, the entire focus is being made on the brittle failure nature of German based reactor pressure vessel material (20MnMoNi 55 steel) at the lower self of DBT region, which is a very important study as far the safety of reactor pressure vessel is concerned. In recent years study on this material dealing with specific topics is not performed. Moreover, the application of Monte Carlo simulation to reduce the burden of performing a huge number of fracture experiments is overcome by this procedure. The author proves the success in the application of the statistical model by matching it with the experimental results in his previous work [29]. In the present work, the author utilizes the statistical model along with FEA to study the effect of strain on brittle fracture through four different strain corrected brittle fracture models. In the end, the fracture toughness predicted from these models is compared with the established ASTM E1921 and master curve results which is a very challenging and interesting part of the work.

2. Material

The material studied is German steel, used in the reactor pressure vessel of Indian PHWR and designated as 20MnMoNi55. The material used in this investigation has received from Bhabha Atomic Research Centre, Mumbai, India. The steel was received in the form of a rectangular block. The specimens were made from this block to determine the fracture toughness of the selected steel using J-integral analysis and the master curve methodology, to understand the fracture behavior of the steel. The RPV material properties during operation are defined by their initial values, material type, chemical composition, and operating stressors, mainly operating temperature and neutron influence. The chemical composition of 20MnMoNi55 is shown in **Table 1**.

Name of element	C	Si	Mn	P	S	Al	Ni	Mo	Cr	Nb
Percentage composition (in weight)	0.20	0.24	1.38	0.011	0.005	0.068	0.52	0.30	0.06	0.032

Table 1.
Chemical composition of 20MnMoNi55.

3. Methods

3.1 Calculation of reference temperature (T_0) and master curve analysis

Brittle fracture probability according to Wallin [1–3], is defined as P_f for a specimen having fracture toughness K_{JC} in the transition region is described by a three-parameter Weibull model as shown by.

$$P_f = 1 - \exp\left[-\left(\frac{K_{JC} - K_{\min}}{K_0 - K_{\min}}\right)^4\right] \tag{1}$$

where,

$$K_{JC} = \sqrt{\frac{J_c.E}{(1-\nu^2)}} \tag{2}$$

Scale parameter K_0 which dependent on the test temperature and specimen thickness, and K_{min} is equal to 20 MPa$\sqrt{m}$ [29].

For single-temperature evaluation, the estimation of the scale parameter K_0, is performed according to Eq. (4).

$$K_0 = \left[\sum_{i=1}^{N} \frac{\left(K_{JC(i)} - K_{\min}\right)^4}{N}\right]^{1/4} + K_{\min} \tag{3}$$

$$K_{JC(median)} = K_{\min} + (K_0 - K_{\min})(\ln 2)^{1/4} \tag{4}$$

Here, T_0 is the temperature at which the value of $K_{JC(median)}$ is 100 MPa$\sqrt{m}$ and is known as the reference temperature. T_0 can be calculated from.

$$T_0 = T_{test} - \frac{1}{0.019} \ln\left[\frac{K_{JC(median)} - 30}{70}\right] \tag{5}$$

3.2 Modified Beremin model

According to the Beremin model [21], the probability of failure is given as,

$$P_f = 1 - \exp\left(-\left(\frac{\sigma_w}{\sigma_u}\right)^m\right) \tag{6}$$

$$\sigma_w = \sqrt[m]{\left(\sum_{j=1}^{n} \sigma_1^j\right)^m \frac{V_j}{V_0}} \tag{7}$$

n is the number of volumes Vj, or elements in a FEM calculation, and σ_1^j the maximal principal stress of the element j and V_j/V_0 is just a scaling based on the assumption that the probability scales with the volume. V_0 is the reference volume which is normally taken as a cubic volume containing about 8 grains i.e., 50 × 50 × 50 μm.

The classical model described above is applicable where plastic strain is negligible or zero for perfectly brittle materials but for ferritic steels where an appreciable amount of plastic strain is observed in the crack tip area this formula cannot capture the failure mechanism perfectly. To impose plastic strain effect on the failure mechanism a correction formulation has been introduced by Beremin [21].

$$\sigma_w = \sqrt[m]{\sum_j \left(\sigma_1^j\right)^m \frac{V_j}{V_0} \exp\left(-\frac{m\varepsilon_1^j}{2}\right)} \tag{8}$$

ε_1^j is the strain in the direction of the maximum principal stress σ_1^j.

Throughout the paper, the Weibull stress is calculated according to Eq. (8).

3.3 Local approach to cleavage fracture incorporating plastic strain effects

This methodology is derived from the work done by Wallin and Laukkanen [30] which is based on the strain effect near the crack tip field. Here the Weibull stress is modified by taking into account a particular volume $^{\delta}$V in the fracture process zone is subjected to a principal stress σ_1 and associated with a plastic strain (ϵ_p). In this mode micro-crack formed by the cracking of brittle particles only participate in the fracture process. It is assumed by Ruggieri and Dodds [30] that a fraction represented by ψ_C of the total number of brittle particles present in FPZ is responsible for nucleating the micro cracks which propagate unstably. This fraction ψ_C is a function of plastic strain but does not depend on microcracks. Based on the weakest link concept limiting distribution for the cleavage fracture stress can be expressed as.

$$P_f(\sigma_1, \varepsilon_p) = 1 - \exp\left[-\frac{1}{v_0}\int_\Omega \psi_c \sigma_1^m d\Omega\right]^{1/m} \tag{9}$$

V_0 represents a reference volume conventionally taken as a unit volume.

Ω is the volume of the near-tip fracture process zone where $\sigma_1 \geq \lambda\sigma_y$. $\lambda = 2$ [30, 31] is normally taken as twice of yield stress for the material.

Now ψ_C is calculated as follows.

$$\psi_c = 1 - \exp\left[-\left(\frac{L}{L_N}\right)^3 \left(\frac{\sigma_{pf}}{\sigma_{prs}}\right)^{\alpha p}\right] \tag{10}$$

L represents the particle size; L_N a reference particle size; σ_{prs} is the particle reference fracture stress; α_p denotes the Weibull modulus shape parameter of particle distribution; and σ_{pf} represents the characteristics of fracture stress.

$$\sigma_{pf} = \sqrt{1.3\sigma_1\varepsilon_p E} \tag{11}$$

where σ_1 represents maximum principal stress, ϵ_p denotes the Maximum plastic strain of those particles whose σ_1 is calculated in the fracture process zone (where $\sigma_1 = 2\,\sigma_y$) and E represents the Youngs Modulus of the particle at different temperatures. Now as assumed by Rugeirri et al. [25] the size of a fracture particle takes the size of a Griffith-like micro-crack of the same size the probability distribution of the fracture stress with increase loading for a cracked solid is given by the following equation.

$$P_f(\sigma_1, \varepsilon_p) = 1 - \exp\left[-\frac{1}{v_0}\int_\Omega \left\{1 - \exp\left[-\left(\frac{\sigma_{pf}}{\sigma_{prs}}\right)^m\right]\right\}\left(\frac{\sigma_1}{\sigma_u}\right) d\Omega\right]^{1/m} \tag{12}$$

As ψ_c is independent of microcrack size so L/L_N is considered to be 1. Therefore, σ_W takes the form.

$$\sigma_W = \left[\frac{1}{V_0} \int_{\Omega} \left\{ 1 - \exp\left[-\left(\frac{\sigma_{pf}}{\sigma_{prs}} \right)^{\alpha_p} \right] \right\} \sigma_1^m d\Omega \right]^{1/m} \tag{13}$$

3.4 Exponential dependence of eligible micro-cracks on ϵ_p

Bordet et al. [31] include plastic strain effects on cleavage fracture in terms of the probability of nucleating a carbide micro crack. The original model considered only freshly nucleated carbides to act as Griffith-like micro-cracks and have the eligibility to propagate unstably take part in the fracture process. But in our work, we considered a simplified model as considered by Bordet et al. [31] and adopt a Poisson distribution by introducing a parameter λ to define ψc given by the following equation.

$$\psi_c = 1 - \exp\left(\lambda \varepsilon_p\right) \tag{14}$$

λ is assumed as the average rate of fracture particles which becomes Griffith-like micro crack with small strain increment. The author has taken the strain increment inconsistency with the quasi-static process. Therefore, the probability of fracture and Weibull stress takes the following form.

$$P_f\left(\sigma_1, \varepsilon_p\right) = 1 - \exp\left[-\frac{1}{v_0} \int_{\Omega} \left\{ 1 - \exp\left(-\lambda \varepsilon_p\right) \right\} \left(\frac{\sigma_1}{\sigma_u} \right) d\Omega \right]^{1/m} \tag{15}$$

$$\sigma_W = \left[\frac{1}{V_0} \int_{\Omega} \left\{ 1 - \exp\left(-\lambda \varepsilon_p\right) \right\} \sigma_1^m d\Omega \right]^{1/m} \tag{16}$$

3.5 Influence of plastic strain on microcrack density

Based upon the work of Brindley and Gurland [32–34] the direct effect of plastic strain on micro-cracking of ferritic steel at different temperatures alter the probability distribution in the FPZ as follows:

$$P_f\left(\sigma_1, \varepsilon_p\right) = 1 - \exp\left[-\frac{1}{v_0} \int_{\Omega} \varepsilon_p^{\beta} \left(\frac{\sigma_1}{\sigma_u} \right)^m d\Omega \right] \tag{17}$$

and the Weibull Stress becomes.

$$\sigma_W = \left[\frac{1}{V_0} \int_{\Omega} \varepsilon_p^{\beta} . \sigma_1^m d\Omega \right]^{1/m} \tag{18}$$

4. Test procedure

4.1 Fatigue pre-cracking

The fracture toughness tests in this investigation were planned on three-point bending (TPB) specimens in L-T orientation. Standard 1T TPB specimens were machined following the guidelines of ASTM E 399-90. The designed dimensions of the specimens were; thickness (B) 25 mm and width (W) = 25 mm which is constant for all the specimen tested and machined notch length (aN) = 10 mm to produce different a/W ratio of 0.5. Fatigue pre-cracking of the TPB specimens was carried out at room temperature at constant ΔK mode as described in ASTM standard E 647 on servo hydraulic INSTRON UTM (Universal Testing Machine) with 8800 controllers having 100 KN grip capacity using a commercial da/dN fatigue crack propagating software supplied by INSTRON Ltd. U.K. The crack lengths were measured by compliance technique using a COD gauge of 10 mm gauge length mounted on the load line of the specimen.

4.2 Fracture test

The estimation of J-integral values of the fabricated specimens was carried out using an INSTRON UTM (Universal Testing Machine) with an 8800 controller with 100 KN grip capacity as described earlier. Tests were done at different temperatures ranging from −100°C to – 140°C. The specimen used is a three-point Bending specimen. The nomenclature along with a picture of the specimen is shown in **Figure 1**.

The Instron FAST TRACK JIC Fracture Toughness Program was used to determine the value of the J integral. This program performs Fracture Toughness on metallic materials following the American Society for Testing and Materials (ASTM) Standard test method E813. The method is applied specifically to specimens that have notches or flaws that are sharpened with fatigue cracks. The loading rate was slow, and cracking caused by environmental factors was considered negligible.

4.3 A result of the tensile test and J_{1C} at different temperatures in the lower self of the DBT region

From the experimental stress-strain results performed at different temperatures for 20MnMoNi55 steel, a clear plastic zone is observed before failure as shown in **Figure 2** [35]. The same plastic strain effect is reflected in the TPB specimen also at the lower self of the DBT region. This provoked us to perform the required strain correction in computing Weibull Stress through different strain correction models as discussed above.

The results of K_{JC} values of TPB specimens at different temperatures are shown in **Figure 3**.

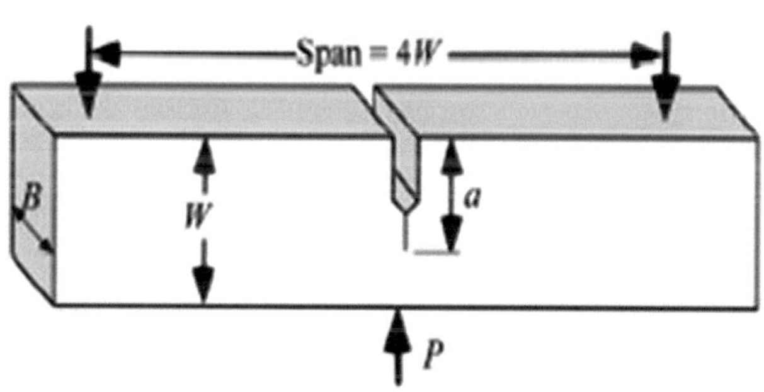

Figure 1.
1 T TPB specimen.

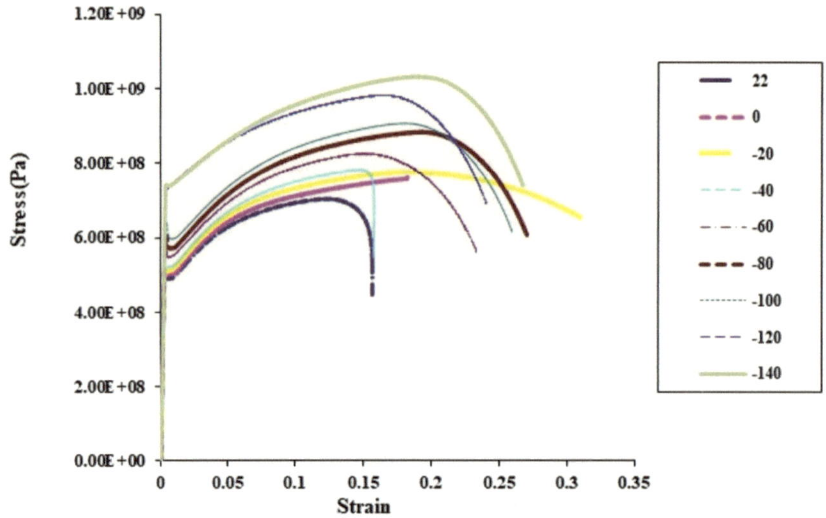

Figure 2.
Stress-strain diagram of 20MnMoNi55 steel at different temperatures [27].

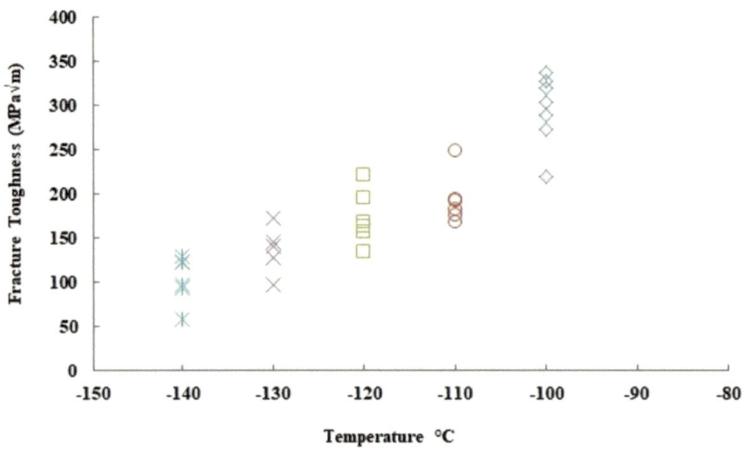

Figure 3.
K_{JC} calculated from fracture toughness test at −110°C.

4.4 Finite element analysis

Finite element analysis of all the fracture tests is performed using ABAQUS 6.13. The material constitutive properties are defined by Young's modulus E, Poisson's ratio ν, and yield stress versus plastic strain obtained from tensile test data performed at different cryogenic temperatures [27]. **Figure 4** shows the stress versus plastic strain diagram at different temperatures and **Table 2** gives the yield stress and ultimate stress versus temperature for 20MnMoNi55 steel at different temperatures in the Brittle Dominated DBT region which is used as a material input parameter for elastoplastic finite element analysis. Isotropic elastic and isotropic hardening plastic material behavior are considered for the material used. 3-D finite element modeling is done for quarter TPB specimen at different temperatures to calculate the Weibull stress for the specimen and hence to calculate T_0 from the Beremin model. The FE model has meshed with 8-node isoparametric hexahedral elements with 8 Gauss points taken for all calculations as referred by

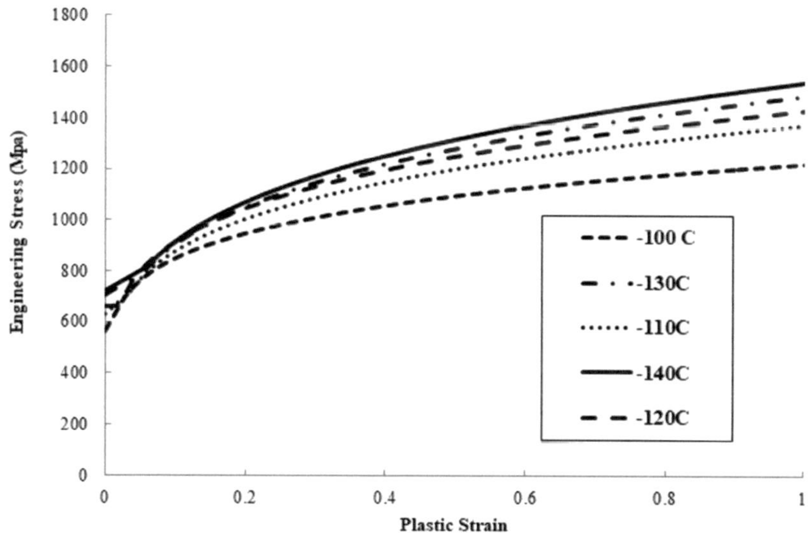

Figure 4.
Engineering stress versus plastic strain for 20MnMoNi55 steel at different temperatures in the brittle dominated DBT region.

Temperature (°C)	Yield strength (MPa)	Ultimate strength (MPa)
−100	593.43	760.49
−110	630.43	786.56
−120	667.06	813.66
−130	701.451	825.054
−140	723.47	856.84

Table 2.
Yield stress and ultimate stress versus temperature for 20MnMoNi55 steel at different temperatures in the brittle dominated DBT region.

IAEA-TECDOC-1631 [36]. Reduced integration with full Newtonian non-linear analysis computation is carried out for all the specimens. In the region ahead of the crack tip the mesh was refined with an element volume of $0.05 \times 0.05 \times 0.05$ mm^3. To facilitate the calculation of Vj, the element size is kept constant near the crack tip [36, 37]. Since a large strain is expected in the crack tip field, a finite strain (large deformation theory) method is used. As the crack extension during the experiment is found to be very small, the crack growth is not simulated in this FE analysis. **Figure 5** shows the boundary conditions and the mesh of the specimen. **Figure 6** shows the region where maximum principal stress exceeded twice the yield stress at that temperature. This region is known as the fracture process zone (FPZ). For such elements, the strain in the direction of the maximum principal stress is also calculated.

4.5 Validation of the FE model and material properties

Figures 7–9 gives a comparison between experimental load versus load line displacement (LLD) of TPB specimen with FE simulated results from Abaqus 6.13 at the −100°C,−110°C and −130°C temperatures The FEA results show a close match with experimental results which validate the used FE model and material

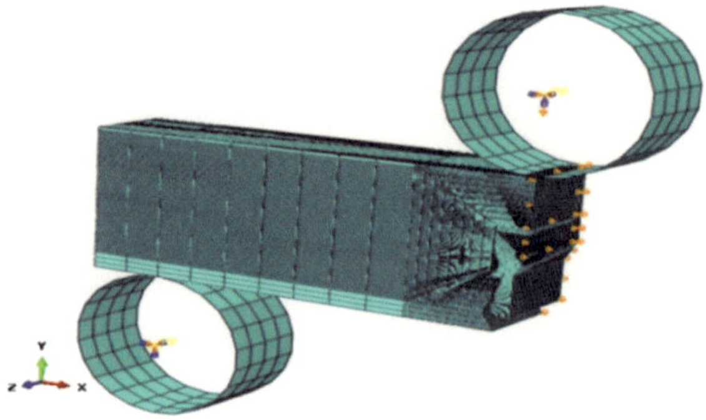

Figure 5.
Quarter TPB specimen model along with boundary conditions.

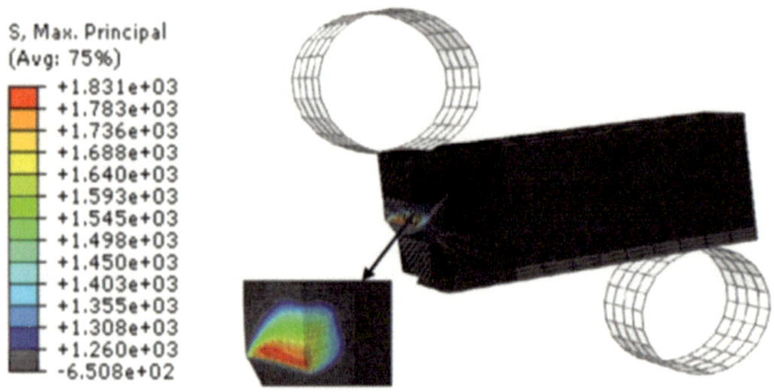

Figure 6.
Maximum principal stress (MPa) distribution in the fracture process zone.

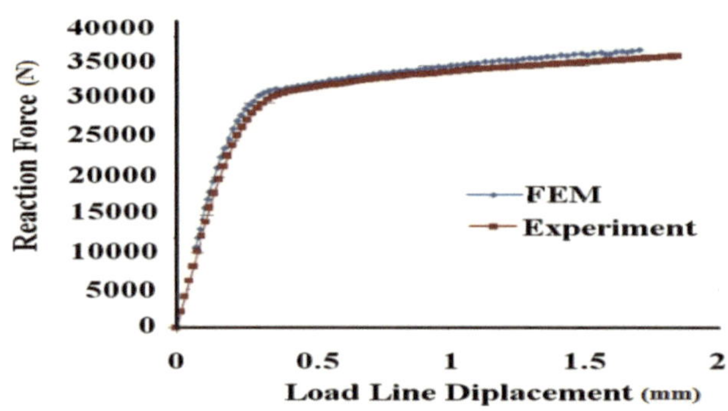

Figure 7.
Comparison of load vs. load line displacement (LLD) −100°C.

parameters. Now for each analysis, the Weibull stress at the failure point can be computed from the FE simulated results.

4.6 Calculation of Weibull stress

Now the maximum principal stress and corresponding strain in the direction of principal stress is known for each element in the fracture process zone, we calculate the Weibull Stress for each model using Eqs. (8), (13), (16), and (18).

The success of the Beremin model for predicting brittle fracture mainly depends on the accuracy of the values of the Beremin material parameters m and σ_u. The

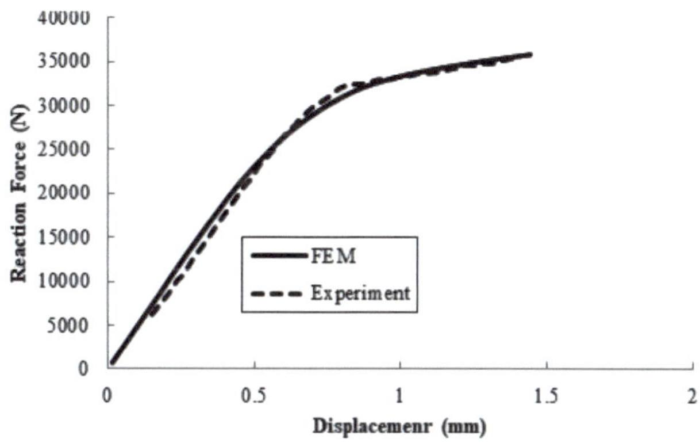

Figure 8.
Comparison of load vs. load line displacement (LLD) −110°C.

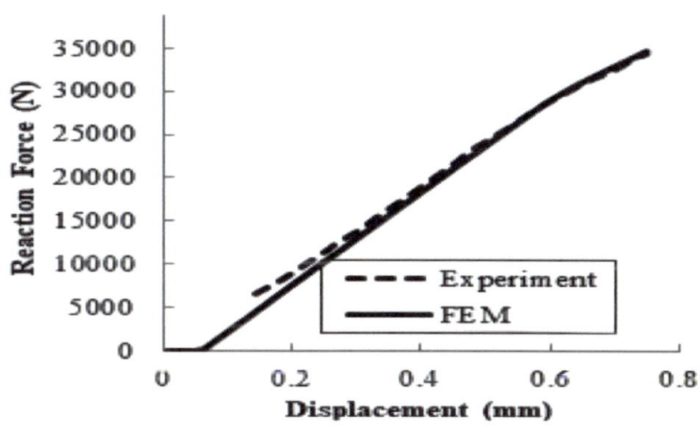

Figure 9.
Comparison of load vs. load line displacement (LLD) −130°C.

Beremin model describes the failure mechanism as an outcome of the distribution of the weakest sites in the statistical material. Hence any material parameters to represent the failure behavior should be determined from a large sample containing variation in candidatures as much as possible. With this in mind, the values of m and σ_u have been determined from the experimental fracture toughness tests at −100°, −110°, −130° which is described as a direct calibration strategy. The process is described vividly for −110°C by K. Bhattacharyya et al. calibration of beremin parameters for 20MnMoNi55 Steel and prediction of reference temperature (T_0) for different thicknesses and a/W ratios [28]. But testing such a huge number of specimens are very expensive so the author used to develop a random number of data with the help of master curve and Monte Carlo simulation [24]. This process is called an Indirect calibration strategy. The entire process is described by the author in their previous work [23], step-wise description only the Step 7 of Article No.4 the Weibull Stress is calculated for different models using Eqs. (8), (13), (16), and (18). The validation of the results simulated from Monte Carlo simulation with the experimental work is also validated by the author in their previous work [24].

4.7 Calibration of $C_{m,n}$

$C_{m,n}$ an important parameter used for the determination of K_{JC} for different models has been calibrated for this material at different temperatures. The process of determination of $C_{m,n}$ for our material is different from that as framed by

Beremin. The entire process is described step by step for −110°C in the previous work done by the author [28]. A similar procedure is used for the determination of $C_{m,n}$ for other temperatures. Once $C_{m,n}$ is calibrated the value of fracture toughness for 5%, 63.2%, and 95% can be determined by Eqs. (6), (9), (12), (15), and (17). These fracture toughness values were then plotted with the experimentally determined master curve methodology as shown in **Figure 12**.

5. Results and discussions

The calibration of Weibull modulus "m" using different models as described above is shown in **Figure 10** and the Weibull scale parameter is shown in **Figure 11**. It is observed that the value of Weibull modulus for the four different models almost coincides at −100°C and −110°C and as the temperature decreases to −120°C to −140°C the variation of in the value of Weibull modulus predicted from the different model is pronounced and it increases with decrease in temperature. As the material moves from the lower self of DBT region to purely cleavage fracture the effect of ductile stretch due to plasticity affect vanishes. As all the four models are functions of plastic strain therefore as it approaches purely brittle failure the strain component almost vanishes therefore prediction capability of the models to some extent becomes biased.

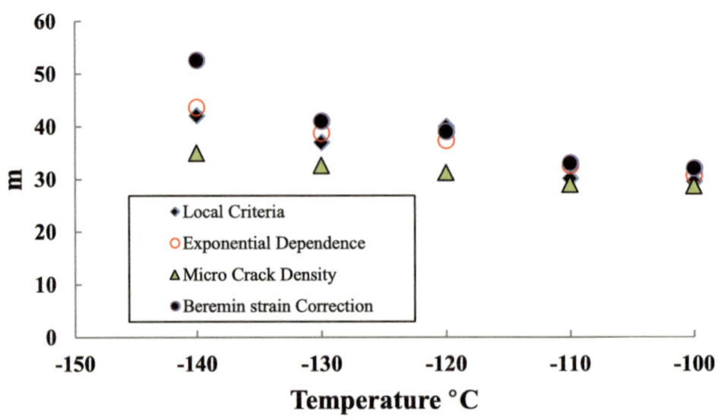

Figure 10.
Variation of Weibull modulus "m" for different temperatures using different strain correction model is shown.

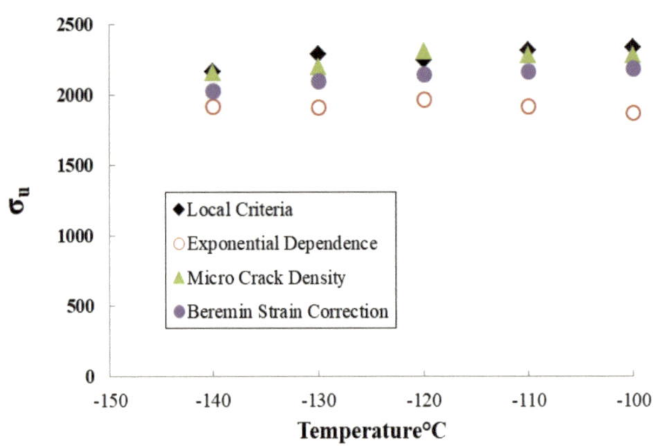

Figure 11.
Variation of Weibull modulus scale parameter "σ_u" for different temperatures using different strain correction model is shown.

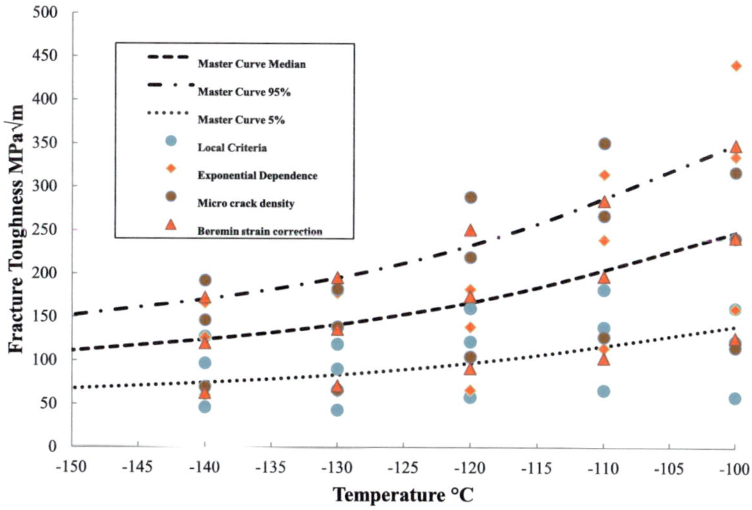

Figure 12.
Master curve determined from experimental results of −120°C compared with the fracture toughness determined from various models.

Weibull modulus "m" and Weibull modulus scale parameter "σ_u" is calculated from different models as explained, now once $C_{m,n}$ is calculated at different temperatures, fracture toughness can be predicted for different probabilities of failure.

It is observed that the prediction capability Beremin strain correction model is much better in comparison to the other three models when validated with the experimental results as shown in **Figure 12**.

Though Claudio Ruggieri and his co-workers in their work [25] showed that fracture toughness predicted from local criteria matches well with the experimental results for A515 Gr 65 pressure vessel steel but the results obtained in this study appeal to be contradictory with their work for the material 20MnMoNi55 steel.

Our study is focussed on the lower self of DBT region starting from −100°C to −140°C where a very small amount of ductile stretch is observed before failure but their work is focussed at −20°C where a huge amount of ductile stretch is observed before cleavage failure for our material. This could be one reason for the deviation of the results with them.

The main aim of this work is to establish the strain affect in the brittle failure-dominated portion of the DBT region, which is observed in the form of ductile stretch in experimental results.

6. Conclusion

1. With the help of finite element analysis, we have a better outlook in fracture process zone and we can bridge the gap between macroscopic observations (like J-integral and fracture toughness) with micro-cracks developed in the fracture process zone.

2. The effect of strain is established in the brittle failure dominated portion of DBT region through different strain correction model.

3. Utilization of statistical model (Monte Carlo simulation) proves to be very useful to reduce the huge cost of performing a large number of experiments at the cryogenic conditions.

4. The values of the Weibull modulus "m" and Weibull modulus scale parameter "σ_u" are calibrated through different brittle fracture models for different temperatures in the brittle failure dominated portion of DBT region for the concerned material.

5. $C_{m,n}$ another important parameter is also calculated for different temperatures for the concerned material.

6. With the help of Weibull modulus "m" and Weibull modulus scale parameter "σ_u" and $C_{m,n}$ the fracture toughness is predicted for different probabilities of failure.

7. The probabilities of failure are then compared with experimentally obtained results.

8. It is observed that the prediction capability Beremin strain correction model is much better in comparison to the other three models when validated with the experimental results.

Whenever fracture mechanics is used from specimen level to component level there is a constrain loss which affects the results. This causes a great lacuna in the application of fracture mechanics to the real engineering problems. This study to some extent put a step forward in overcoming the lacuna by using extensive finite element analysis and different brittle fracture models on specimen level and tried to predict the results in comparison with experimental counterpart. With the hope that in future application of fracture mechanics will not be limited to specimen level. This study will propel more research work in this field and the development of new models.

Author details

Kushal Bhattacharyya
Department of Mechanical Engineering, Netaji Subhash Engineering College, Kolkata, India

*Address all correspondence to: bhattacharyyakushal3@gmail.com

References

[1] Wallin K. The scatter in KIC result. Engineering Fracture Mechanics. 1984; **19**:1085-1093

[2] Wallin K. The master curve method: a new concept for brittle fracture. International Journal of Materials and Product Technology. 1999;**14**(2–4): 342-354

[3] Wallin. K, Master Curve of ductile to brittle transition region fracture toughness round-robin data. The "EURO" Fracture toughness curve. VTT Report 1998. ISBN 951-38-5345-4

[4] Bhattacharyya K et al. Study of constraint effect on reference temperature (T0) of reactor pressure vessel material (20mnmoni55 Steel) in the ductile to brittle transition region. Procedia Engineering. 2014;**86**:264-271

[5] Ruggieri C et al. Transferability of elastic-plastic fracture toughness using the Weibull stress approach: Significance of parameter calibration. Engineering Fracture Mechanics. 2000; **67**:101-117

[6] Gao X. An engineering approach to assess constraint effects on cleavage fracture toughness. Engineering Fracture Mechanics. 2001;**68**:263-283

[7] Wallin K. The size effect in ***Klc*** results. Engineering Fracture Mechanics. 1985;**22**(I):149-163

[8] Smith JA. The effect of crack depth (a) and crack depth to width ratio (a/W) on the fracture toughness of A533-B steel. Journal of Pressure Vessel Technology. May 1994; **116**(2): 115-121

[9] Dodds RH et al. A framework to correlate a/W ratio effects on elastic-plastic fracture toughness (Jc) International Journal of Fracture. 1991; **48**:1-22

[10] Minnebo P. Constraint-Based Master Curve Analysis of a Nuclear Reactor Pressure Vessel Steel Results from an Experimental Program Carried Out within the IAEA CRP-8 Project. EUR 24092 EN-2009

[11] Gupta M et al. A review of T-stress and its effects in fracture mechanics. Engineering Fracture Mechanics. Jan 2015;**134**:218-241. DOI: 10.1016/j.engfracmech.2014.10.013

[12] Ayatollahi MR et al. Determination of T -stress from finite element analysis for mode I and mixed-mode I/II loading. International Journal of Fracture. 1998; **91**:283-298

[13] Castro JTP et al. Comparing Improved Plastic Zone Estimates Considering Corrections based on T-Stress and a Complete Stress Fields. Proceedings of "First IJFatigue & FFEMS Joint Workshop" Forni di Sopra (UD), Italy; 2011. pp. 58-65

[14] Wallin K. Quantifying Tstress controlled constraint by the master curve transition temperature T0. Engineering Fracture Mechanics. 2001; **68**:303-328

[15] Henry BS. The stress triaxiality constraint and the Q-value as a ductile fracture parameter. Engineering Fracture Mechanics. 1997;**57**(4):375-390

[16] Moattari M. Modification of fracture toughness master curve considering the crack-tip Q-constraint. Theoretical and Applied Fracture Mechanics. Aug 2017, **90**:43-52. DOI: 10.1016/ j.tafmec.2017.02.012

[17] Cravero S. A Two-Parameter Framework to Describe Effects of Constraint Loss on Cleavage Fracture and Implications for Failure Assessments of Cracked Components J.

of the Braz. Soc. of Mech. Sci. & Eng. Apr 2019;**141**:021401-7.

[18] Graba M. The influence of material properties and crack length on the *Q*-stress value near the crack tip for elastic-plastic materials for centrally cracked plate in tension. Journal of Theoretical and Applied Mechanics. 2012;**50**(1):23-46

[19] Bhowmik S et al. Evaluation, and effect of loss of constraint on a master curve reference temperature of 20MnMoNi55 steel. Engineering Fracture Mechanics. 2015;**136**:142-157

[20] Bhattacharyya K et al. Modelling the constraint effect on reference temperature with finite element parameters for reactor pressure vessel material 20MnMoNi55 steel. Defence Science Journal. 2020;**70**(3):323-328. DOI: 10.14429/dsj.70.12886

[21] Beremin FM. A local criterion for cleavage fracture of a nuclear pressure vessel steel. Metallurgical and Materials Transactions A. 1983;**14A**:2277-2287

[22] Khalili A, Kromp K. Statistical properties of Weibull estimators. Journal of Materials Science. 1991;**26**: 6741-6752

[23] Bhattacharyya K et al. Variation of Beremin model parameters with temperature by Monte Carlo simulation. Journal of Pressure Vessel Technology. 2019;**141**:021401. DOI: 10.1115/1.4042121

[24] Ruggieri C, Dodds RH Jr. An engineering methodology for constraint corrections of elastic–plastic fracture toughness—Part I: A review on probabilistic models and exploration of plastic strain effects. Engineering Fracture Mechanics. 2015;**134**:368-390. DOI: 10.1016/j.engfracmech.2014. 12.015

[25] Ruggieri C, Savioli RG, Dodds Jr RH. An engineering methodology for constraint corrections of elastic–plastic fracture toughness—Part II: Effects of specimen geometry and plastic strain on cleavage fracture predictions. Engineering Fracture Mechanics. 2015; **146**:185-209. DOI: 10.1016/j.engfrac mech.2015.06.087

[26] Ruggieri C, Dodds RH Jr. A local approach to cleavage fracture modeling: An overview of progress and challenges for engineering applications. Engineering Fracture Mechanics. Jan 2018;**187**:381-403. DOI: 10.1016/j.engfracmech.2017. 12.021

[27] Ruggieri C. A modified local approach including plastic strain effects to predict cleavage fracture toughness from subsize precracked charpy specimens. Theoretical and Applied Fracture Mechanics. Feb 2020;**105**: 102421. DOI: 10.1016/j.tafmec.2019. 102421

[28] Bhattacharyya K et al. Calibration of Beremin parameters for 20MnMoNi55 steel and prediction of reference temperature (T0) for different thicknesses and a/W ratios. Ratios Journal of Failure Analysis and Prevention. 2018;**18**:1534–1547. DOI: 10.1007/s11668-018-0549-7

[29] Bhattacharyya K, Acharyya S. Validation of Monte Carlo simulation technique for calibration of cleavage fracture model parameters, with the calibrated values from experimental results for reactor pressure vessel material 20MnMoNi55 steel in the lower self of ductile-to-brittle transition region. Journal of Pressure Vessel Technology. Dec 2021;**143**: 021401-1 to 021401-7. DOI: 10.1115/ 1.4051021

[30] Wallin K, Laukkanen A. New developments of the Wallin, Saario, Törrönen cleavage fracture model. Engineering Fracture Mechanics. 2008; **75**:3367-3377

[31] Bordet SR, Karstensen AD, Knowles DM, Wiesner CS. A new statistical local criterion for cleavage fracture in steel. Part I: Model presentation. Engineering Fracture Mechanics. 2005;**72**:435-452

[32] Brindley BJ. The effect of dynamic strain-aging on the ductile fracture process in mild steel. Acta Metallurgica. 1970;**18**:325-329

[33] Lindley TC, Oates G, Richards CE. A critical appraisal of carbide cracking mechanism in ferritic/carbide aggregates. Acta Metallurgica. 1970;**18**: 1127-1136

[34] Gurland J. Observations on the fracture of cementite particles in a spheroidized 1.05% C steel deformed at room temperature. Acta Metallurgica. 1972;**20**:735-741

[35] Bhowmik S et al. Application and comparative study of master curve methodology for fracture toughness characterization of 20MnMoNi55 steel. Materials and Design. 2012;**39**:309-317

[36] International Atomic Energy Agency. Master Curve Approach to Monitor Fracture Toughness of Reactor Pressure Vessels in Nuclear Power Plant, IAEA-TECDOC-1631,Vienna; 2009b

[37] Tiwari A, Avinash G, Sunil S, Singh RN, Per Stahle Chattopadhyay J, et al. "Determination of Reference Transition Temperature of In-RAFMS in Ductile Brittle Transition Regime Using Numerically Corrected Master Curve Approach," Engineering Fracture Mechanics. 2015;**142**:79-92

Chapter 9

Simulation Model of Fragmentation Risk

Mirko Djelosevic and Goran Tepic

Abstract

In this chapter, a simulation model for fragmentation risk assessment due to a cylindrical tank explosion is presented. The proposed fragmentation methodology is based on the application of Monte Carlo simulation and probabilistic mass method. The probabilities of generating fragments during the explosion of the tank were estimated regardless of the available accident data. Aleatoric and epistemic uncertainty due to tank fragmentation has been identified. Generating only one fragment is accompanied by aleatoric uncertainty. The maximum fragmentation probability corresponds to the generation of two fragments with a total mass between 1200 *kg* and 2400 *kg* and is 17%. The fragment shape was assessed on the basis of these data and fracture lines. Fragmentation mechanics has shown that kinematic parameters are accompanied by epistemic uncertainty. The range of the fragments in the explosion of the tank has a Weibull distribution with an average value of 638 *m*. It is not justified to assume the initial launch angle with a uniform distribution, since its direction is defined by the shape of the fragment. The presented methodology is generally applicable to fragmentation problems in the process industry.

Keywords: Simulation, fragmentation, explosion, cylindrical tank, risk assessment

1. Introduction

The most common accidents with dangerous substances involve leaks, fires and explosions [1]. If these events are an integral part of the accident chain, then they are manifested through a domino effect [2, 3]. The main reason for fires in process plants is the presence of flammable vapors [4]. Fires lead to heating of process installations and increase of pressure in them creating conditions for explosions due to BLEVE effect [5]. Explosions of process equipment due to the domino effect imply a pronounced fragmentation effect [6]. Fragmentation action in the accident chain (domino effect) is characterized by the fact that it is both a cause and a consequence of explosions [7]. The fragmentation effect during the explosion of the tank is accompanied by a very pronounced uncertainty of geometric and kinematic parameters of the fragments [8]. Large-scale accidents that have occurred in recent times are the result of progressive technological developments, and the typical examples are Toulouse [9] and Neyshabur [10]. Fragmentation barriers are used as a form of protection against the fragmentation effect [11, 12]. The fragmentation effect and the mechanism of formation during the explosion of the tank are presented in [13]. The literature recognizes the basic geometric characteristics as the number and shape of fragments, and defines the kinematic parameters through the trajectory and velocity of the fragments [14]. An initial procedure for estimating the number and mass of a fragment of a LPG storage tank was proposed by Baker et al. [15]. This research served as a starting point for some recent studies in the field of tank fragmentation [16, 17]. Tank fragmentation analysis should identify potential hazards and risks in terms of protecting process equipment from accident escalation [18]. Fragmentation barriers are used as protection against the fragmentation effect and were originally used in nuclear plants [19]. There are several models in the literature according to which the impact energy of a fragment into the target zone is estimated based on the fragment velocity [20–22]. Accident data show that two or three fragments are usually formed during tank fragmentation [23]. It has been determined that the distinct presence of fire during the fragmentation of the tank usually gives one fragment [24]. Previous research for the number of generated fragments estimation has exclusively used the maximum entropy model [25]. This model is based on accident data and shows that explosions with more than five generated fragments are very rare accidents [26, 27].

Tank explosions with the BLEVE effect very rarely provide more than three generated fragments [28]. The implementation of the maximum entropy model is possible only if there are available accident data for this type of process equipment [29]. Fragmentation mechanics analyzes the flight of a fragment and for that purpose the literature sources state a simplified mathematical model [30]. This model is represented in all recent research and was originally proposed by Mannan [31]. Greater mass of the fragment corresponded to the higher initial kinetic energy or the initial velocity [32]. Risk assessment due to the action of fragments is very often estimated in the literature on the basis of kinetic energy of fragments [33]. This energy is usually defined by the percentage share of expansion energy [34]. Some recommendations suggest that this percentage ranges between 5% and 20% [35]. This procedure of determining the initial velocities of the fragments has significant deviations from the real values, so it can only be used for general estimation. The aerodynamic properties of the fragments have a great influence on its kinematic parameters and are reflected in the uncertainty of the shape [36]. Djelosevic and Tepic conducted a complex study of cylindrical tank fragmentation in terms of identifying aleatoric and epistemic uncertainty [37, 38]. The same authors established a correlation of geometric and kinematic parameters of fragments, stating the importance of simulation technique in fragmentation analysis. This chapter will present a general methodological framework for the study of tank fragmentation under the conditions of the BLEVE effect, which is characteristic of the gas industry [39]. The focus of this research is on the analysis of types of uncertainty that follow fragmentation parameters with special reference to fracture probabilities and elimination of conditions that introduce influential fragment sizes into the zone of epistemic uncertainty [40].

2. Fragmentation methodology

Tank fragmentation implies physical separation of fragments from the tank construction itself. The basic feature of fragments is the kinetic energy they have just before hitting a target. Greater kinetic energy of the impact creates greater potential hazard as a result of the fragmentation effect of the explosion. Assessing the kinetic energy of fragments is a complex task which requires identification of geometric and kinematic parameters. The basic geometric parameters are the shape and mass of fragments, whereas the most important kinematic parameter is the initial velocity of a fragment. Geometric and kinematic parameters are not independent since the shape of a fragment affects its initial velocity and initial launch direction. Literature resources on the assessment of geometric and kinematic parameters of fragments are scarce. Therefore, the assessment of these parameters in this paper was conducted using probabilistic and simulation techniques. The probabilistic approach will first be presented and thereafter the simulation analysis of fragmentation.

2.1 Probabilistic approach

The probabilistic approach is used for the assessment of fragment shapes, number of generated fragments as well as for the identifaction of aleatoric and epistemic uncertainty. The effect of uncertainty in assessing geometric and kinematic parameters is extremely important because if a parameter has the same or approximately the same probability with the change of influential factors, then it is definitely accompanied by aleatoric uncertainty. Aleatoric uncertainty is typical of those parameters whose uncertainty cannot be eliminated. On the other hand, there is epistemic uncertainty. If a parameter has epistemic uncertainty, that kind of uncertainty can be eliminated by additional research procedures and the parameter can be subjected to deterministic principles. A typical example of epistemic uncertainty is the initial launch angle of a fragment. Literature resources show that any angle value has the same probability of occurence and thus this parameter is introduced into the zone of aleatoric uncertainty. We will show that this is not justified and that the intitial launch angle of a fragment is defined by the shape of the fragment, i.e. by potential fracture lines of a tank.

Probabilistic analysis of influential fragmentation factors is an efficient way of distinguishing between epistemic and aleatoric uncertainty. The probabilistic approach is based on the use of the probabilistic mass method which was originally developed by Djelosevic and Tepic [37, 38]. The main purpose of this method is the assessment of tank fragmentation probabilities on the basis of ideal values and the mass factor. Ideal fragmentation probabilities are assessed using statistical simulation on a sufficient number of samples. The precondition for a sufficient number of statistical samples is clear convergence of results, which, in this case, is achieved with more than 100,000 samples. Ideal fragmentation probabilities were obtained under the assumption of uniform stress state and strength (homogeneity) of the material. Ideal fragmentation probabilities are those that correspond to the explosion of a tank with a uniform stress state, and the values that refer to the specific number of generated fragments are presented in **Table 1**.

Since the actual stress state of a tank is not uniform, the ideal fragmentation probabilities have to be corrected. The correction factor used is the so-called mass factor (f_{mass}) which represents the ratio between the mass of the part of the tank with the non-uniform stress state and the total mass of the tank. Mass factor values for typical cylindrical tanks with torispherical end caps range between 0.55 and 0.75. Greater values indicate greater uniformity of the stress state and vice versa. The effect of fire contributes to greater non-uniformity so the mass factor in this case has lower values. The assessment of mass factors requires division of the tank into segments. Cylindrical tanks should be devided into three segments irrespective of their construction type and size. The first segment comprises the cylindrical part between the supports and it is defined by length L (S1) according to **Figure 1**. The other two segments (S2 and S3) comprise the parts of the tank outside the supports

Number of fragments	1	2	3	4	5	6	≥ 7
The probability of ideal fragmentation	1/2	1/3	1/8	1/30	1/150	1/800	5/12000

Table 1.
Ideal tank fragmentation probabilities.

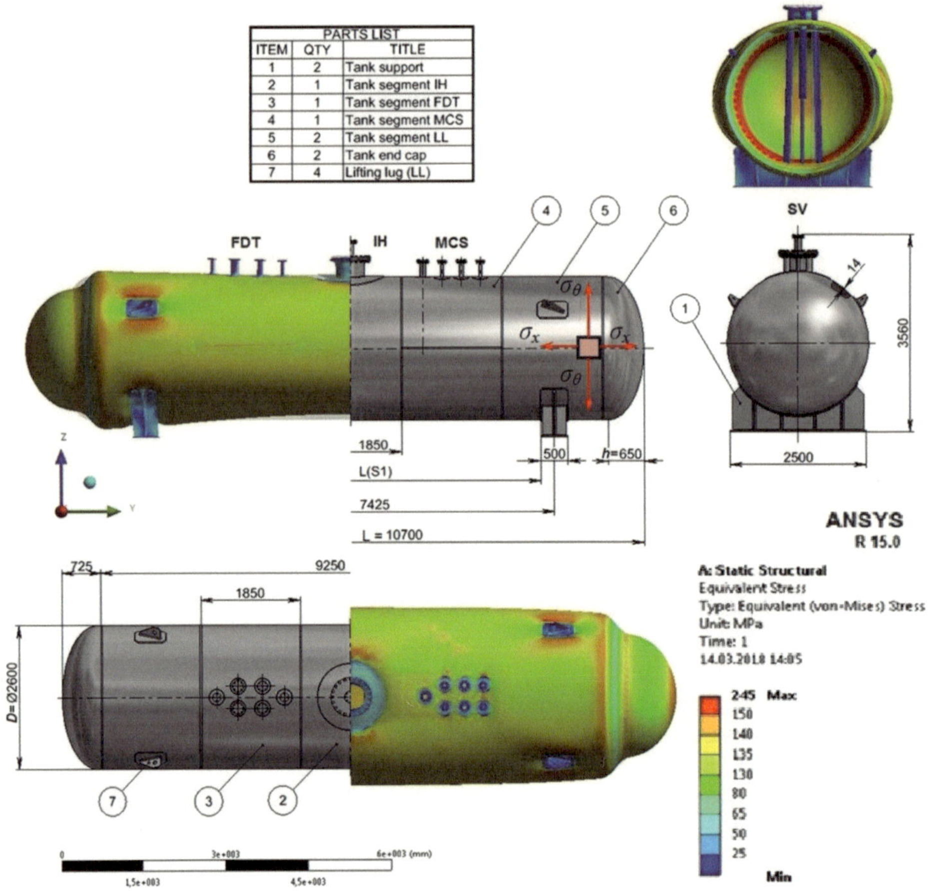

Figure 1.
Type of the tank and critical stress zones.

and their lengths amount to [L – L(S1)]/2. Tank fracture can occur in individual segments (either S1 or S2 or S3) or in at least two segments (S1 and S2 or S1 and S3 or S2 and S3) or in all three segments (S1 and S2 and S3). Depending on the mentioned scenarios, fracture probabilities and conditional fracture probabilities resulting from tank fragmentation are derived. Fracture probabilities are associated with tank fragmentation within only one segment (either S1 or S2 or S3). Conditional fracture probabilities comprise fragments which were generated from at least two segments.

Fracture probabilities and conditional fracture probabilities according to the number of fragments are given in **Table 2**.

Conditional fracture probabilities for individual fragmentation scenarios (Sc1 ... Sc8) were obtained on the basis of fracture probabilities values for segments S1, S2 and S3. This way the probabilistic mass method enabled easy assessment of fracture probabilities according to the number of generated fragments. The literature on the subject usually mentions the entropic model for this purpose where deviation of about 50% for the probability of the third fragment is observed. The entropic model is based on accident data fitting, whereas the probabilistic mass method is independent of accident data. A comparative analysis of fracture probabilities according to the number of generated fragments for the entropic model, the probabilistic mass method and accident data is presented in **Figure 2**.

Fragment number	Fracture probability for segments [%]			Conditional probability of fracture S1 if there is damage to segments S2 and/or S3 [%]							
	S1	S2	S3	Sc1	Sc2	Sc3	Sc4	Sc5	Sc6	Sc7	Sc8
ONE (1)	35.500	6.250	6.250	32.959	2.197	2.197	0.146	54.932	3.662	3.662	0.244
	27.500	11.250	11.250	21.661	2.746	2.746	0.348	57.105	7.239	7.239	0.918
TWO (2)	25.000	4.167	4.167	22.960	0.998	0.998	0.043	68.880	2.995	2.995	0.130
	18.333	7.500	7.500	15.686	1.272	1.272	0.103	69.876	5.666	5.666	0.459
THREE (3)	9.375	1.563	1.563	9.084	0.144	0.144	0.002	87.815	1.394	1.394	0.022
	6.875	2.813	2.813	6.494	0.188	0.188	0.005	87.960	2.545	2.545	0.074
FOUR (4)	2.500	0.417	0.417	2.479	0.010	0.010	0.000	96.689	0.405	0.405	0.002
	1.833	0.750	0.750	1.806	0.014	0.014	0.000	96.700	0.731	0.731	0.006
FIVE (5)	0.500	0.083	0.083	0.499	0.000	0.000	0.000	99.334	0.083	0.083	0.000
	0.367	0.150	0.150	0.366	0.001	0.001	0.000	99.335	0.149	0.149	0.000
SIX (6)	0.094	0.016	0.016	0.094	0.000	0.000	0.000	99.875	0.016	0.016	0.000
	0.069	0.028	0.028	0.069	0.000	0.000	0.000	99.875	0.028	0.028	0.000
SEVEN (7)	0.013	0.002	0.002	0.012	0.000	0.000	0.000	99.983	0.002	0.002	0.000
	0.009	0.004	0.004	0.009	0.000	0.000	0.000	99.983	0.004	0.004	0.000
≥ EIGHT (8)	0.019	0.003	0.003	0.019	0.000	0.000	0.000	99.975	0.003	0.003	0.000
	0.014	0.006	0.006	0.014	0.000	0.000	0.000	99.975	0.006	0.006	0.000

Table 2.
Fracture probabilities and conditional fracture probabilities of tank fragmentation.

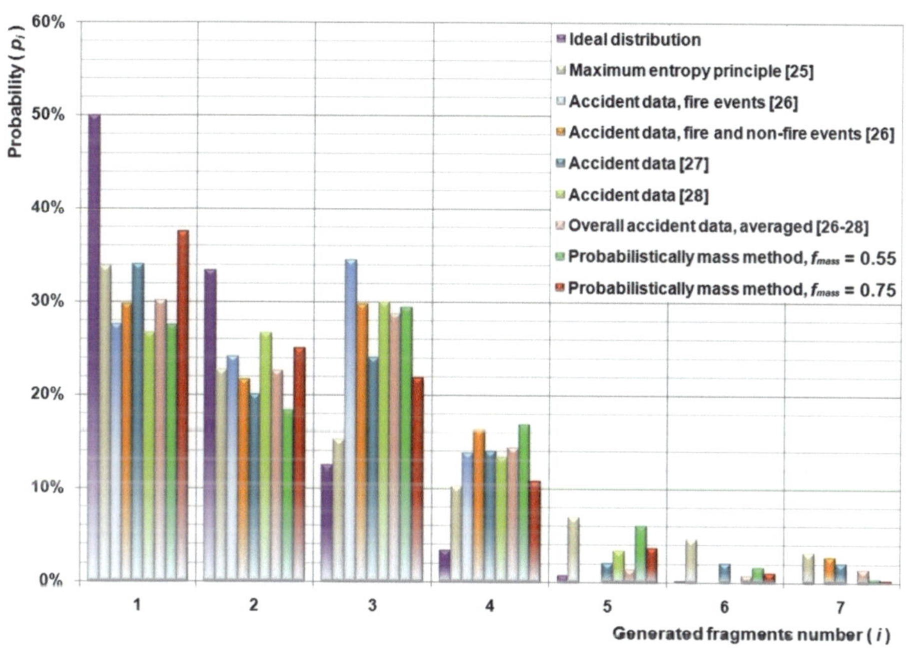

Figure 2.
Comparative analysis of fracture probabilities.

2.2 Fracture scenarios

Fracture scenarios of fragmentation in tank explosions imply a qualitative and quantitative analysis which is performed using Failure Tree Analysis (FTA). The triggering event that leads to tank fragmentation is reaching the critical pressure. The major central events are fractures of segments 1, 2 and 3. Qualitative analysis offers eight potential scenarios (Sc1 … Sc8). All the scenarios imply tank fragmentation apart from scenario Sc5 which excludes the possibility of tank fracture when the critical pressure is reached. Quantitative analysis implies the assessment of fracture probabilities and conditional fracture probabilities. These probabilities were assessed by means of Monte Carlo simulation using the probabilistic mass method, and they are presented in **Table 2**.

A significant part of this research deals with the assessment of the mass of fragments generated during tank explosion. For that purpose, the mass of fragments is expressed via the percent share of the empty tank mass which amounts to 12.3 *t*. Simulation results with over 100,000 samples show that fragments whose sum of masses ranges between 10% and 20% of the empty tank have the greatest generation probability. In this concrete case, the mass of fragments is between 1,230 *kg* and 2,460 *kg*. The maximum fragmentation probability is observed when two fragments are generated in an explosion and it amounts to around 17%.

Generation of fragments with small mass (smaller than 1% of the total mass of the tank) and generation of more than six fragments are extremely rare. The distribution of probabilities for the generated fragments depending on their mass is presented in **Figure 3**.

The probabilities of generation of only one fragment for different masses have uniform distribution, which points to aleatoric uncertainty. This means that the shape and mass of only one fragment generated in an explosion cannot be predicted

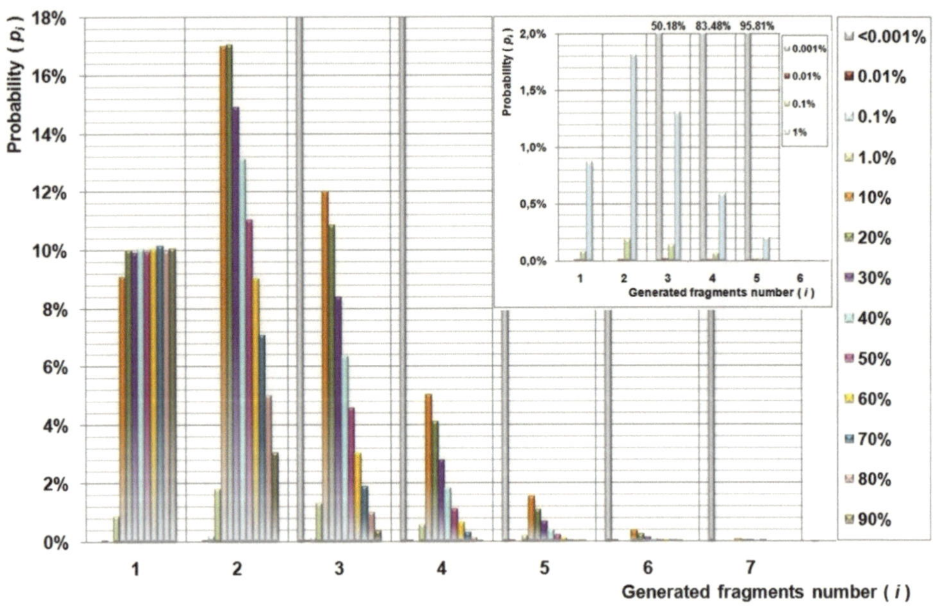

Figure 3.
Distribution of probabilities for fragments with different mass.

with certainty. On the other hand, generation of two or more fragments is accompanied by epistemic uncertainty. This means that with adequately used methodology, the mass and shape of two or more generated fragments can be predicted with certainty with the probabilities shown in **Figure 3**.

2.3 The assessment of shape and mass of fragments

The assessment of mass and shapes of fragments requires defining fracture lines, i.e. the lines along which tank fracture occurs. Fracture lines are zones of pronounced stress of the material and are the result of different construction conditions. Definition of fracture lines implies prior pressure analysis. In this case, the analysis was performed using the ANSYS software for the operating pressure of 16.7 *bar*. Fracture lines can spread in those zones whose pressure exceeds 130 *MPa*.

Accordingly, the investigated cylindrical tank has a total of 13 typical areas from which tank fragments can be generated (**Figure 4**). The most pronounced stress of the tank is on the transition from the cylinder to the end caps of the tank. This is the reason why during an explosion a fragment containing a larger or smaller part of the end cap is almost always generated.

Tank fragmentation is most often accompanied by generation of one, two or sometimes three fragments. Their typical shapes, mass, fracture zones and generation probabilities are given in **Table 3**. The most probable scenario in tank explosion is the one with the generation of two fragments with the sum of masses of around 1,200 *kg* or 2,255 *kg*.

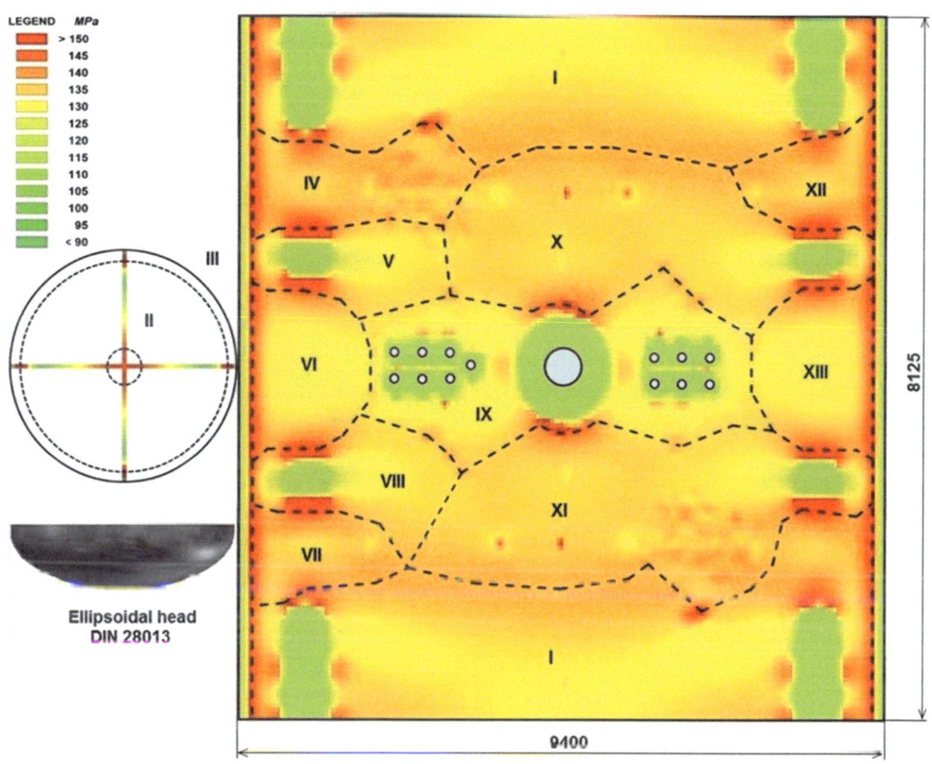

Figure 4.
Stress state of the tank with fracture lines.

Number of fragments	Fragmentation configuration Shapes of fragments	Fragment mass [kg]	Fragmentation probability [%]	Fracture zone
2	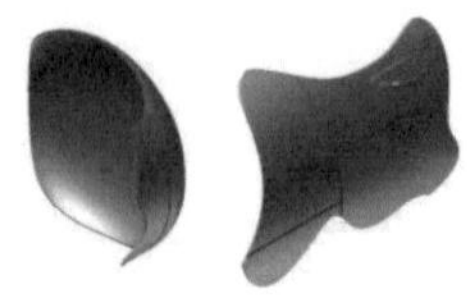	820 480	17.04	II + III IV + V
2	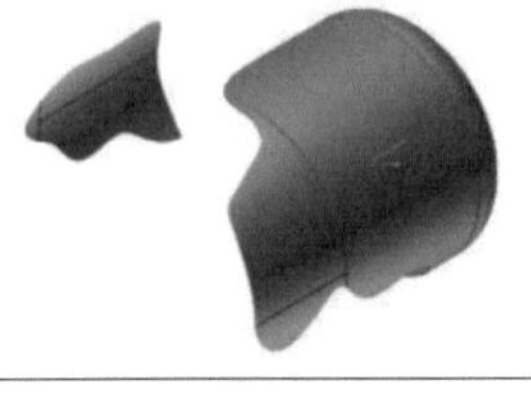	205 2050	17.06	VII + VIII II + III + IV + V + VI
3		190 805 235	12.00	VII II + III V

Table 3.
Characteristic fragmentation forms of the tank.

3. Fragmentation mechanics

Fragmentation mechanics involves modeling the flight of fragments created by a tank explosion. The basic characteristics of the fragments include geometric and kinematic parameters. We identified the geometric parameters in Section 2 of this chapter and they include number, mass and shape of the generated fragments. The main kinematic parameters include the initial velocity and the initial direction of the fragment launch. The literature does not provide information on the distribution of the range of fragments.

Therefore, it is not justified to assume that the range of the fragments is accompanied by a random distribution. Also, it is not justified to introduce assumptions about the uniformity of kinematic parameters of the fragments, just because we do not have enough available information about their behavior. Assuming a uniform distribution for some of the kinematic parameters, we enter an area of aleatoric uncertainty that does not allow an adequate assessment of fragmentation risk. The authors of this chapter start from the assumption that the generation of fragments does not follow a stochastic process, thus putting epistemic uncertainty in the foreground.

3.1 Fragment flight model

The flight of the fragment takes place under the influence of inertial, gravitational and aerodynamic forces (air resistance and lift force). The trajectory of the fragment uniquely determines the vector form of the equation of motion:

$$m_{fr} \cdot \vec{a}_{fr} = \vec{W}_D + \vec{W}_L + \vec{G} \tag{1}$$

The air resistance force (W_D) and the thrust force (W_L) are defined by:

$$\vec{W}_D = -\left(\frac{1}{2}\rho_v C_D A_D v_{fr}\right) \cdot \vec{v}_{fr} \tag{2}$$

$$\vec{W}_L = -\left(\frac{1}{2}\rho_v C_L A_L v_{fr}\right) \cdot \vec{v}_{fr} \tag{3}$$

The flight of each of the fragments should be observed in the local coordinate system Oxyz and then the projections of the vector differential Eq. (1) read:

$$a_{x,fr} = \dot{v}_{x,fr} = \left(-k_D v_{x,fr} - k_L v_{z,fr}\right)\sqrt{v_{x,fr}^2 + v_{z,fr}^2} \tag{4}$$

$$a_{y,fr} = \dot{v}_{y,fr} = 0 \tag{5}$$

$$a_{z,fr} = \dot{v}_{y,fr} = \left(k_L v_{x,fr} - k_D v_{z,fr}\right)\sqrt{v_{x,fr}^2 + v_{z,fr}^2} - g \tag{6}$$

Practically, the flight of the fragment is completely described by (4), (5) and (6), since there is no motion in the direction of the y-axis (that is why we observed the motion in the local coordinate system). The Taylor's series method was used to solve the coupled system of nonlinear differential equations. The ratio of the velocity components initially defines the initial launch angle of the fragment (**Figure 5**). This simply proves that the initial launch angle is not accompanied by a stochastic distribution.

3.2 Initial conditions

The initial conditions define the kinematic parameters at the initial moment (at the moment of the tank explosion). These conditions are necessary for initiating the procedure of numerical solution of differential equations and read:

$$x_{fr}(t_0) = x_0 \wedge z_{fr}(t_0) = z_0 \tag{7}$$

$$v_{x,fr}(t_0) = v_{xo} \wedge v_{z,fr}(t_0) = v_{zo} \tag{8}$$

$$\dot{v}_{x,fr}(t_0) = a_{xo} \wedge \dot{v}_{z,fr}(t_0) = a_{zo} \tag{9}$$

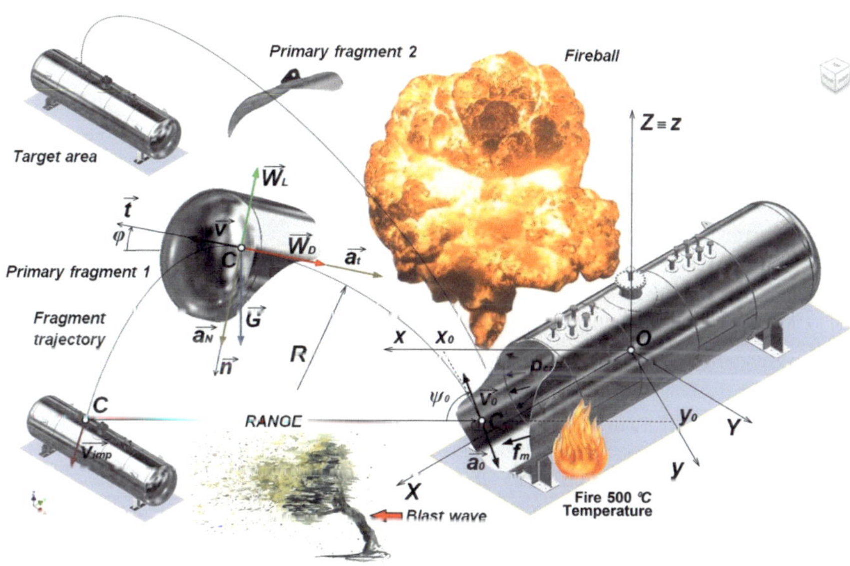

Figure 5.
Kinematic parameters due to tank fragmentation.

These values are used as the first step in the numerical procedure and they are of unknown magnitude at the moment. At this level we know that the initial velocity is not independent of the initial acceleration of the fragments. The components of velocity at any moment t read as follows:

$$v_{x,fr} = \sqrt{\frac{(-a_x)}{k_D + k_L \cdot tg\phi} \cdot \frac{1}{\sqrt[4]{1 + tg^2\phi}}} \tag{10}$$

$$v_{z,fr} = \sqrt{\frac{(-a_z)}{k_D + k_L \cdot tg\phi} \cdot \frac{tg\phi}{\sqrt[4]{1 + tg^2\phi}}} \tag{11}$$

The direction of the fragment velocity can be determined at any time during the flight of the fragment, including the initial moment using the expression:

$$\phi = arctg\left(\frac{v_{z,fr}}{v_{x,fr}}\right) = arctg\frac{\left(\frac{k_L}{k_D}\right) \cdot (-a_x) + (-a_z) - g}{\left[(-a_x) + \left(\frac{k_L}{k_D}\right) \cdot a_z + g\right]} \tag{12}$$

In order to obtain the initial launch angle of a fragment, it is necessary to know the components of the initial acceleration. In the continuation of this chapter, we show how the initial acceleration occurs. It is important to point out that (12) shows an unjustified assumption about the uniformity of the initial acceleration. Thus, the initial launch angle as a kinematic parameter is classified in the category of epistemic uncertainty. Literature sources in this area do not use the initial acceleration parameter, although it is a way to remove uncertainty regarding a reliable fragmentation risk assessment.

3.3 Defining the initial acceleration

We have previously came to the conclusion that in order to define the initial velocity and the initial launch angle of a fragment, it is necessary to know the initial acceleration. Hence the idea and justification for introducing this kinematic parameter into fragmentation analysis. The initial acceleration is proportional to the force of pressure on the fragment. This force is created by the critical pressure p_{cr} acting on the inner surface of the fragment. The proportion of explosive energy transferred to the fragment is limited by the action of critical pressure. This means that the proportion of explosive energy of the fragment transferred to the fragment depends on the tensile strength of the material.

The procedure for determining the initial acceleration is based on this assumption. The lower tensile strength gives less initial kinetic energy of the fragment. Fragment generation occurs when the von Mises's stress reaches a critical value under the action of internal tank pressure and is defined as:

$$\sigma_{cr} = \sqrt{\sigma_x^2 + \sigma_\theta^2 - \sigma_x\sigma_\theta + \frac{3}{2}(\sigma_x - \sigma_\theta)^2} = 102 \cdot p_{cr} \tag{13}$$

Where the corresponding components of the von Mises's stress are given by:

$$\begin{aligned} \sigma_x &= 101 \cdot p_{cr} \\ \sigma_\theta &= 105 \cdot p_{cr} \end{aligned} \tag{14}$$

Separation of fragments from the tank as a whole occurs when the critical stress reaches the value of tensile strength of the material f_m, so the critical pressure is determined as $p_{cr} = f_m/102$. Accordingly, the initial acceleration of a fragment can be defined by:

$$a_o = \frac{F}{m_{fr}} = \frac{p_{cr}A_{fr}}{\rho\delta A_{fr}} = \frac{f_m}{102\rho\delta} = const \tag{15}$$

This proves that the initial acceleration has a constant value for a certain type of steel from which the tank is made. LPG transport and storage tanks have a constant wall thickness of 14 mm. The density of steel ρ is also constant and amounts to 7850 kg/m^3.

Thus, by knowing the initial acceleration, we are able to define the initial velocity and the initial launch angle, as well as all other kinematic parameters at any time during the flight of the fragment. It should be borne in mind that the initial acceleration depends on the tensile strength of the material whose values are subject to variation with temperature changes. The change in the influence values on the range of the fragment is given in **Table 4**.

3.4 Fragments range

Based on a mathematical model that describes the flight of the fragment and the initial acceleration, we are able to determine the range of the fragments. For this purpose, all geometric and kinematic parameters will be classified into two groups. The first group consists of invariant parameters and their value is fixed. The second group consists of variable parameters, ie those whose value changes during the flight of the fragment. This group includes the coefficients of aerodynamic and thrust acceleration k_L and k_D. The invariant parameters are fully defined and their value is not subject to variation during the flight of the fragment (tank wall thickness, tensile strength and specific weight of the material). Variable parameters change during the flight of the fragment due to the rotation of the fragment or some other effects. Variable parameters are defined via the air resistance coefficient C_D and the thrust coefficient C_L.

The literature gives approximate values of these coefficients which depend on the shape of the fragment. Smaller fragments whose mass does not exceed a few hundred kilograms generally have the shape of a shell (shells are aerodynamic shapes), so they have a pronounced thrust effect. In order to estimate the range of the fragments, the fluctuation of variable parameters is performed, whereby different trajectories of the fragments are obtained. These trajectories enable the definition of the limit values of the coefficients k_L and k_D. Parabolic trajectories of

Parameter	**Temperature [^oC]**						
	20	**200**	**400**	**500**	**600**	**800**	**900**
Tensile strength R_m [MPa]	500	469	373	252	117	23	11
Critical pressure $p_{cr}[bar]$	48.5	45.5	36.2	24.5	11.4	2.2	1.1
Initial acceleration $a_0[m/s^2]$	43,746	41,034	32,635	22,048	10,237	2,012	962
Initial velocity $v_0[m/s]$	1,985	1,923	1,715	1,409	960	426	294
Average range $R_{avg}[m]$	395.6	394.3	389.3	380.0	359.1	303.1	272.2

Table 4.
Temperature influence on the influence values of the fragment range.

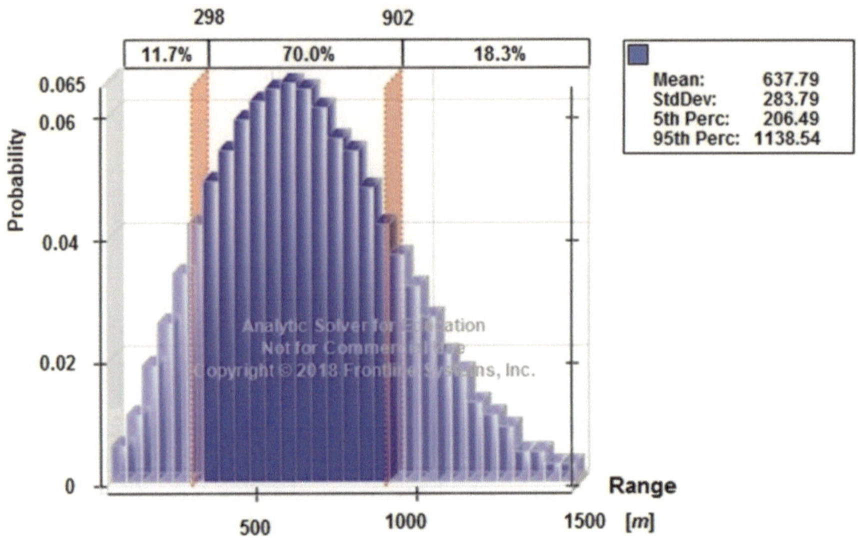

Figure 6.
Distribution of the range of typical shaped fragments.

small height (fragment reaches small range and small height) as well as pointed trajectories (fragments reach large height and relatively small range) are very rare and represent boundary cases for the selection of coefficients k_L and k_D. Fragment range estimation was realized by Monte Carlo simulation by processing 240 different samples for different fragmentation parameters. Fragments of up to a few hundred kilograms launched at an angle of up to 35^0 can be adequately represented by the Weibull's distribution with parameters 2.3 and 723.8 as shown in **Figure 6**.

The maximum probability of 6.5% corresponds to fragments with a range between 613 m and 663 m. This means that 6.5% of the total number of generated fragments will fall to the target between these distances.

4. Fragmentation simulation model

The basic questions that are asked in the simulation model of fragmentation are related to the assessment of fragmentation density and sector angle. Fragmentation density refers to the number of fragments whose range will correspond to an area. The sector angle refers to the most common angle in the horizontal plane at which the fragments will burst due to the explosion of the tank. The simulation model of fragmentation is based on the research given in Sections 2 and 3.

4.1 Assessment of fragmentation density

The density of fragments is estimated based on the results of fragmentation mechanics for different trajectories, masses and shapes of fragments. Assessment of fragment density requires the definition of appropriate distribution functions depending on the mass, the initial launch angle of the fragment and the limit values of the coefficients k_L i k_D. The considered masses of fragments are in the range from 200 kg to 200 kg, while the initial angles take values from 5^0 to 35^0. The characteristic density functions for the defined parameters are given in **Table 5**.

Mass	Init. ang.	Coeff. of drag/lift accel. $\times 10^{-4}$ $[m^{-1}]$				Type of probability density function (pdf)	Parameters of pdf	
$m_{fr}[kg]$	$\psi_0\ [^0]$	$k_{D,min}$	$k_{D,max}$	$k_{L,min}$	k_{Lmax}		a	b
200	15	60	150	0	30	Rayleigh	264.436	—
500	15	50	80	0	22	Weibull	4.150	667.458
500	25	50	80	0	20	Gamma	10.978	51.535
800	5	40	80	0	25	Rayleigh	486.687	—
800	15	40	70	0	21	Log Normal	671.079	212.938
1350	5	39	60	0	16	Gamma	11.678	64.910
1350	15	39	60	0	15	Log Normal	802.359	228.446
1350	35	35	59	0	13	Weibull	2.984	773.223
2000	15	31	55	0	16	Weibull	2.910	890.345

Table 5.
Fragmentation densities for characteristic fragments.

4.2 Sector angle assessment

The sector angle is the angle in the horizontal plane under which the fragment is launched. The explosion of the cylindrical tanks is accompanied by the generation of fragments from segments 1, 2 and 3 according to **Figure 1**. If the fragments belong only to segment 1, then their bursting is done exclusively in the axial direction. If the fragments belong only to segments 2 and 3, then the scattering of the fragments takes place in the action direction. In practice, the most common cases are when we have the generation of fragments from segments 1 and 2 or 1 and 3. Therefore, the first step in assessment the sector angle is to define the fragmentation probabilities by tank segments. For this purpose, the results on fracture and conditional fracture probabilities presented in Section 2 are used. Limit values of fragmentation probabilities by the number of generated fragments are given in **Table 6**.

The sector angles α and β are determined on the basis of the following formula:

$$p_f\left(1 - \frac{90^0 - \beta}{\alpha}\right) = \frac{11}{26} \tag{16}$$

The condition should be added to the previous formula: $\alpha + \beta = \pi/2$, where p_f is the fragmentation probability of the tank corresponding to the first segment (S1).

Segment	Value	Fragmentation probability (p_f) for the number of generated fragments [%]							
		1	2	3	4	5	6	7	≥ 8
1 | (2 and/or 3)	max	83.21	80.33	76.94	75.48	75.04	74.60	75.00	76.00
	min	64.11	60.86	57.11	55.54	55.26	55.20	52.94	53.85
Only 2 or 3	min	16.25	19.25	22.88	24.46	24.96	25.40	25.00	24.00
	max	33.75	37.62	42.28	44.28	44.74	44.80	47.06	46.15
Only 2 and 3	min	0.54	0.42	0.18	0.06	0.00	0.00	0.00	0.00
	max	2.14	1.52	0.61	0.18	0.00	0.00	0.00	0.00

Table 6.
Fragmentation probabilities by number of generated fragments.

Sectoral angle	Value	Sectoral angle (2α and 2β) for the number of generated fragments							
		1	2	3	4	5	6	7	≥ 8
2α [^{0}]	max	275.4	287.8	283.0	285.0	285.6	286.2	285.8	284.4
	min	303.4	310.0	318.4	322.2	323.0	323.2	329.2	326.6
2β [^{0}]	max	84.6	81.2	77.0	75.0	74.4	73.8	74.2	75.6
	min	56.6	50.0	41.6	37.8	37.0	36.8	30.8	33.4

Table 7.
Sector angles by number of fragments generated.

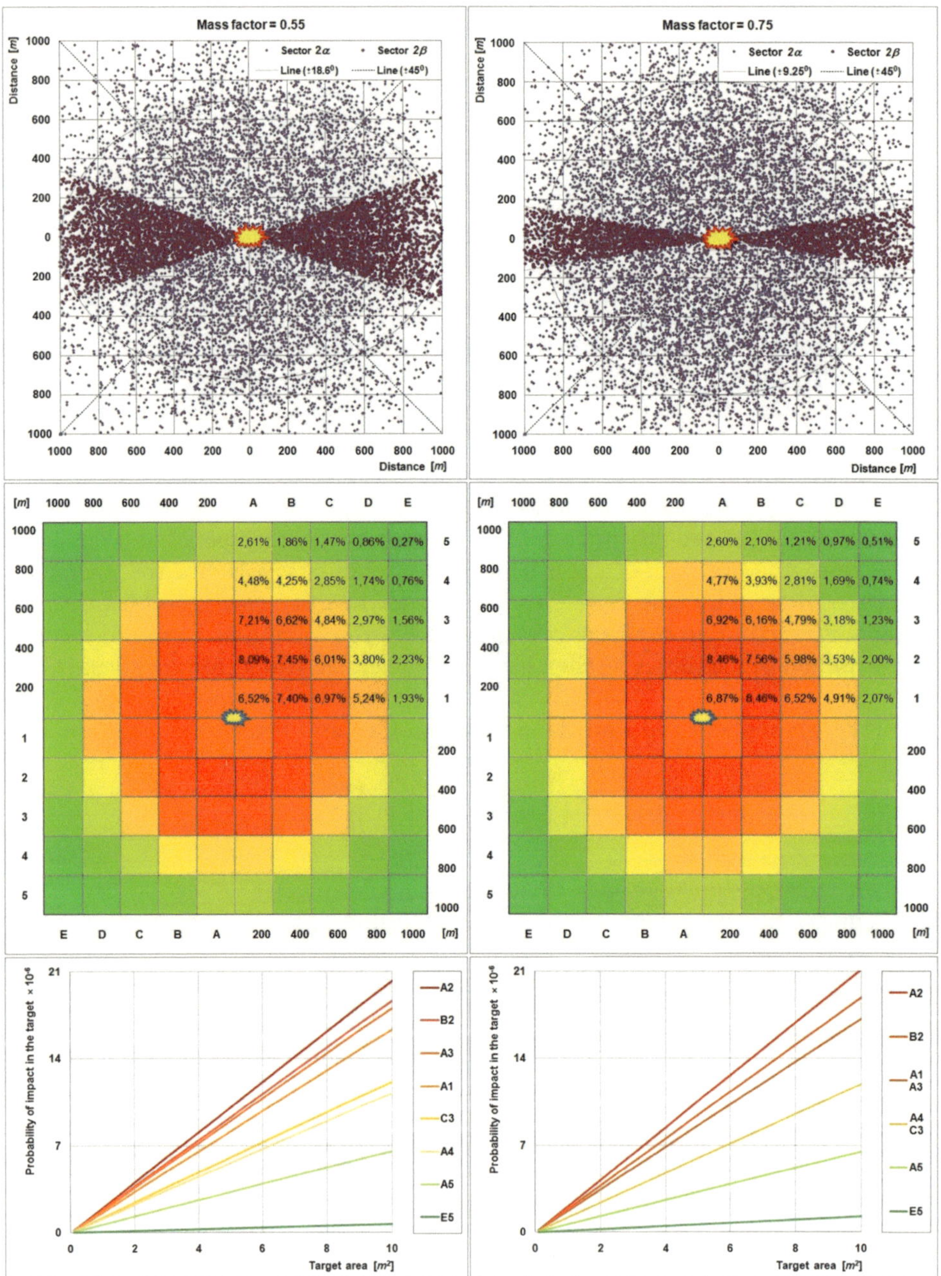

Figure 7.
Burst simulation with risk matrix.

Sector area 2α covers fragmentation zones in which fragments generated predominantly from the cylindrical part of the tank are located (S1). Sector angle 2β covers the area corresponding to the fragments generated from segments 2 and 3 (S2 i S3). Sector angles by number of generated fragments are given in **Table** 7.

5. Risk assessment

The risk assessment is the final phase of the fragmentation analysis due to the explosion of the LPG tank. Defining fracture probabilities, shape and mass of fragments, as well as kinematic parameters are the basis for defining fragmentation density and sector angles. The fragmentation risk assessment is performed based on the results given in Section 4. Research has shown that the fragmentation effect of a 50,000-liter tank can endanger objects and people at distances greater than 1,000 meters. Therefore, fragmentation stands out as the dominant hazard versus shock wave and thermal effect during tank explosion. Fragmentation risk assessment is carried out by dividing the area around the focus of the accident into quadrants measuring 200 × 200 meters within which the number of fragments is observed.

A larger number of fragments gives a higher fragmentation risk and vice versa. The fragmentation risk analysis is given for the limit values of the mass factor (0.55 and 0.75), so we obtain the limit values of the fragmentation risk. For example, for quadrant D1 we have that the limit values of fragmentation risk are 5.24% and 4.91%, which gives an average value of 5.08%. Although there is a significant deviation between the limit values of the mass factor (0.55 and 0.75), based on the appropriate risk matrices, we can conclude that the deviation of the limit values of fragmentation risk is only a few percent. This indicates that the presented methodology based on the identification of uncertainties provides convergent and reliable solutions in the assessment of fragmentation risk. Simulations of fragment bursting as well as risk matrices are shown in **Figure** 7.

6. Conclusions

In this chapter, a fragmentation simulation model for risk assessment due to the explosion of a cylindrical tank is presented. According to the literature, fragmentation models exclusively use accident data. In addition, parameters for which there is insufficient information available are assumed with a uniform distribution. These are the main shortcomings of existing fragmentation models. These shortcomings have been remedied by applying the proposed fragmentation model. The simulation of fragment scattering is considered through the issue of uncertainty in the estimation of geometric and kinematic parameters. Fracture probabilities were estimated without available accident data for the considered tank type. Fragmentation mechanics enabled the definition of characteristic trajectories and the definition of limit values for coefficients k_L i k_D. Introducing the initial acceleration into the analysis, we came to know about the correlation of certain geometric and kinematic parameters of the fragment. Most of the influential parameters are accompanied by epistemic uncertainty, so the initial velocity cannot be estimated on the basis of the explosive energy of the tank, nor can the initial launch angle of the fragment be assumed by a stochastic distribution. Fragments weighing up to a few hundred kilograms are best represented by the Weibull distribution, and the most probable range of the fragments is between 670 m and 680 m. The risk matrix is given per square meter for an area of 4 km^2.The probability of impact of the fragment in the base target of 10 m^2 is from $1.6{\cdot}10^{-5}$ to $2.1{\cdot}10^{-5}$.

Acknowledgements

This work has been partially supported by the Ministry of Education and Science of the Republic of Serbia within the Project No. 34014 and by the project "Naučni i pedagoški rad na doktorskim studijama", University of Novi Sad, Faculty of Technical Sciences.

Nomenclature

D	External diameter of the tank (D = 2600 mm)
h	Height of the elliptic head (h = 650 mm)
δ	Wall thickness of the tank (δ = 14 mm)
p	Operating pressure of the tank (p = 16.7 MPa)
σ_x	Longitudinal stress of the tank
σ_θ	Circumference stress of the tank
a_0	Initial acceleration of the fragment
F	The inertial force of the fragment
ρ	Density of steel S355J2G3 (ρ = 7850 kg/m^3)
p_{cr}	Critical tank pressure
f_m	Tensile strength of steel S355J2G3
m_{fr}	Mass of the fragment
W_D	Force of air resistance during flight of the fragment
W_L	Lifting force of the fragment
G	Gravitational force ($G = m_{fr} \cdot g$, g = 9,81 m/s^2)
ρ_{air}	Air density (ρ_{air} = 1.20 kg/m^3)
C_D	Coefficient of force of air resistance
C_L	Coefficient of lift force
A_D	The area of the frontal projection of the fragment
A_L	The area of the lateral projection of the fragment
v_{fr}	Velocity of the fragment
a_{fr}	Acceleration of the fragment
k_D	Coefficient of drag acceleration
k_L	Coefficient of lift acceleration
x	Horizontal coordinate
y	Vertical coordinate
Δt	Time interval

Author details

Mirko Djelosevic* and Goran Tepic
Faculty of Technical Sciences, University of Novi Sad, Novi Sad, Serbia

*Address all correspondence to: djelosevic.m@uns.ac.rs

References

[1] Hemmatian B, Abdolhamidzadeh B, Dabra R.M, Casal J: The significance of domino effect in chemical accidents. J. Loss Prev. Process Ind. 2014;29:30-38. Doi:10.1016/j.jlp.2014.01.003

[2] Abdolhamidzadeh B, Abbasi T, Rashtchian D, Abbasi S.A: Domino effect in process industry accidents – An inventory of past events and identification of some patterns. J. Loss Prev. Process Ind. 2011;24:575-593. Doi: 10.1016/j.jlp.2010.06.013

[3] Dabra R.M, Palacios A, Casal J: Domino effect in chemical accidents: Main features and accident sequences. J. Hazard. Mater. 2010;183:565-573. Doi: 10.1016/j.jhazmat.2010.07.061

[4] Bariha N, Mishra I.M, Srivastava V. C. Fire and explosion hazard analysis during surface transport of liquefied petroleum gas (LPG): a case study of LPG truck tanker accident in Kannur, Kerala, India. J. Loss Prev. Process Ind. 2016; 40:449-460. Doi:10.1016/j.jlp.2016.01.020

[5] Eckhoff R.K: Boiling liquid expanding vapor explosions (BLEVEs): A brief review, J. Loss Prev. Process Ind. 2014;32:30-43. Doi: 10.1016/j.jlp.2014.06.008

[6] Sun D, Jiang J, Zhang M, Wang Z: Influence of the source size on domino effect risk caused by fragments. J. Loss Prev. Process Ind. 2015;35:211-223. Doi: 10.1016/j.jlp.2015.05.005

[7] Khan F.I, Abbasi S.A: Major accidents in process industries and an analysis of causes and consequences. J. Loss Prev. Process Ind. 1999;12:361-378. Doi: 10.1016/S0950 4230(98)00062-X

[8] Khakzad N, Amyotte P, Cozzani V, Reniers G, Pasman H. How to address model uncertainty in the escalation of domino effects? J. Loss Prev. Process Ind. 2018; 54:9-56. Doi:10.1016/j.jlp.2018.03.001.

[9] Lang T, Schwoebel V, Diène E, Bauvin E, Garrigue E, Lapierre-Duval K, Guinard A, Cassadou S. Assessing post-disaster consequences for health at the population level: experience from the AZF factory explosion in Toulouse. J Epidemiol Community Health. 2007; 61 (2):103–107. Doi:10.1136/jech.2005.043331

[10] Jahangiri K, Ghodsi H, Khodadadizadeh A, Yousef Nezhad S. Pattern and nature of Neyshabur train explosion blast injuries. World J. Emerg. Surg. 2018;13(3). Doi:10.1186/s13017-018-0164-7.

[11] Landucci G, Argenti F, Spadoni G, Cozzani V. Domino effect frequency assessment: The role of safety barriers. J. Loss Prev. Process Ind. 2016;44:706-717. Doi:10.1016/j.jlp.2016.03.006

[12] Kang J, Zhang J, Gao J. Analysis of the safety barrier function: Accidents caused by the failure of safety barriers and quantitative evaulation of their performance. J. Loss Prev. Process Ind. 2016; 43:361-371. Doi:10.1016/j.jlp.2016.06.010

[13] Baker W.E, Cox P.A, Westine P.S, Kulesz J.J, Strehlow R.A. Explosion Hazards and Evaluation, Elsevier, Amsterdam, 1983.

[14] CCPS, Guidelines for Evaluating the Characteristics of Vapor Cloud Explosions, Flash Fires and BLEVE's, Center for Chemical Process Safety, American Institute of Chemical Engineers, New York, 1994.

[15] Baker W.E, Kulesz J.J, Ricker R.E, Bessey Westine P.S, Parr V.B. Workbook for Predicting Pressure Wave and Fragment Effects of Exploding Propellant Tanks and Gas

Storage Vessels. NASA CR-134906. NASA Scientific and Technical Information Office, 1997, Washington.

[16] Hauptmanns U. A Monte-Carlo based procedure for treating the flight of missiles from tank explosions. Prob. Eng. Mech. 2001;16:307-312. Doi: 10.1016/S0266-8920(01)00023-6

[17] Hauptmanns U. A procedure for analyzing the flight of missiles from explosions of cylindrical vessels. J. Loss Prev. Process Ind. 2001;21:395-402. Doi: 10.1016/S0950-4230(01)00011-0

[18] Sun D, Jiang J, Zhang M, Wang Z, Zang Y, Yan L, Zhang H, Du X, Zou Y. Investigation on the approachof intercepting fragments generated by vessel explosion using barrier net. J. Loss Prev. Process Ind. 2017;49:989-996.

[19] Moore C.V. The design of barricades for hazardous pressure systems. Nucl. Eng. Des.1967;5:81-97. Doi:10.1016/0029-5493(67)90081-7

[20] Brode H.L. Blast wave from a spherical charge. Phys. Fluids. 1959; 2: 217-229. Doi:10.1063/1.1705911

[21] Baker W.E. Explosions in Air, University of Texas Press, Austin, 1973.

[22] Baum M.R. The velocity of missiles generated by the disintegration of gas pressurized vessels and pipes. Trans. ASME. 1984;106:362-368. Doi:10.1115/1.3264365

[23] Nguyen Q.B, Mébarki A.M, Saada R. A, Mercier F, Reimeringer M. Integrated probabilistic framework for domino effect and risk analysis. J. Loss Prev. Process Ind. 2009;40:892-901. Doi: 10.1016/j.advengsoft.2009.01.002

[24] Mishra K.B. Multiple BLEVEs and fireballs of gas bottles: Case of a Russian road carrier accident. J. Loss Prev. Process Ind. 2016;41:60-67. Doi: 10.1016/j.jlp.2016.03.003

[25] Mébarki A, Mercier F, Nguyen Q.B, Saada R.A. Structural fragments and explosions in industrial facilities. Part I: Probabilistic description of the source terms. J. Loss Prev. Process Ind. 2009;22: 408-416. Doi:10.1016/j.jlp.2009.02.006

[26] Holden P.L, Reeves A.B, Fragment hazards from failures of pressurised liquefied gas vessels. IchemE Symposium Series No. 93. 1985:205.

[27] Holden P.L. Assessment of missile hazards: Review of incident experience relevant to major hazard plant. Safety Reliab. Directorate, Health Safety Directorate, 1988.

[28] Nguyen Q.B, Mebarki A, Mercier F., Saada R.A, Reimeringer M. The domino effect and integrated probabilistic approaches for risk analysis. In: proceedings of the Eight International Conference on Computational Structures Technology, Sep. 2006, Las Palmas, Spain. 2006. p. 27-34. <hal-00719771>

[29] Sun D, Jiang J, Zhang M, Wang Z, Huang G, Qiao J. Parametric approach of the domino effect for structural fragments. J. Loss Prev. Process Ind. 2012;25:114-126. Doi:10.1016/j.jlp.2011.06.029

[30] Gubinelli G, Zanelli S, Cozzani V. A simplified model for the assessment of the impact probability of fragments. J. Hazard. Mater. 2016; A116:175-187. Doi: 10.1016/j.jhazmat.2004.09.002

[31] Mannan S. Lees' Loss Prevention in the Process Industries, fourth ed., Elsevier, Oxford, 2012.

[32] Baum M.R. The velocity of large missiles resulting from axial rupture of gas pressurised cylindrical vessels. J. Loss Prev. Process Ind. 2001;14:199-203. Doi:10.1016/S0950-4230(00)00039-5

[33] Mébarki A, Nguyen Q.B, Mercier F. Structural fragments and explosions in

industrial facilities. Part II: Projectile trajectory and probability of impact. J. Loss Prev. Process Ind. 2009;22:417-425. Doi:10.1016/j.jlp.2009.02.005

[34] Center for Chemical Process Safety (CCPS). Guidelines for evaluating the characteristics of vapor cloud explosions, flash fires, and BLEVEs. New York: American Institute of Chemical Engineers, 1994.

[35] Tugnoli A, Gubinelli G, Lamducci G, Cozzani V. Assessment of fragment projection hazard: Probability distributions for the initial direction of fragments. J. Hazard. Mater. 2014;279: 418-427. Doi:10.1016/j.jhazmat.2014.07.034

[36] Gubinelli G, Cozzani V. Assessment of missile hazards: Evaluation of the fragment number and drag factors. J. Hazard. Mater. 2009;161:439-449. Doi: 10.1016/j.jhazmat.2008.03.116

[37] Djelosevic M, Tepic G. Identification of fragmentation mechanism and risk analysis due to explosion of cylindrical tank. J. Hazard. Mater. 2019; 362: 17-35. Doi:10.1016/j.jhazmat.2018.09.013

[38] Djelosevic M, Tepic G. Probabilistic simulation model of fragmentation risk. J. Loss Prev. Process Ind. 2019; 60:53-75. Doi:10.1016/j.jlp.2019.04.003

[39] Manescau B, Chetehouna K, Sellami I, Nait-Said R, Zidani F. BLEVE Fireball Effects in a Gas Industry: A Numerical Modeling Applied to the Case of an Algeria Gas Industry. Fire Safety and Management Awareness, July 16th 2020: Fahmina Zafar and Anujit Ghosal, IntechOpen, Doi:10.5772/intechopen.92990.

[40] Nannapaneni S, Mahadevan S. Reliability analysis under epistemic uncertainty. Rel. Eng. & Sys. Safety. 2016;155:9-20. Doi:10.1016/j.ress.2016.06.005

Section 5

Electric Power Systems

Chapter 10

Simulation Modeling of Integrated Multi-Carrier Energy Systems

Nikolai Voropai, Ekaterina Serdyukova, Dmitry Gerasimov and Konstantin Suslov

Abstract

Integrated multi-carrier energy systems give good possibilities to have high effectiveness of energy supply to consumers. Transformation of energy systems under the impact of internal and external factors remarkably strengthens the technological integration of those systems and supports development of integrated multi-carrier energy systems. The concept of energy hub is developed for modeling and simulation of integrated multi-carrier energy systems. Based on previous research, a simulation model of the energy hub is being developed. The basic principles of building a simulation model of an energy hub concept are discussed. Realization of simulation model using Matlab/Simulink is proposed. Simulation results for the integrated electricity and heat systems are explained to demonstrate the capabilities of the simulation energy hub model. A case study for application of the simulation model is discussed.

Keywords: integrated multi-carrier energy systems, simulation modeling, energy hub, energy converters, energy storage, energy consumption optimization, Matlab/Simulink software

1. Introduction

Modern energy supply systems, primarily electricity, heat and gas systems represent a developed energy infrastructure that provides consumers in the economic and social sectors with various energy types with the required reliability, the required quality and at an affordable price. The development and opposition of these energy systems is under the influence of a new paradigm of customer-oriented energy supply. Recently, the requirements for the reliability of power supply and the quality of the types of energy supplied to consumers have significantly increased due to computerization and digitalization of consumer production processes and the expansion in the use of "high" production technologies by the consumer.

The design and operation of these energy systems tend to consider them independently of each other. The systems under discussion however interact quite closely with each other, for example, when electricity and heat is generated using gas as fuel at cogeneration under normal and emergency conditions, when electric heaters are used by consumers in the case of accidents in the heating system, etc.

The new conditions for the development and computerization of the infrastructure energy systems contribute to the expansion of interaction between them as many new actors appear that can provide ancillary services. The consumers with controlled load, managing their energy load, can have self-generation sources and energy storage units, and simultaneously, depending on the current conditions, be involved in conversion, storage and generation of the required type of energy; electric vehicles can deliver stored electricity to the power supply system during peak hours, etc.

The development of information and telecommunication technologies bring about additional opportunities for joint coordinated management of the expansion and operation of the energy systems under consideration.

All the above features increase significantly the interest in the research of virtually new facilities, i.e., integrated energy systems (IESs) [1, 2]. The primary basic problem here is the technology of modeling the sophisticated IESs. This chapter focuses on the main principles of the IES simulation technology relying on the capabilities of the Matlab/Simulink system and the energy hub concept.

The further presentation is structured as follows. Section 2 presents basic information about the features of IES and the history of research in this strand. Section 3 provides an overview of the energy hub concept. Section 4 contains a description of the nature of mathematical models of IES based on the integration of traditional models of the components of the considered IES of power supply systems. Section 5 discusses the principles of energy hub modeling used in most of the studies conducted, and the advantages and disadvantages of the models. Section 6 analyzes the capabilities of the Matlab/Simulink system for IES modeling. Section 7 presents a new approach to building a simulation model of an energy hub developed by the authors. Section 8 discusses the proposed technology for constructing an IES simulation model. Section 9 contains a description of one of the problems solved using the developed simulation model. The conclusion to this Chapter summarizes the results of the studies performed.

2. Integrated multi-carrier energy systems

Objective trends in energy systems development (electric power, heat, gas, oil, oil products supply systems, etc.) lead to creation of integrated multi-carrier energy systems. These tendencies are determined by strengthening of technological integration not only during production of energy (for example, electric power and heat on the co-generation plants (CGP) by using gas as the fuel), but also under energy consumption based on implementing different kinds of energy for the same objectives. For example, it is possible to use heat from centralized heating system based on CDP or from individual electric boilers, electric or gas individual furnaces, and so on. In these cases individual energy systems (electric power, gas and heat supply systems) acquire the interdependences not only between production plants and consumption of individual systems, but also between load flows in networks of these systems. Particularly significant interrelations between individual energy systems we can meet in emergency conditions. Taking into account above mentioned peculiarities we have to consider joint operation and expansion of individual energy systems [1, 2].

In [2], the authors explain the elements of the concept of integrated energy systems as a three-layer structure in three dimensions, similar to Rubik's cube (see **Figure 1**). The groups of layers can be defined as follows:

- Layers of systems - power systems, heating/cooling systems, gas systems;
- Layers of scale - super-systems, mini-systems, micro-systems;
- Layers of functions - energy, communication and management, decision making.

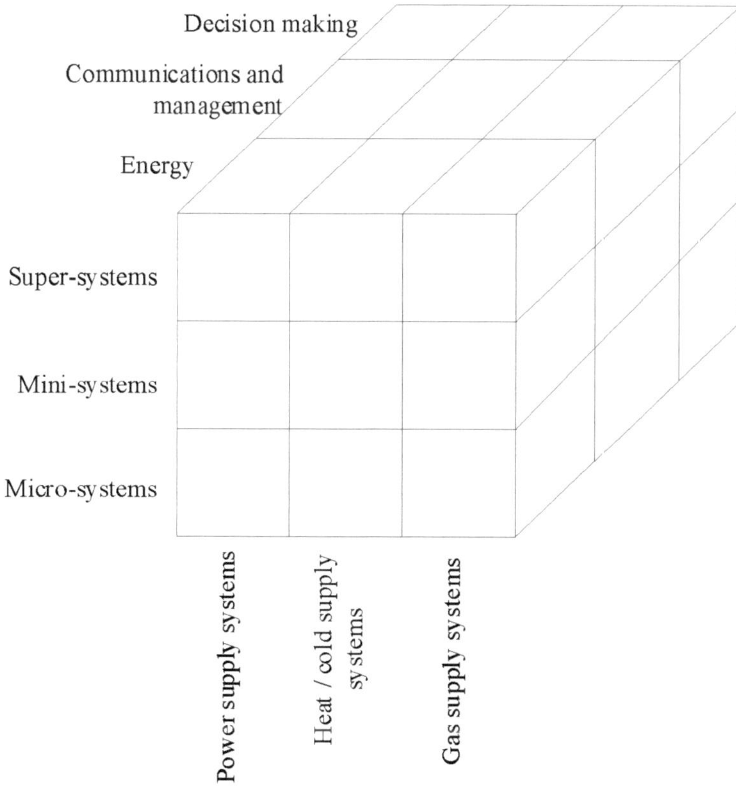

Figure 1.
Three-layer structure of integrated energy systems in three dimensions.

Integrated multi-carrier energy systems, as well as their individual energy supply systems, especially electric power, heat and gas supply systems, have important infrastructural role in the enhancement of optimal operation of different economy sectors and acceptable life of citizens in any country. There are concrete requirements to necessary level of power supply reliability to consumers and high quality of supplied energy, and also to effectiveness of operation and development of above mentioned infrastructural energy systems. It is necessary to note, that the requirements to increase reliability and quality of energy supply first of all are forming under the influence of digitalization and computerization in technological processes of consumers [3, 4].

In 1999 actually the first research project started concerning energy delivery systems from production of different kinds of energy to retail markets [5]. End use energies included electricity and heat. Such kinds of energy were studied, as electric power, gas, oil, as well as conversion between different kinds of fuel (gas power plants, hydro power plants, co-generations, heating pumps, plants for production of liquid natural gas, and so on). The possibilities of alternative storages were studied, for example hydro accumulating plants and liquid natural gas storages. This project was as the stimuli for preparation of methodology of comprehensive analysis of complicate energy delivery systems with several kinds of energy including technological, economic and ecology aspects. It was planned, that such methodology will be very flexible and will allow the integrated energy companies to make comprehensive analysis their investments and general optimization of their energy supply systems.

The project "Vision of Future Energy Networks (VFEN)" was proposed by group of authors and supported by industry [6, 7]. Horizon of planning is since 30 up to 50 years. Economic, ecology and technological aspects localize the research conditions. General hybrid approach includes different kinds of energy, which consider the synergy between electrical, chemical and heat energies (it is possible, between the other kinds of energy).

An integration of different energy systems into technologically joint body gives new functional possibility, using complex innovative technologies for integrated energy system operation and creation of smart integrated multi-carrier energy systems (SIES). Such systems have many dimensional structures of functional possibilities and development properties. They consider big number of factors: intelligence, effectiveness, reliability, controllability, flexible use of technologies for energy transformation, transportation and preservation, active demand. Protection and control systems have to react to emergency and unreal behavior and to ensure SIES after such events. It is important to develop the models and software for on-line decision making, especially in the conditions of large disturbances [8–10].

3. Energy hub concept

Tendency towards technological integration of energy supply systems gave birth to the notion of an energy hub [1, 8], that implies an integrated facility with multiple inputs and outputs, which represent different types of energy. This facility has internal elements for the support of some functions, i.e., transformation, conversion and storage of different kinds of energy. It is necessary to note [10], that the energy hub concept can be used rather wide – from representing some individual transmission element to a building or a part of the city.

Following [7], we will consider an example of the energy hub shown in **Figure 2**. The Figure shows the inputs and outputs of the energy hub, as well as its internal components and their interconnections (electric transformer, electric battery, micro-turbine, heat exchanger, furnace, cooler and hot water storage).

In [11], an overview of the main provisions of the energy hub concept is presented. Four main functionalities of the energy hub concept are identified, including the input, conversion, storage and output of the considered types of energy. At the same time, most of the studies discussed in the overview, use electric and gas networks as the studied facilities of the energy hub. Various types of power plants especially those based on renewable energy resources, and those relying on promising innovative technologies, such as, fuel cells, for example, were studied as sources of generation.

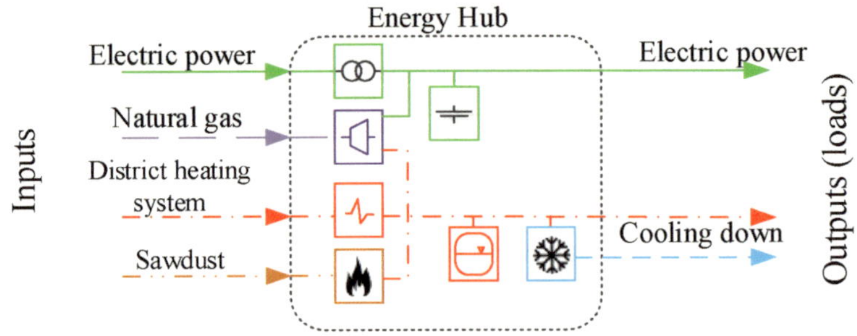

Figure 2.
Example of a specific energy hub containing a transformer, microturbine, heat exchanger, furnace, cooler, battery, and hot water storage.

4. Conventional modeling of integrated multi-carrier energy systems

It is necessary to take into account, that technological and market strengthening of individual energy systems requires more intensive studies of modeling integrated multi-carrier energy systems for the investigation and control of their operating conditions and expansion planning. There are two basically different approaches for modeling integrated multi-carrier energy systems: based on conventional mathematical models of individual energy systems [12, 13] and to use energy hub concept [14, 15].

Let us represent as the example conventional mathematical model of integrated multi-carrier energy system, including electric power and heat supply systems, in following form (1)–(6) [13]:

$$F_{obj} \to \min \tag{1}$$

subject to:

$$E_{k\,\min} \le E_k^t \le E_{k\,\max},\ k \in N_{par}^e,\ t = \overline{1,T}, \tag{2}$$

$$H_{k\,\min} \le H_k^t \le H_{k\,\max},\ k \in N_{par}^h,\ t = \overline{1,T}, \tag{3}$$

$$0 \le P_i^t \le E_{i\,\max},\ i = \overline{1,N},\ t = \overline{1,T}, \tag{4}$$

$$F_{w\,\max}^{it} \ge F_w^{it},\ F_{q\,\max}^{it} \ge F_q^{it}, \tag{5}$$

and balance between electricity and heat production is:

$$\sum_{t=1}^{T}(W_t + Q_t) = \sum_{i=1}^{N}\sum_{t=1}^{T}(W_i^t + Q_i^t) = \sum_{t=1}^{T} P_i^t \Delta t, \tag{6}$$

where F_{obj} is objective function, its structure depends on the sense of solved problem for example, active power; F_{q_i} is volumes of fuel used at source i for heat production; F_{q_i} is volumes of fuel used at source i for electricity production; P_i is used (installed) capacity of source i; W_i is supply of electricity from source i; Q_i is supply of heat from source i; W is is electric power output total value in the system; Q is heat output total value in the system; E_k is current state parameter of electric network; $E_{k\,\min}$ and $E_{k\,\max}$ are technically admissible current state operating parameters limits of the electric network; H_k is current state parameter of heat network; $H_{k\,\min}$ and $H_{k\,\max}$ are technically admissible current state operating parameters limits of the heat network; P_i is used (installed) capacity of source i; $P_{i\,\max}$ is maximum (installed) capacity of source i.

5. Main current principles of the energy hub modeling

References [14, 16, 17] present a system of algebraic equations that relate input variables of the energy hub into output variables. Both variables present different kinds of energy:

$$\begin{pmatrix} L_\alpha \\ L_\beta \\ \vdots \\ L_\gamma \end{pmatrix} = \begin{pmatrix} C_{\alpha\alpha} & C_{\beta\alpha} & \cdots & C_{\gamma\alpha} \\ C_{\alpha\beta} & C_{\beta\beta} & \cdots & C_{\gamma\beta} \\ \vdots & \vdots & \ddots & \vdots \\ C_{\alpha\gamma} & C_{\beta\gamma} & \cdots & C_{\gamma\gamma} \end{pmatrix} \begin{pmatrix} E_\alpha \\ E_\beta \\ \vdots \\ E_\gamma \end{pmatrix} \tag{7}$$

or, in matrix presentation,

$$L = C \cdot E \tag{8}$$

Energy in the input and output ports is represented by vector-columns $E = [E_\alpha, E_\beta \dots E_\gamma]$ and $L = [L_\alpha, L_\beta \dots L_\gamma]$, C is a matrix of direct relations, that describes conversion of energy forms from input to output. Each member of the matrix relates one specific input to a certain output.

In case of solving the inverse problem, a matrix of inverse conversions is introduced

$$\begin{pmatrix} E_\alpha \\ \vdots \\ E_\gamma \end{pmatrix} = \begin{pmatrix} d_{\alpha\alpha} & \cdots & d_{\gamma\alpha} \\ \vdots & \ddots & \vdots \\ d_{\alpha\gamma} & \cdots & d_{\gamma\gamma} \end{pmatrix} \begin{pmatrix} L_\alpha \\ \vdots \\ L_\gamma \end{pmatrix} \tag{9}$$

Relations between coefficients of inverse and direct transformations have a unique form:

$$d_{\beta\alpha} = \begin{cases} c_{\alpha\beta}^{-1} \; if \; c_{\alpha\beta} \neq 0 \\ 0 \; else \end{cases} \tag{10}$$

Should there be N output ports and one input port, the energy through each output channel would be distributed following the equation:

$$E_{im} = \sum_{n=1}^{N} d_{im} L_{in} \tag{11}$$

It is necessary to note, that the most part of references, which deal with the energy hub modeling, including dissertations [18, 19] for different problems investigations concerning integrated multi-carrier energy systems, are using linear energy hub models. These studied problems include calculation and optimization of power flow in integrated multi-carrier energy systems, reliability of electric power and heat supply to consumers, optimization of integrated energy system expansion, and some others [1, 10, 12, 14–17].

Above mentioned studies showed potentials of considered approach to use the linear energy hub model and at the same time the problems of its application. The matter is, that it is necessary to determine the matrix coefficients in (7), which relate inputs and outputs of the energy hub. But this determination faces some difficulties even for linear case. Really these coefficients can have complicate structure including non-linearities. Moreover, above mentioned energy hub models allow to solve only stationary problems in integrated multi-carrier energy systems. Dynamic problems consideration based on energy hub concept had not been studied yet, what had noted as the favorite direction of further investigations [18].

It is necessary to draw the attention on the first known results of dynamic problems study in [20] using conventional mathematical model of integrated energy system based on technique of the theory of singular perturbations (small parameters). This technique was used for presentation of individual energy systems in the integrated multi-carrier energy system.

The above mentioned peculiarities of energy hub modeling stimulate to search the other possibilities to solve these problems. Next Section allows such possibilities.

6. Matlab/Simulink capabilities and simulation model construction

The simulation modeling approach can be as the basic technology for construction of integrated multi-carrier energy system model. Let us use the capabilities of Matlab/Simulink software for suggested technology development. The following components of necessary simulation model construction procedure of integrated multi-carrier energy systems we will have to take into account:

- The initial information about modeled integrated energy system includes the topology and parameters of different kinds of elements (objects) of individual energy systems. We have to note the initial element in every individual energy system, which will as the start point for topological model creation of every individual energy system.

- Current versions of Matlab/Simulink software include rather developed library of models for elements of different technological systems – electric power, pneumatic, hydraulic ones, and the others. These models of elements are presented using transfer functions and can be implemented for dynamic processes study in different technological systems. We deal with steady state conditions, therefore it is necessary to convert initial dynamic models into the static form.

- The above mentioned library of Matlab/Simulink software does not contain complicated elements with multi-input and multi-output structure. Such elements are energy hubs. The co-generation plant can be as the example of such complicated element (object) with one input (gas) and two outputs (electric power and heat). At the level of consumption such elements of integrated multi-carrier energy system include conversion function of one kind of energy into the other. The energy hub models are forming the specific additional library. These models also implement such functions as energy storages and summation of different kinds of energy.

- It is necessary to note, that there are two kinds of energy conversion elements: 1) they change the characteristics of the energy channel without conversion of energy form into the other one (for example, electrical transformer, heat exchanger, and so on); 2) they change not only characteristics of the energy channel, but also convert one kind of energy into the other one.

- Different kinds of energy in integrated multi-carrier energy system have different measurement units (kWh, Gcal, etc.). Therefore Joule (J, W.s) is considered as a basic unit of measurement. The transformation function of different unit to the basic one was implemented using Matlab/Simulink software.

- After the creation of topological models of different individual energy systems networks using above mentioned procedures it is necessary to connect them each other. Energy hub models of energy production plants (co-generation plants, heating plants, etc.) and complex consumers with several kinds of consumed energy (electric power, heat, etc.) play the role of such connectors.

- Constructed simulation model of integrated multi-carrier energy system really is the basic part of any full simulation model for solving some concrete problem of integrated energy system. The statement of concrete solved problem requires additional procedures for problem formalization and results interpretation. For example, the solving loss minimization problem in electrical network requires load flow calculation as the basic procedure and optimization algorithm formalization for solving full necessary problem.

After above mentioned procedures the simulation model of integrated multi-carrier energy system is ready to usage for solving different problems.

7. Elements of technology for modeling of the energy hub

Figure 3 shows general structure of the energy hub simulation model, which was constructed by Matlab/Simulink software capabilities [21]. This structural scheme presents three energy supply channels: 1 - electric power; 2 - heat; 3 - gas. The model implements the functions of transformation, conversion and storage of energy, and an additional summation function whose concept is understandable from **Figure 3**.

Figure 4 presents detail structure of the energy hub simulation model for the electric power supply channel using representation of elements by images of Matlab/Simulink software. Here 1 and 6 present direct and inverse transformations of state variables; 2 and 4 present the electricity transfer; 3 presents the transformer sub-station model; 5 presents the energy storage device model.

Figure 4 takes into account the peculiarities of simulation modeling procedures of Matlab/Simulink software, including propagation and conversion of presented system.

Figure 5 gives an example of an integrated scheme based on two energy supply channels. A black line here denotes a channel of a heat network; gray one denotes a

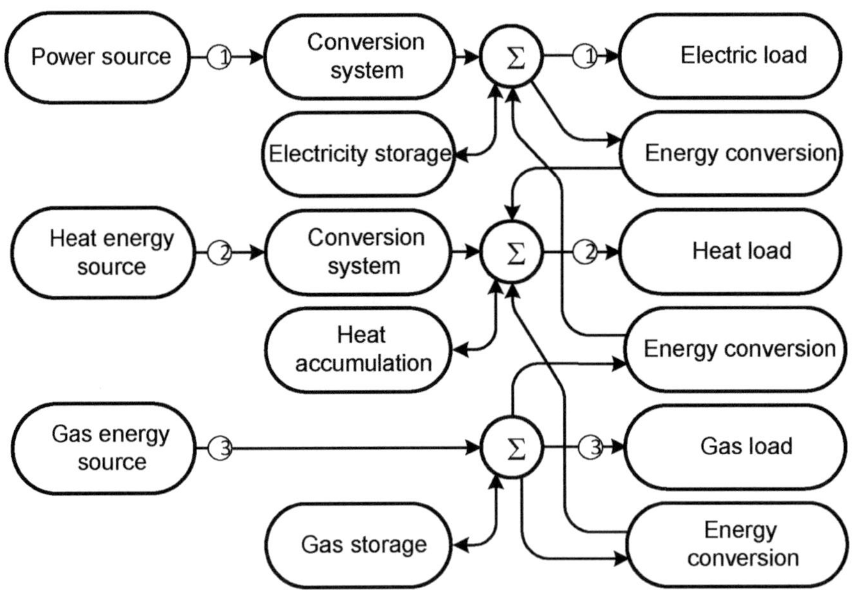

Figure 3.
General structure of the energy hub simulation model. (channels: 1 - electric power supply; 2 - heat supply; 3 - gas supply).

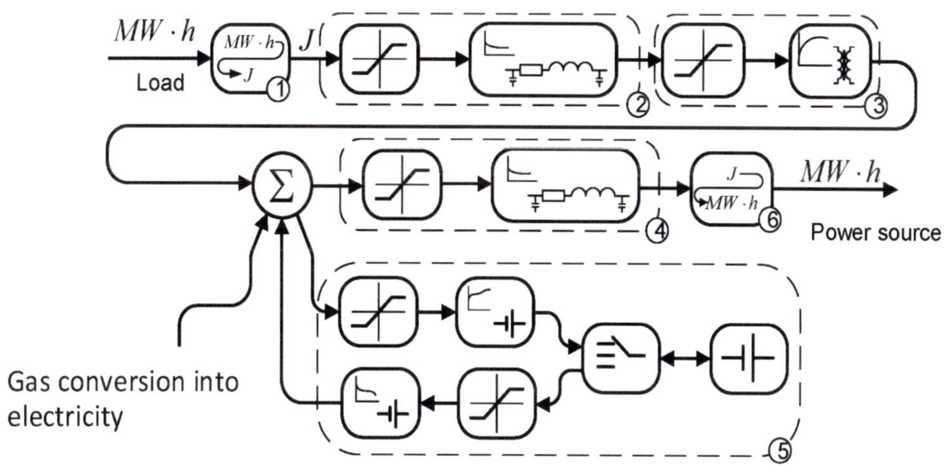

Figure 4.
Flow chart of electric power supply channel for the energy hub simulation model.

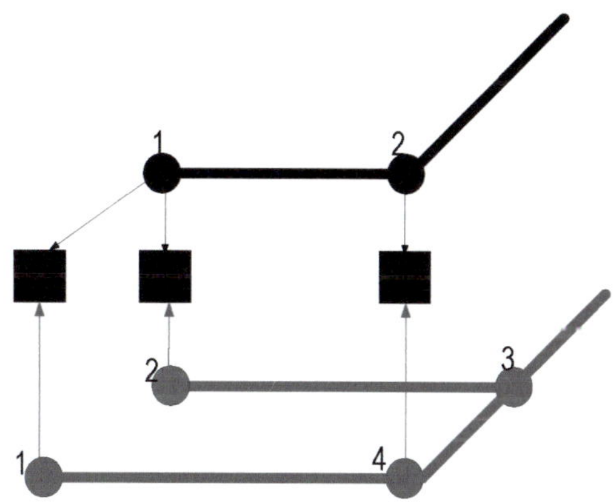

Figure 5.
An integrated scheme based on two energy supply channels.

channel of an electric network. A squares denote hubs locations, which represent electricity and heat consumers.

The example in **Figure 5** shows, that two energy channels (electric power and heat) go to one consumer. Taking into account storage systems and systems for conversion of electric power into heat we will have complex energy hub based on presented consumer.

Rather simple elements of electric power and heat supply systems can be presented by simulation models from Matlab/Simulink library. According to above noted approach, it is necessary to create additionally the library of energy hubs simulation models. **Figure 6** presents the integrated simulation model of energy supply systems (electric power and heat) including energy hub with two energy supply channels.

The simple elements of individual energy supply systems are used from the basic library of Sim Power Systems which is sub-system of Matlab/Simulink.

WEI and WS elements represent energy consumption and storage, which connected by electric power supply channel. WH is the energy consumption by heat supply channel. KP1 - KP4 are the elements, which works taking into account efficiency of the energy conversion. WE-WH represent electricity converted into the heat energy.

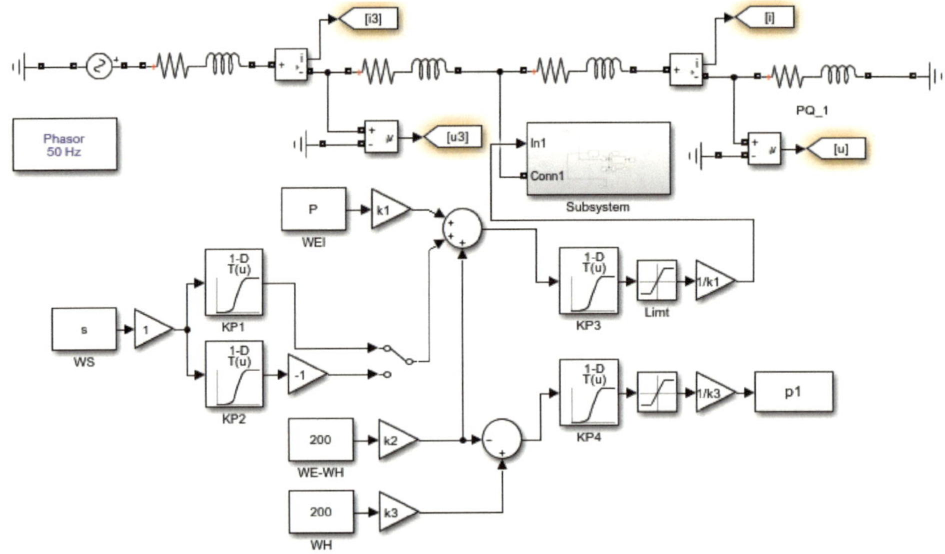

Figure 6.
Construction of integrated simulation model with two energy supply channels in Matlab/Simulink.

8. An algorithm of simulation model construction for integrated multi-carrier energy systems

A general approach to constructing a simulation model of an integrated multi-carrier energy system and to solution of different problems with its help can be represented as follows [21, 22] (see **Figure 7**).

Input data about the studied integrated multi-carrier energy system is prepared including the matrices of parameters of individual energy systems (their network topologies, electric and hydraulic resistances of electric lines and pipelines), as well as vectors of nodes parameters (electric power and heat generations, loads, storages, etc.).

The necessity to use of two libraries of integrated energy system elements was noted earlier in Section 6. An algorithm for simulation model construction of integrated multi-carrier energy system selects required model of the next element from the point of view of individual energy system topology (depending on the element type) either from library of typical elements in Matlab/Simulink software or from additional library, which includes the energy hubs models. After that required model attaches to necessary node (nodes for energy hub model) of integrated energy system. As it is noted in Section 6, the energy hub model has several inputs and several outputs, which connect different individual energy systems into integrated multi-carrier energy system.

As we said in Section 6, above mentioned procedure creates so called basic part of integrated energy system simulation model. It is necessary to work out an additional part for simulation model, which represents the specifics of concrete calculated problem (see Section 6).

Matlab/Simulink software contains the object-oriented programming language, which has used for construction of integrated multi-carrier energy system simulation model. **Figure 7** represents simplified flow chart of the basic part of discussed algorithm taking into account three individual energy systems: electric power, heat and gas supply systems.

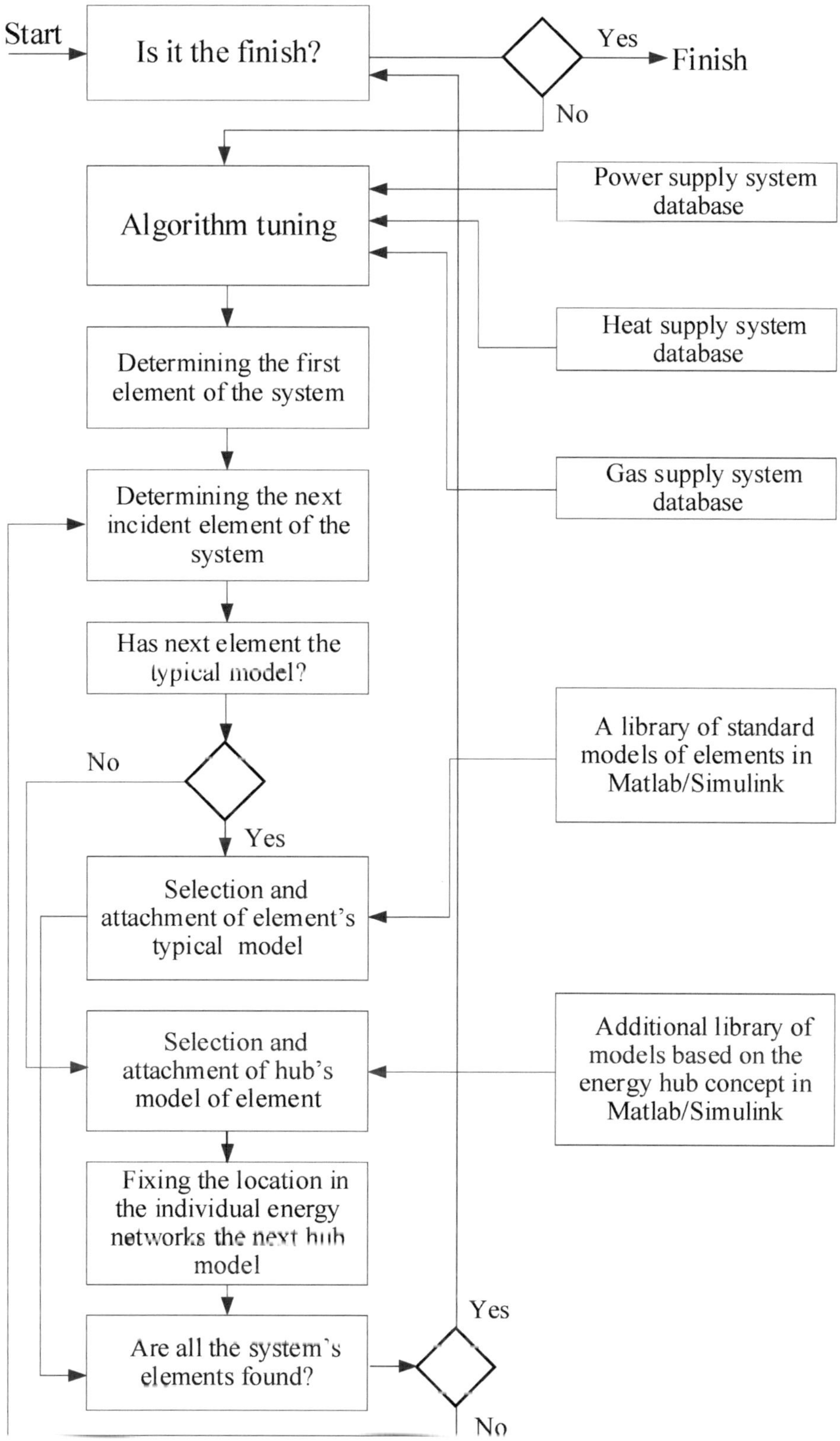

Figure 7.
Flow chart of algorithm for constructing basic part of integrated energy system simulation model.

9. Illustrative case study

An integrated energy system is considered including the electricity and heat supply systems of a block of 9 dormitories of a University campus. The diagram of the electric network of the integrated energy system is shown in **Figure 8**, the diagram of the heat network is topologically about similar, since each dormitory is a consumer of both electricity and heat. The diagram of the heat network is not given, since the load of heat pipelines in the problem solved does not including, but, on the contrary, decreases, i.e., there are no network constraints on heat transfer.

In **Figure 8** FS is feeding substation, the nodes 11, 12, 13, 14, 15 are transformer substations 6/0.4 kV.

Figures 9 and **10** indicate the total annual electricity and heat consumption curves for the entire block of dormitories, respectively. We assume that thermal energy is consumed only for heating. The daily heat load curve is uniform. The irregularity factor of daily electrical load curve is 0.4 (the ratio of the load value during the night minimum period from 23:00 to 7:00 to the peak load value). Daily curves of heat and electrical load are the same for all dormitories.

We consider the conditions for preventing overload of the electrical network. To this end, the total load power during the night minimum of the daily load curve, including its power level plus the amount of power consumed to convert electricity into heat, should not exceed the daily maximum load. In this case, the load flow in the electrical network will not change and there will be no overloads.

Table 1 shows monthly data on the parameters of electricity supply to consumers on the University campus.

The values of daily maximum load are used to calculate the values of conventional maximum possible electricity consumption of the campus per month with the formula:

$$E_{\text{mon. max}} = \left(P_{\text{day. max}} \cdot 24\right) \cdot 30, \tag{12}$$

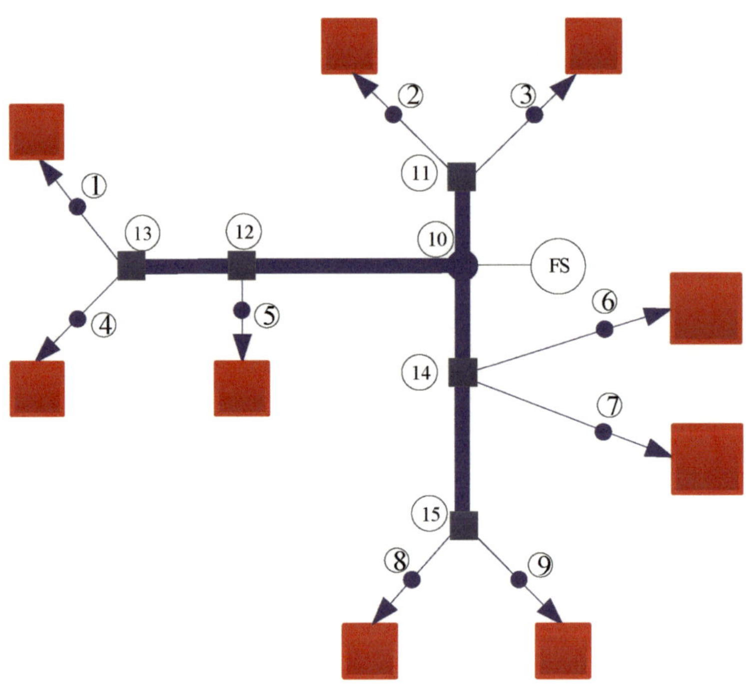

Figure 8.
Diagram of the electrical supply system.

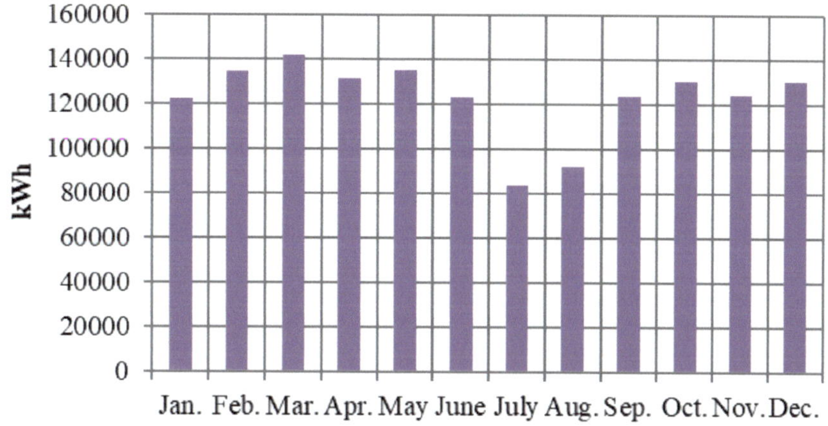

Figure 9.
Electricity consumption of 9 dormitories.

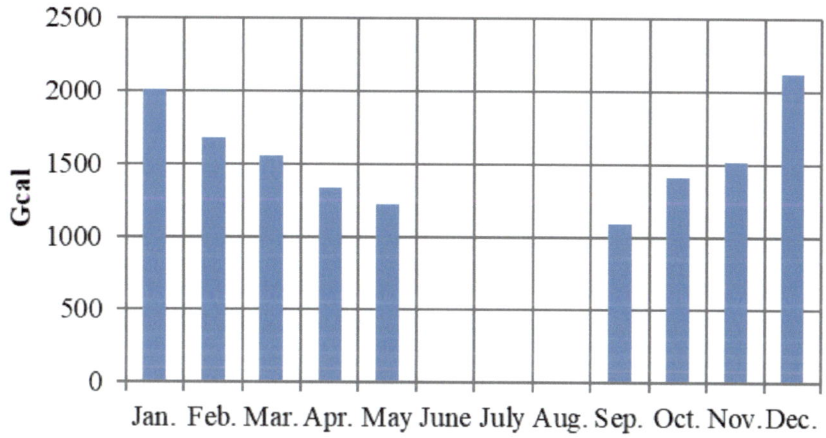

Figure 10.
Heat consumption of 9 dormitories.

where $E_{mon.\max}$ is the maximum possible conventional value of electricity consumption per month; $P_{day.\max}$ is a daily maximum load.

The amount of electricity that can be converted into heat (conversion potential) is determined by:

$$E_p = E_{mon.\max} \cdot 0,6 \cdot 0,33, \tag{13}$$

where E_p is the potential for converting electricity into heat per month; coefficient 0.6 reflects the share of free power within the night minimum load; coefficient 0.33 determines the share of duration of the night minimum daily load curve (8 hours), during which electricity is paid for at the minimum night rate.

Conversion of electricity into heat is carried out according to the relationship:

$$1\,\text{kWh} = 0,00086\,\text{Gcal}.$$

In **Table 1**, the last two columns indicate two options for the amount of electricity to be converted to heat: the entire (100%) conversion potential and 50% of this potential.

	Power consumption of 9 dormitories, kWh	Night zone (from 23 to 7), kWh	Payment for electricity consumption at night without conversion, $	Daily peak load, kW	Maximum electricity consumption per month, kWh	The amount of electricity (potential) to convert to heat per month, kWh	50% of electricity for conversion to heat per month, kWh
Jan.	122000	47000	519	476	343000	67900	34000
Feb.	134000	50000	547	459	331000	65500	33000
Mar.	142000	53000	583	418	301000	59600	30000
Apr.	131000	48000	526	392	282000	55900	28000
May	135000	50000	552	489	352000	69700	35000
June	123000	48000	525	441	318000	62900	31000
July	83600	33000	361	324	233000	46200	23000
Aug.	92300	36000	401	344	248000	49100	25000
Sep.	123600	48000	533	443	319000	63100	32000
Oct.	130000	51000	561	410	295000	58500	29000
Nov.	125000	49000	536	450	324000	64200	32000
Dec.	130000	51000	559	489	352000	69700	35000

Table 1.
University campus power consumption data.

Following the current pricing system for electricity and heat, electricity rates are differentiated throughout the day: a preferential night rate from 23:00 to 7:00 is $ 0.011 per kWh. Heat rate is $ 20.6 per Gkal.

In general terms, the following relations are valid:

$$C_e = E_p \cdot t_3, \tag{14}$$

C_e is the cost of electricity before conversion into heat.

$$C_{tn} = E_t \cdot t, \tag{15}$$

t is heat tariff; E_t is thermal energy.

The following relations are valid:

$$C_e = C_{en} + C_{ed}, \tag{16}$$

$$C'_e = C_{en} + C'_{en} + C_{ed}, \tag{17}$$

$$C_t = C_{tn} + C_{td}, \tag{18}$$

$$C'_t = C_{tn} - C'_{tn} + C_{td}, \tag{19}$$

$$C'_{en} < C'_{tn}, \tag{20}$$

where C_{en} is cost of electricity before conversion at night; C_{ed} is cost of electricity before conversion to daytime; C_t is cost of thermal energy before conversion; C_{tn} is the cost of thermal energy at night; C_{td} is the cost of thermal energy in the daytime; C'_e is cost of electricity after conversion; C'_{en} is cost of electricity at night after conversion; C'_t is cost of thermal energy after conversion; C'_{tn} is the cost of thermal energy after conversion at night.

The results of the calculations of the considered options are presented in **Figures 11** and **12**.

Let us return to the condition of preventing the electrical network overloads, formulated above. An analysis of the transfer capability and loading of individual ties lines in the case of electricity conversion into heat, according to the condition assumed, shows that this loading is not the same (see **Table 2**).

It is important to estimate some limiting volume of conversion of electricity into heat at night taking into account the possibilities of electrical network. These possibilities depend on free transfer capabilities of ties and permissible loading of transformers on feeding substation. Required parameters of electrical network and basic load flow calculation without consideration of active losses you can see on the **Table 2**. Consumption load at night which found on the previous stage (basic load flow) for each consumer is 490 kW. The permissible loading of transformers on

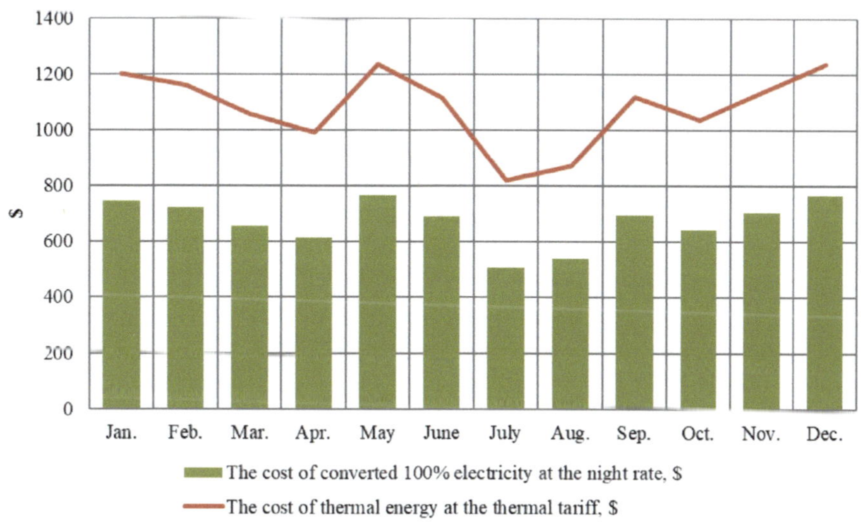

Figure 11.
Comparison of payment with electricity conversion into heat factored in 100%.

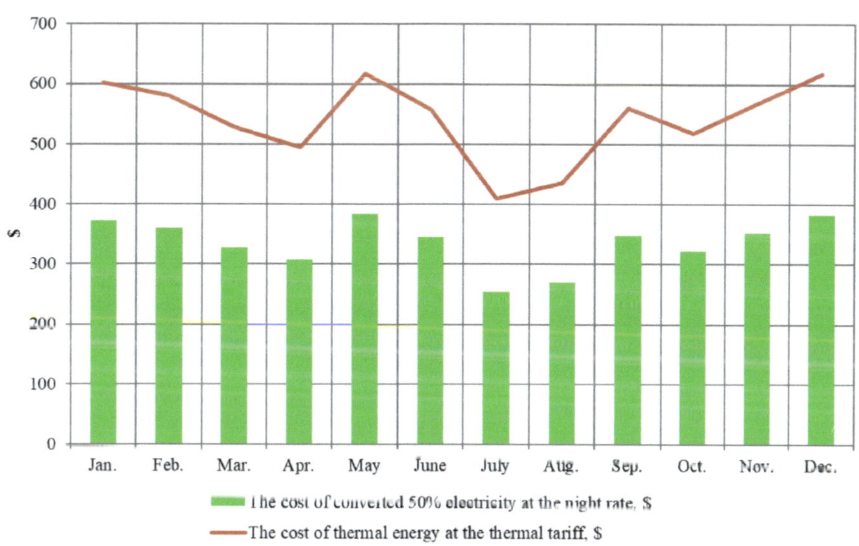

Figure 12.
Comparison of payment with electricity conversion into heat factored in 50%.

Ties numbers	10–11	10–12	10–14	12–13	14–15
Transfer capabilities of ties, kW	2100	2100	2100	2100	2100
Load flow, kW	980	1470	1960	980	980

Table 2.
Required parameters of electrical network and basic load flow.

feeding substation is 6000 kW. Let us consider, that cable lines from transformer substations 6/0.4 kV to import of electricity into building do not have the limits of transfer capabilities. As for limiting volume of heat supply for each consumer, let us to consider 380 kW after re-calculation into converted electricity.

Let us formalize optimization problem as following:

Objective function:

$$\Delta P_{FS} \rightarrow \max, \tag{21}$$

Subject to:

$$\Delta P_{FS} \leq \Delta P_{FS\lim}, \tag{22}$$

$$P_{ij} \leq P_{ij\lim}, \tag{23}$$

$$P_{kheat} \leq P_{kheat\lim}, \tag{24}$$

$$\Delta P_{FS}^{l+1} = \Delta P_{FS}^{l} + h\frac{\Delta P_{FS}^{l}}{1/\Delta P_{ij}}, \tag{25}$$

$$\Delta P_{ij} = P_{ij\lim} - P_{ij}, \tag{26}$$

where ΔP_{FS} is additional power for conversion into heat; $P_{ij\lim}$ is transfer capability of tie ij, i, j = 1, 2, 3; $P_{kheat\lim}$ is top re-calculated to electricity level of heat for consumer.

k, k = 1–9; h is the step of optimization; l is the number of iteration. The second member in right part of (25) is the similar to gradient of objective function.

Several beginning steps of optimization are along the ray 10–11 to use the possibility for additional conversion of electricity into heat. The results of these iterations are 380 kW for consumer 2 and 380 kW for consumer 3 as the additional converted volumes of electricity. These volumes along the ray 10–11 are top volumes for additional conversion. One next iteration deals with the ray 10–12, where it is possible to use 380 kW for consumer 5 and the rest on this ray 250 kW (2100–1470 – 380 = 250) for consumers 1 or 4. The iteration along the ray 10–14 allows to use 140 kW additional converted electricity (2100–1960 = 140) for consumers 6 or 7 or 8 or 9.

It is possible to see, that we could use more electricity for additional conversion into heat, but the problem is in hhe electrical network limitation.

10. Conclusion

Creation of the integrated multi-carrier energy systems is progressive trend in development of energy supply systems. Joint expansion of individual energy systems leads to enhancement of economic efficiency and reliability of energy supply to consumers. It is necessary to have the efficient tools for expansion planning and operation management and control of integrated multi-carrier energy systems.

Energy hub concept is progressive way for modeling and simulation of integrated energy systems, but there are some problems in determination of the coefficients of connection of each individual input and each individual output of the energy hub simulation model.

This Chapter represents new approach to solve above mentioned problems based on the possibilities of Matlab/Simulink software taking into account the elements of energy hub concept. The main idea of suggested approach deals with the construction of simulation model of integrated multi-carrier energy system considering the models of simple typical elements from the Matlab/Simulink library and complicate energy hub models from additional library, which is created based on Matlab/Simulink software possibilities.

Illustrative case study shows the efficiency of suggested approach.

Acknowledgements

This study was performed according to project # FWEU-2021-0002 of State Assignment of Fundamental Investigation Program of Russian Federation for 2021-2030.

Author details

Nikolai Voropai[1]*, Ekaterina Serdyukova[1], Dmitry Gerasimov[2] and Konstantin Suslov[2]

1 Melentiev Energy Systems Institute of Siberian Branch of the Russian Academy of Sciences, Irkutsk National Research Technical University, Irkutsk, Russian Federation

2 Irkutsk National Research Technical University, Irkutsk, Russian Federation

*Address all correspondence to: ni.voropai@yandex.ru

References

[1] Arnold M, Andersson G. Decomposed electricity and natural gas optimal power flow. In: 16th Power System Computation Conference. Glasgow, Scottland, UK, July 26 – 30; 2008; 6 p.

[2] Voropai NI, Stennikov VA. Mint: Integrated intelligent energy systems. Izvestiya RAN, Energetika; 2014; No. 1; pp. 64 – 73. (in Russian)

[3] Jin Wei, Kundur D. Two-tier hierarchical cyber-physical security analysis framework for smart grid. In: IEEE PES General Meeting; San Diego; USA; July 22 – 27, 2012; 5 p. doi:10.1109/pesgm.2012.6345633

[4] Voropai NI, Goubko MV, Kovalev SP, Massel LV, Novikov DA, Raikov AN, Senderov SM, Stennikov VA. Mint: Development problems of digital energetics in Russia. Problemy Upravleniya; 2019; No. 1, pp. 2 – 14. (in Russian)

[5] Bakken BH, Haugstad A, Hornnes KS, Vist S, Gustavsen B, Roynstrand J, Simulation and optimization of systems with multiple energy carriers. In: 1999 Conference of the Scandinavian Simulation Society (SIMS); Linko Eping; Sweden; August 11 – 15; 1999; 7 p.

[6] Geidl M, Favre-Perrod P, Klockl B, Koeppel G, A greenfield approach for future power systems. In: CIGRE 2006 General Session; Paris; France; August 22 – 27; 8 p.

[7] Geidl M, Koeppel G, Favre-Perrod P, Klockl B, Andersson G, Frohlich K, The energy hub – a powerful concept for future power systems. In: Third Annual Carnegie Mellon Conference on the Electricity Industry; 2007; Vol. 13; p. 14.

[8] Geidl M, Koeppel G, Favre-Perrod P, Andersson G. e.a. Energy hubs for the future: A powerful approach for next-generation energy systems. In: IEEE Power and Energy Magazine; 2007; Vol.5; No.1; pp.24–30. DOI:10.1109/MPAE.2007.264850

[9] Voropai NI, Stennikov VA, Barakhtenko EA. Mint: Integrated energy systems: Challenges, trends, ideology. Problemy Prognozirovaniya; 2017; No. 5; pp. 39 – 49. (in Russian) DOI: 10.1134/S107570071705015X

[10] Koeppel G, Andersson G. Mint: Reliability modeling of multi-carrier energy systems. Energy; 2009; Vol. 34; No. 3; pp. 235 – 244. DOI: 10.1016 / j.energy.2008.04.012

[11] Mohammadi M, Noorollahi Y, Mohammadi-Ivatloo B, Yousefi H. Mint: Energy hub: From a model to a concept – a review. Renewable and Sustainable Energy Reviews; 2017; Vol. 80; pp. 1512 – 1527. DOI:10.1016/j.rser.2017.07.030

[12] Chaudry M, Jenkins N, Strbac G. Mint: Multi-time period combined gas and electricity networks optimization, Electric Power System Research; 2008; Vol. 78; No. 5; pp. 1265 – 1279. DOI: 10.1016 / j.epsr.2007.11.002

[13] Voropai NI, Stennikov VA, Barakhtenko EA, Voitov ON, e.a. Mint: A model for control of a steady-state of intelligent integrated energy system, Energy Systems Research; 2018; Vol. 1; No. 1; pp. 57 – 66. DOI: 10.25729/esr.2018.01.0007

[14] Geidl M. Optimal power flow of multiple energy carriers. In: IEEE Transactions on Power Systems; 2007; Vol. 22; No. 1; pp. 145 – 155. doi:10.1109/tpwrs.2006.888988

[15] Almassalkhi M, Hiskens I. Optimization framework for the analysis of large-scale networks of energy hubs. In: 17th Power System

Computation Conference; Stockholm, Sweden; August 22 – 26, 7 p.

[16] Geidl M, Andersson G. Optimal coupling of energy infrastructures; In: 2007 IEEE Lausanne Power Tech, Lausanne; Switzerland; July 17 – 21; 2007; 6 p. DOI: 10.1109 / PCT.2007.4538520

[17] Zhang X, Shahidehpour M, Alabdulwahab A, Abusorrah A, Optimal expansion planning of energy hub with multiple energy infrastructures. In: IEEE Transactions on Smart Grid; 2015; Vol. 6; No. 5; pp. 2302 – 2311. DOI: 10.1109 / TSG.2015.2390640

[18] Geidl M. Integrated modeling and optimization of multi-carrier energy systems. PhD Dissertation. Swiss Federal Institute of Technology; Zurich, Switzerland; 2007; 125 p.

[19] Koeppel GA. Reliability considerations of future energy systems: Multi-carrier systems and the effect of energy storage. PhD Dissertation. Swiss Federal Institute of Technology, Zurich, Switzerland; 2007; 139 p.

[20] Fu Shen, Ping Ju, Shahidehpour M, e. a. Singular perturbation for the dynamic modeling of integrated energy systems. In: IEEE Transactions on Power Systems, 2020; Vol. 35; No. 3; pp. 1718 – 1728. DOI:10.1109/ TPWRS.2019.2953672

[21] Voropai N, Gerasimov D, Ukolova Ek, Suslov K, e. a. Simulation approach to integrated energy systems study based on the energy hub concept. In: 2019 IEEE Power Tech; Milan, Italy; June 23–27;2019;5 p. DOI:10.1109/ PTC.2019.8810666

[22] Voropai NI, Gerasimov DO, Serdyukova EV, Suslov KV. Designing a simulation model of integrated multi-carrier energy system using the energy hub concept. In: Methodological Problems of Large Energy Systems Reliability Study. Int. Conf. Proceedings. Kasan, Russia; September 21 – 25, 2020; Issue 2; pp. 333 – 342. (in Russian).

Section 6

Economy

Chapter 11

Using Simulation Modeling for Finding the Limits of Economic Development Lending without a Financial Crisis

Yuriy V. Vasylenko

Abstract

The only existing approach to analyze the impact of excessive credit on the economy is based on statistics. Its main drawback is small intervals of changes in countries' indicators, limited by current values. So researchers cannot notice how too much credit causes a financial crisis. To eliminate this and other shortcomings of the statistical approach, the author proposes a different approach: to use for such an analysis an economic model in which one can change credit levels. The most adequate model is a causal simulation model that reflects the main types of legal and shadow economic activity in their relationship. The author has developed such a model. This model showed that the level of loans 25% of output (51.8% of GDP), could create Ukraine's financial crisis. Since loans are mainly used for investment, the author introduced the concept of the technical productivity of investment to link them with the technical progress, and with GDP growth. The technical productivity of investment measures their ability to reduce the rate of material or labor costs. Besides, the introduction of an indicator of technical productivity of investment made it possible to obtain an analytical dependence of the rate of economic growth on the level of loans and technical productivity of investment.

Keywords: financial crisis, limits of secure lending, simulation model, legal and shadow economic activities, technical productivity of investment

1. Introduction

Loans have always been considered the driving force of the economy. Even the financial crisis of 2008–2009 did not change the indisputability of this view in many researchers. For example, Lorenzo Cappiello, Arjan Kadareja, Christoffer Kok Sørensen and Marco Protopapa [1] only regret that "the Financial crisis erupting in mid-2007 which led to the need for banks to deliver their balance sheets and possibly to reduce their loan supply".

Manaresi and Pierri [2] insist that "To grow and thrive, firms need reliable access to external funding".

Sadaf Majeed, Syed Faizan Iftikhar, Zeeshan Atiq found a difference in the effects of loans to household and enterprise in Pakistan: the former does not cause economic growth, while the latter has a positive impact [3].

At the same time, more balanced thoughts emerged. Thus, Dirk Bezemer, Anna Samarina, and Lu Zhang already distinguish the nature of the impact of loans in the short and medium-term: "The impact is positive in the short term and negative in the medium term" [4].

Leading scholars such as Stiglitz, Paul Krugman, and others have pointed to excessive credit as the immediate cause of the 2008–2009 financial crises [5].

After the crisis of 2008–2009, there were works to create early warning systems for the financial crisis [6–8].

All the above-mentioned analytical studies of financial crises and other works are based on a single basis - statistical. The advantage of this approach is the coverage of a wide range of countries or firms. But there are drawbacks. The first is small intervals of changes in countries' indicators, limited by actual values. So researchers cannot notice how too much credit causes a financial crisis. The second drawback is the heterogeneity of the samples for analysis - they mix countries with large differences in institutional characteristics and development levels. The third one is the lack of accurate estimates of the results because the probability distributions of economic indicators are mostly far from normal (Gaussian).

2. Methodology of work

We propose a different approach to analyzing the impact of excessive credit on the economy, free from these shortcomings. We apply the author dynamic model [9], adequate to the Ukrainian economy (see below).

In this model, we simulated Ukraine's economy's smooth development over 5 or 10 years (results were similar) at different lending levels in different economic situations. After that, in the sixth or eleventh year, we stopped lending and watched how much real GDP would fall than the base year. This is a very approximate financial crisis model; however, it gave estimates in the first approximation.

For this approach to work, the adequacy of the model to the real economy of a country or a homogeneous group of countries must be very high. The next section lists the model features that ensure its adequacy.

One uses loans primarily on active investment. So one has to investigate investment more deeply and link them with technical progress, and GDP growth.

Investments in the literature are mostly explored for their payback, which is mainly a microeconomic approach. Nobody analyzed their impact on an economy in general. To assess them on a national scope, one has to characterize investments in macroeconomic terms.

The creators of economic dynamics, Robert Solow, Roy Harrod [10, 11], did not set themselves to create such a characteristic. It is not clear why they did not link investment, on the one hand, with the growth of capital, and on the other hand, with technical progress.

One of the technical progress indicators is reducing material or labor costs due to active investment. But no one measured the ability of an investment to reduce the rate of material or labor costs. This characteristic is the direct technical result of active investment and a direct factor in increasing the enterprise's competitiveness. An economist, having such a characteristic of investment, can quantify the potential of the totality of any investment, regardless of their specific nature. That is, he will be able to move from the microeconomic level to macroeconomic analysis. Moreover, he can no longer derive identity (like in Roy Harrod), but the equation of economic dynamics, which links the GDP growth rate with the endogenous factor of technical progress - investment, a source of which is income.

To fill this gap, move to the macro level, and link investment to GDP growth, we have introduced such an indicator and called it the **technical productivity of**

investment. We accepted that the technical productivity of investment is equal to 1 if the rate of decrease in the rate of expenditure (per unit of output) is equal to the increase of the active part of fixed assets through investment. If technical productivity is 2, then the reduction is twice as much.

These two points are central to our approach.

Else we propose to expand the measurement of investment economic efficiency - to measure it not only in terms of profit but also in all value-added. We believe that this indicator reflects the efficiency for the state, and secondly, the more far-sighted interests of the business owner in minimizing the turnover of skilled workers. Therefore, simultaneously with the payback, our model calculates the efficiency of investment by GDP, i.e., the ratio of real GDP growth obtained in one or more subsequent years-cycles to the real amount of investment in one or more previous years-cycles. In some sense, this indicator is the inverse of Roy Harrod's capital ratio [11], but it uses real indicators rather than nominal ones.

Besides, the model considers that if the bank issued loans *k* times more than it received deposits *D*; thereby, it created an emission of (*k*-1) x *D*. In Ukraine, over the past 15 years, banks issued more loans (for example, in 2005–2008, 2011, and 2013) and less (in 2009, 2010, 2012, and 2014) than they received deposits. We believe, together with monetarists, that prices for all goods will increase accordingly. Rising prices for intermediate goods will increase the cost of all goods, thus reducing gross profit; rising prices of final consumption goods and housing reduce households' purchasing power, higher investment goods' prices reduce real investment. Therefore, the increase in loans has positive and negative results. Let us see which wins lower.

Researchers of the financial crisis, as well as creators of systems of early warning crisis, measured the level of credit as a share of credit in GDP (for example, [6–8]). But GDP is used for purchasing consumer and investment goods only, and loans are taken else to produce intermediate consumption goods included in the output.

We believe that it is methodologically correct to take loans' share, not GDP, but output. GDP is not proportional to output. A more efficient economy produces more GDP per unit of intermediate consumption, a less efficient economy - less.

3. A brief history of economic modeling and an adequate model of economic dynamics of Ukraine

To analyze the impact of excessive credit on the economy by the proposed method, we must have a dynamic model with a very high level of adequacy to the Ukrainian economy.

At that time, when scientists began to build economic science on physics principles, they discuss its nature, roles of mathematic, and the very possibility of such a building. Here is what J. von Neumann and O. Morgenstern wrote in 1944:

> *"To illuminate the concepts that we will apply to economics, we present and will continue to present some illustrations from physics. Many sociologists object to drawing such parallels for 48 different reasons, among which they usually cite the statement that economic theories cannot be modeled on the model of physical ones, since economic theories take into account social, economic phenomena, since they have to take into account psychological factors, etc. Such claims are immature, to say the least. Undoubtedly, it seems reasonable to uncover what led to progress in other sciences and investigate why the application of these principles cannot lead to progress in economics. If there really is a need to apply some other principles to*

economics, this can only be revealed in economic theory's actual development. This, in itself, will be a revolution in science. But since, almost certainly, we have not yet reached such a state and it is in no way clear that there is a need to use completely new scientific principles, it would be unreasonable to consider anything other than the interpretation of problems in the way that has already led to the creation of physical science." [12].

So far, no consistent economic theory based on the principles of physics has been created. Moreover, many economists have spoken of the crisis of economics. Paul Samuelson in 1947 wrote:

"The economist comforts himself ... with the thought that he is forging tools that will eventually lead to results. This promise is always for the future; we are like well-trained athletes who do not participate in competitions and therefore lose shape." ([13], p. 4).

Moreover, the very definition of economics is changing. In the first half of the 19 century, economics was seen as a study of the "nature and causes of the wealth of nations" (Smith), "the laws governing the distribution of what is produced on earth" (Ricardo), and "the laws of the movement of capitalism" (Marx). After 1870, it was believed that the economy analyzed human behavior in different markets. Mark Blaug wittily remarked that many early studies then lost the right to be called economics [14].

As a result of this approach, Paul Krugman has subjected destructive criticism of economic models that cannot predict anything [15].

One of the fathers of economic dynamics, Roy Harrod [11], submitted the fundamental equation of economic dynamics which determines the growth rate: the amount of savings, expressed as a share of net income divided by the capital ratio. The capital ratio is the ratio of net (rather than gross) investment to the increase in output or income over the same period.

Note that the output growth (which stands in the denominator of the Harrod's capital ratio) to its base value - this is the rate of economic growth. Thus, Harrod's fundamental equation of economic dynamics is not an equation - it is nothing more than some identity where the output indicator is determined by itself. If we solve it, we get the known identity: savings equal the capital increase.

Roy Harrod himself understood this perfectly; moreover, he emphasized that this equation is a truism because it is easily deduced from the standard definitions of macroeconomic variables included in this equation.

However, he called this identity an equation. On its varieties - equation of guaranteed and natural growth rate, recommendations were calculated to the governments of different countries to forecast and regulate the rate of economic growth in the 60s of the twentieth century. But only in 6 of the 88 countries where they were used were the expected results obtained [16]. Attempts to predict economic growth based on the Harrod-Domar model have failed. The researchers concluded that the model does not explain the main determinants of economic growth. And how could it explain if it is not an **equation** but **identity**? **The rate of output growth does not follow from this identity.** What the increase in output in the capital ratio we set, the same increase in output we get at the output of the Harrod fundamental equation of economic dynamics.

Robert Solow [10] turned this identity into the equation by replacing the capital ratio with a labor efficiency indicator.

For this he received the Prize behalf of Nobel in 1987. This was, in my opinion, a rather strange decision, because:

1. His labor efficiency was exogenous and not related to investment.

2. It is the Solow exogenousness of economic growth was the object of criticism №1, and less than 10 years later the Ramsey-Cass-Koopmans model appeared, in which the saving rate changed in each period [17, 18], due to which this model became considered endogenous. However, it was later recognized as exogenous because, as in the Solow model, scientific and technological progress in the Ramsey-Kass-Kupmans model is not the result of decision-making by economic agents but is set exogenously [19, 20]. The very problem of determining the norm of conservation F. Ramsey [21] studied in the 20s with differential equations.

   That is, R. Solow took two steps back at once: in time and in terms of abandoning the endogenous nature of economic growth.

3. In choosing the model, R. Solow was behind K. Marx who divided the economy into two sectors: the production of consumer goods and means of production.

4. Solow's model is not his own model, but the Cobb–Douglas model, the adequacy of which to the real economy is in great doubt - deviations of the actual data of the production of the USA manufacturing industry from the Cobb–Douglas function was more than 15 percent, which does not provide sufficient reason to believe this power function adequate the existing economy. And this is for one homogeneous industry, where companies have approximately the same levels of capital intensity. Which will be the mistake for the whole economy, where there are capital-intensive and labor-intensive industries? Mark Blaug ([14], p. 653) expressed a more general idea:

   "In itself, the concept of production function - a set of all known production technologies - is so general that it cannot be called meaningful".

5. So all studies of R. Solow of various modifications of the Cobb–Douglas model (with constant elasticity of substitution (CES), with constant return on a scale, with decreasing return on a scale, etc.) are, in fact, the study of mathematical properties of the power function, but whether this function reflects real economy - a big question.

6. In the Solow model, investments make up a fixed share of production. In fact, all production is not a source of investment. There are three inner investment sources: gross profit, wages, and taxes (in public sector investment), which together account for GDP. The volume of production also contains material costs that are not proportional to GDP: in industries and countries with higher production efficiency, they are smaller, with lower efficiency – higher (this is why it is necessary to take the output as a result indicator of the economy, not GDP). Therefore, an investment cannot be taken as a share of output that includes anything that has nothing to do with investment.

They cannot be taken as a share of GDP because the shares of savings in gross profit, wages, and the share of investment in the budget, formed from taxes, are significantly different. Non-financial corporations in Ukraine direct all gross adjusted disposable income to gross savings. Financial corporations of Ukraine, for example, in 2018–2099.4%. The general government sector directed 8.1% of gross adjusted disposable income to gross savings. Households are unlikely to use social benefits to save as cash, much less in kind. They save from three sources: gross profit (mixed-income), which amounted to 2018 25.7% of the gross balance of primary incomes, wages of employees (71.9% of the gross balance of primary incomes), and property income (2.5%). The share of gross savings in the gross balance of primary household income was 3.9%. Non-profit institutions serving households directed 8.6% of their gross adjusted disposable income to gross savings [22].

Thus, the average share of gross savings in gross adjusted disposable income in Ukraine of 14.4% does not contain any useful information. This is the average temperature of patients in the hospital, possibly except for the morgue. Because if we determine the optimal share of gross savings in the country's gross adjusted disposable income, and even more so in production, then there is no economic agent who can use it, there is no source of investment from which to take this share. From this optimal share, it is impossible to move to the optimal shares for agents who own or manage investment sources: for households, for non-profit organizations serving households, and for the general government sector.

Note also that to determine the exact optimal share of savings is not simple. As R. Pindyck [23] stated, there is great uncertainty in Nordhaus models' input indicators and the like, particularly regarding the discount factor, small changes of which strongly affect the value of the optimum, and there is no justification for these small differences. W. Nordhaus and E. Moffat [24] themselves note the poor reproducibility of their results.

Modern economists rightly believe that economic development should be carried out at the expense of some endogenous resources. Thus, in the Agion-Howitt model [25], economic growth is a consequence of individuals' decisions, not an exogenously given variable. But for some reason, they focused on the norm of saving and did not study the sources of development, i.e., sources of investment.

In all these and similar models, one of the main determinants of growth is the uninformative **amount of capital,** regardless of its **productivity.** The latter depends on the nature of the investment. Economic growth models did not address either the nature of investment or investment effectiveness. But this needs to be taken into account because the pace of technological progress and economic growth depends on it.

None of the economic dynamics models reflects the shadow economy, although, in many countries, it significantly affects economic dynamics. For Ukraine, we showed this in [26].

This analysis's general conclusion is as follows: **it is impractical to study the economic dynamics on the highly aggregated models, such how mentioned above, where the main determinant of growth is the uninformative amount of capital, regardless of its productivity.** They are inadequate to the real economy, especially in those countries where the shadow economy is significant. Deviations from the real economy are so great that **any economic dynamics in them become indistinguishable**. It is impossible to draw constructive conclusions appropriate for a particular economy or group of similar economies on such models. The conclusions that have been obtained have either already turned out to be wrong (Harrod-Domar model) or will turn out to be such, or they cannot be applied in a particular economy (Solow model).

4. An adequate model of economic dynamics of Ukraine

When creating our model of economic dynamics of Ukraine, we sought **to achieve the greatest adequacy**. We are more inclined to believe Alexander Gray:

"Economic science if this is different from other sciences that there is no imminent transition from least to most credibility, there is no inexorable desire to go all the way, the truth, which, as once revealed, will be true for all time, until the complete eradication of any opposite teaching, "(quoted by [14], p. 3).

Duc to the limited scope, we present only the model features that ensure its adequacy and several basic equations of the model (the whole system of equations is given in [9, Chapter 2]).

1. The model is causal. The causal model is better suited to solving various analytical problems than the regression one.

2. The model is imitative. Back in the early 60-s of the XX century, von Bertalanffy wrote: "UTS opens new horizons for us, but its compliance with the empirical data remains scant." Therefore the author tried to stay away from theories and to be closer to life. He simulated real-world economic processes and mechanisms which undoubtedly exist in the economy, not forcing them into the Procrustean bed of theories and macroeconomic hypothesis (hypothesis of monetarists, Keynes, equilibrium) that a priori rigidly predefine the economic behavior and make a model inadequate to the real economy. In the language of control theory, all these hypotheses relate to the economy's management, but not to the economy itself. Introducing them into the model of an object means to mix object and a control system so one cannot analyze the "pure" economy (economy as a system with internal positive feedback is unstable) and synthesize a control system correctly. We used a priori more appropriate path (long known in the theory of control systems, but have never been used in economics): to display only an object of control, but not control actions: the real economy, not the ideas.

   In this respect, it is closer to the ASPEN model [27]; true, the author found out about it in 2018. Of course, our model differs in the specifics of the Ukrainian economy.

3. The model is systemic, i.e., it displays the complete system of essential micro- and macroeconomic mechanisms (formation of prices, cost, producers' and state income, taxes, emission, bank rates, transfers, etc.; for example, most models do not show the interrelation between the increase in salary and taxes and an increase in cost price whereas our model reflects this and all other interconnections.

4. The model reflects seven basic types of shadow economic activities in conjunction with the legal. Any model of economic dynamics (and statics) in the world has no such conjunction, they reflect the shadow economy separately from the legal one (as, for example, in [28–30].

   Significant shadow types machinations that exist in Ukraine:

   In private companies:

   a. In addition to legal exists the shadow production of each product;

   b. Part of the salary is paid illegally; taxes on all illegal amounts are not paid;

   c. One overstates material costs to avoid paying VAT and income tax.

   d. Both private and public companies include the following areas:

e. Prices of public procurement exaggerated;

f. The state returns VAT for sham sales both domestically and for export,

g. VAT is not returned on time or in full;

h. Companies give bribes to officials for not "noticing" shadow machinations. The absolute size of the bribe is proportional to the size of the shadow equipment.

i. We accepted the overall level of the shadow economy for 2018–2027 32%, as determined by the Ministry of Economic Development and Trade of Ukraine. Using the general level of the shadow economy, the model determines the levels of the shadow parts in different goods. They were found to be different and, in some ways, consistent with the data of the Ministry of Economic Development and Trade: the highest level of the shadow is in the financial sector (after exports, in which the Ministry of Economic Development did not evaluate the level of shadow). This confirms the model's adequacy to both the shadow and the legal sectors of the Ukrainian economy.

5. We tried to choose the optimal level of aggregation of goods in this model, in the sense that we abandoned complete aggregation, as in Solow [10] but did not introduce excessive specification by industry, as in V. Leontief [31]. We chose the principle of the system of national accounts (SNA): we formed product groups, manufacturers, and consumers so that they behave equally when devaluation and inflation: final consumption goods, intermediate consumption, investment, exports, and imports. The model reflects the production of raw materials (intermediate goods), lacking in most models. Without this, it is impossible to accurately reflect changes in all goods' cost (including the intermediate goods themselves) and the value-added in its production.

So, we increased the number of goods from three in the dependent economy model [32] or eight in the ASPEN model [27] to 19, the number of producers to 29 (14 private, 14 state-owned enterprises and the general government sector (GGS)). Each pair companies: private and public, produces one of 14 products:

a. non-tradable consumer goods and services[1] together with distribution and retail (index *1* in below equations);

b. Tradable goods of final consumption (only production without distribution and retail) sold domestically (index *2* in below equations);

c. Tradable goods of intermediate consumption (production together with distribution and retail) sold domestically (index *3*);

d. Tradable goods of investment consumption (without housing) sold domestically together with distribution and retail (index *3 N*);

e. Housing sold domestically together with distribution and retail (index *3H*);

[1] Next, we will say "goods", referring to "goods and services".

f. distribution and retail of import of these four goods; we included these services in domestic output and GDP (indexes *4, 5, 5 N, 5H*).

g. Consumer export together with distribution (index *2E*),

h. Intermediate export together with distribution (index *3E*),

i. Consumer goods' distribution (index *2D*),

j. Consumer goods' retail (index *2R*),

k. Financial services (index *F*).

The GGS produces government services (index *DU*).

Imports are, of course, not made in Ukraine, but four types of import (points b-e) increase the number of goods to 19.

In this way, we linked all systemic relations to each other: production (the cost of production reflects material costs (different in each cycle through different investments in previous years), wages, payment for banking services, depreciation (also different due to different investments) and taxes), consumption (products are purchased at the expense of salaries, profits, credits, pensions, and taxes), capital accumulation (at the expense of profits, salaries, taxes, and credits).

6. Based on the availability of 14 types of private and state-owned enterprises which produce 14 types of goods, and GGS, we divided households into 86 groups: 28 aggregated employees of all 14 types of private and 14 state-owned enterprises, which produce the above-mentioned goods; 14 business owners; 14 directors of state-owned enterprises (in Ukraine, they act like owners); pensioners; public servants; 28 aggregated officials, each of whom receives a bribe from one of 28 enterprises.

7. We took output (instead of GDP) as the primary outcome of production. This eliminates a lot of inaccuracies and methodological incommensurability, for example:

a. The neoclassical model of demand for money takes into account only trade agreements that are linked to the GDP, but the sale of intermediate products does not include, and in Ukraine, it is more than 60%;

b. in the traditional regression models, export demand is mainly determined by the GDP of the importing country, but this is true only for end-use products, but the need for exports of intermediate consumption goods, which is in the export of Ukraine 80–90%, depends precisely on the output of the importing country;

c. GDP, which measures only the value-added contained in the goods, sometimes is compared with indicators that measure the full value: exports, imports, consumption, supply, and so on. For example, in the computation of openness of the country to the outside world, this leads to the fact that the two countries with different levels of the cost of producing a unit of output (hence, with varying levels of GDP per unit of output) for which this indicator has the same value, according to the existing methodology are considered equally open. In contrast, the degree of openness of an economy is more at the country in which the level of GDP per unit of output above. Openness to the outside world should be measured by the ratio of export value to the output. For example, a traditional indicator showed that the degree of openness of Ukraine in 1993 increased from 24 to 26%, whereas, according to our indicator, it dropped from 11 to 10%. In 1996 the opposite was true.

d. The use of GDP as the model output is automatically (though implicitly) introduces the hypothesis of the constancy of economic efficiency (GDP per unit of intermediate consumption), while the simulation of production and costs reflects its changes.

e. When identifying the empirical data model, primary values in the physical dimension may be probabilistically distributed according to the normal (Gaussian) law for which statistical evaluation of the significance and accuracy of the model coefficients is derived. Nonlinear conversion of primary values (multiplication of prices on volumes) changes the normal probability distribution to the other, for which statistical estimates are incorrect.

Besides, the author believes that the hypothesis of a balance between supply and demand (which has never been proved either theoretically or practically) has not been disproved because all models used GDP as a result indicator instead of output. Namely, using output allowed us to detect an imbalance between consumed and produced GDP.

At different levels of efficiency per unit of intermediate consumption, our model produces different amounts of GDP. In particular cases, the hypothesis of equilibrium between supply and demand, between consumed and produced GDP, is performed. Now imagine that the efficiency decreased. Each product unit contains less added value; therefore, the whole output has less GDP. Now households will be able to buy at the new GDP only a part of the goods produced; thus, the balance between supply and demand is broken. It follows that the balance is carried out only for one particular subset of the values of economic efficiency. So, a much more powerful set of unbalanced economy hardly comes into the view of economists. Our model takes them into account.

Our dynamic model is a set of models of successive economic cycles[2]. The model of each cycle is a model of the Ukrainian economy's transition from the previous cycle to the next one under the influence of exogenous and endogenous factors; all three sources of investment: gross income, wages, and taxes, are results of the previous cycle (endogenous factors for it) and input (exogenous) factors for the next cycle.

Each cycle's model is a direct description of the real economic activity of all 28 aggregated enterprises and GGS (**Figure 1**). Each company produces its own product, sends it to the market and receives revenues. From it, the company pays taxes, returns debts with interest, restores spent fixed assets, buys working capital, pays salaries. The owner of the enterprise spends part of the remaining profit on his personal needs, and sends the other part to investments.

---

[2] For the convenience of applying the existing statistical accounting, it is accepted that one cycle is equal to one year.

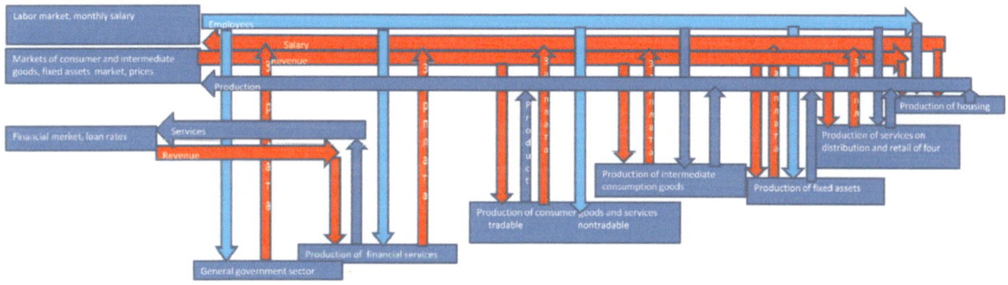

Figure 1.
Block diagram of the model of one economic cycle.

The output of each consumer product is initially equal to the total demand of all 86 of its consumers (listed above), the volume of each intermediate goods (domestic, import, and bank services) - to the total demand for it by all 28 enterprises and the GGS. Afterward one can set any over- or underproduction for any product. Since amounts of intermediate goods' demand also include the demand of the enterprise which produces intermediate goods itself, the systems of Equations [9] are recurrent. We solved them by the sequential approximation method implemented in Excel using the macros developed by the author. In the next economic cycle, enterprises purchase material and labor resources at perhaps new prices and at norms that may have decreased in proportion to investment in previous periods. The aggregated investment enterprise increases (decreases) its production following the increase (decrease) in all enterprises' total gross profit, including this enterprise itself.

Thus, the model calculates each resultative indicator by taking into account all the main direct, feedback, and cross-links, all the main economic mechanisms existing in the economy of Ukraine. Therefore, all known multipliers are determined not once and for all, as traditionally, but for each economic situation newly. For each factor, we make several alternative calculations to ensure the consistency of the regularities found. Therefore, an adequate model is the best justification of conclusions, more conclusive than theoretical considerations, which, at all their correctness, cannot take into account all problem aspects and all factors affecting the development of the economy, especially if the consequences are multiple, opposite, and the outcome depends on which one is stronger.

Here are the basic equations of the model (all equations see in [9]).

The price model consists of non-devaluation I_{nj} and devaluation $a_j \cdot (I\text{-}1)$ components of inflation. The *j*-th product owner raises the price more or less than devaluation index I (coefficient a_j and index I_{nj}):

$$I_j = \left(I_{nj} + a_j \cdot (I-1)\right), j = 1, 2, 2_D, 2_R, 2_E, 3, 3_N, 3_H, 3_E, 4, 5, 5_N, 5_H \tag{1}$$

The owner has to rise salary under devaluation, but he sets wages (I_{vj}) usually less than price (I_j ($b_j < 1$)):

$$I_{vj} = 1 + b_j\left(I_j - 1\right), j = 1, 2, 2_D, 2_R \tag{2}$$

Cost of *j*-th product unit is modeled as a sum of conditionally variable and constant (the last summand) expenses for domestic and imported materials, wages, contributions to pension and social insurance funds with norm c (being changed due to devaluation), amortization am_j and taxes which has been aggregated in groups with homogeneous devaluation behavior of their bases (value added - t_{DWj}, natural resources - t_{Rj}, excise and import duties - t_{IM} and others taxes - t_{INj}):

$$s_j = I_3 \cdot z_{30} \cdot n_{3,j} + I_5 \cdot z_{50} \cdot n_{5,j} + (I_{vj} + c) \cdot w_{j0} \cdot p_j + am_3 + t_{DWj0} \cdot I_{DWj}$$
$$-t_{Rj0} - t_{INj0} \cdot I_{Wje1} + S_{jc0} \cdot K_{j0}/K_j, \tag{3}$$

where n_{3j}, n_{5j}, p_j - amount of domestic and imported materials and person-years required for j-th product unit production;

w_{j0} - average annual salary;

Index $_0$ corresponds to base period (before devaluation).

Net profit per unit of product which remains at the disposal of the company's owner depends on prices (1) and cost (3), on tax on profit t_{Dj} and on interest on loans for working capital t_{krj}:

$$d_3 = z_3 - s_3 - t_{FOj0} \cdot I_{vj} - t_{krj} \tag{4}$$

Predesigns on the model have shown that devaluation and inflation **always** reduce real GDP rather more than existing theories showed. To be assured in it, we have entered the best variant into the model: 1) production output is equal to sum of purchases of all consumers; 2) the model does not take into account imbalances in transition process during which the "invisible hand" of Adam Smith balances market, not immediately, but after many underproductions and overproductions of goods that will never find a buyer. This is done to maximize confidence in the negative effect of devaluation. If such "best" model shows a negative effect, in fact it could be only worse, but a positive result may be the same or substantially weakened, or not happen.

Model of j-th good purchase by k-th consumer K_{kj} is oriented on the price (1) and on specific k-th consumer income I_{Vk} (not on average CPI): the consumer demand curve shifts from the nominal price on specified income change (size of shift f_k can be varied).

On the supply side production is limited to the growth of interest on loans for working capital t_{krj} and its shortages, which caused by increasing cost I_{sj} due to devaluation, but supply is growing due to part d_{Kj} of emission EM_3 for the development of production:

$$K_{kj} = l_{kj} \cdot (z_k/(1 + f_k \cdot (I_{Vk} - 1)))^{mk} \cdot (1 - h_j \cdot t_{krj}/t_{krj0}) / \left(1 + g_j \cdot (I_{sj} - 1)\right)$$
$$\cdot (1 + d_{EM} \cdot EM_3 \cdot d_{Kj} \cdot ob/S_{j0}), \tag{5}$$
$$j = 1, 2_R, 4, k = 1, 2, 2_D, 2_R, 2_E, 3, 3_N, 3_H, 3_E, 4, 5, 5_N, 5_H, BG, PN, BD, DU$$

where ob - velocity of money.

First, we took the sedate demand functions, then - linear. Results were similar.

For tradable goods *2* the effect of substitution of import Im_{40} by output W_{20} in part r_2 is considered:

$$K_{2j}^{IM} = K_{2j} \cdot (1 + r_2 \cdot (1 - K_4/K_{40})) \cdot Im_{40}/W_{20} \tag{6}$$

The production output is made of j-th product sales (5) to all consumers, the whole economy output is made of all products.

Production of intermediate goods *3* and their import *5* are determined by demand of all 11 producers, export - of the outside world. When manufacturer buys his own goods, there is a vicious circle: the volume of purchase depends on his salary, but it – from manufacture volume equal to purchases volume. In the model there are many such recurrent equations' systems. They are solved iteratively: on the first step any values of all unknown variables are substituted in the equations and the first results are defined. On the second step these first results are substituted in the equations and their second approximations are defined etc. The author did not investigate convergence of recurrent procedure, but already on 5-6th step the error did not exceed 0.001%.

We also have applied the same iterative process to the decision of equations' systems that is not recurrent and difficult for the decision; it has converged to the correct decision.

From this, it follows a proposal for the lazy: even if the method of solution of any system of equations is known but is hard enough, you cannot spend time on it, but easy to implement this iterative algorithm in Excel and get a solution quickly and without errors, that often encountered in the "manual" decision in the "*quadrature*". Even for experienced mathematicians it may be easier to use iteration than 1) to identify the appropriate method's suitability for this problem (i.e., to check the scope of application), and 2) to use it. If a method is not yet developed, the more so. The convergence of the process is also easier to identify every time than to prove it.

CPI is determined not for goods' "basket", but for all consumer goods (difference makes no more than 8–10%):

$$I_{sz} = (I_1 \cdot W_1 + I_2 \cdot W_2 + I_4 \cdot Im_{4I})/(W_1 + W_2 + Im_{4I}) \tag{7}$$

State revenues consist of tax and non-tax revenues and contributions to social insurance funds. Devaluation adds to these a variable part of emission:

$$\begin{aligned} D_{DU} = I_V \cdot T_{FO0} + I_D \cdot T_{D0} + I_{DW} \cdot T_{DW0} + T_{R0} + I_W \cdot \cdot T_{IN0} + I_{IM} \\ \cdot T_{IM0} + u_D \cdot EM_D + u_2 \cdot EM_2 + EM_{3DU}, \end{aligned} \tag{8}$$

where I_{DW}, I_{ChD}, I_W - indices of added value, net income and production throughout the country;

T_{FO} – income tax;

T_{IM} - excise and import duties.

Whereas inflation has been almost always an occurrence in Ukraine, devaluation happened only sometimes (moreover, a revaluation took place in 2001–2006). Therefore it is expediently to divide total emission on such parts:

1. The first one causes the devaluation *I*:

$$EM_D = (I - 1) \cdot IM_0/ob \tag{9}$$

Devaluation in turn causes inflation:

$$I_{szD} = 1 + EM_D \cdot ob/(W_0 + IM_0 - Ex_0) \tag{10}$$

2. the second one is related to non-devaluation inflation:

$$EM_2 = (I_{sz} - 1) \cdot (W_0 + IM_0 - Ex_0)/ob - EM_D \tag{11}$$

3. and the third one (as a part q_3 of *M2*) which government directs to social sector or to production as "short" investments that allow rapid growth according to a latest factor in (5):

$$EM_3 = q_3 \cdot M_2 \tag{12}$$

Trivial equations like $D_j = K_j \cdot d_j$ or $VVP = (V + D + AM + T)$ are not given.

All parameters of the model are made variable. It gives the opportunity to investigate not only dependences, but their characters and also to define ranges of their invariance.

The model allows optimizing strategies of the 14 manufacturers (each or all of them) using major (maximum added value) and supplementary criteria (maximum gross profit, market expansion and many others) by the algorithm of the multi criteria compromise [33].

The model allows to study the devaluation dynamics of nominal and real indices for different goods and across the country: the cost, the gross and net profit, rate and the amount of wages, value added, production output, GDP, the share of wages in the cost, unemployment, pensions, tax and non-tax revenues of budget, salaries of budget employees, foreign exchange earnings from intermediate and consumer exports, physical volume of exports, physical volumes and currency expenses on intermediate and consumer imports, trade balance, the structure of export and import, the share of import in intermediate and final consumption, producer and consumer prices, manufacturing GDP per unit of intermediate consumption, which characterizes the ultimate operational efficiency of economy from the national point of view, efficiency of production of some goods from terms of other economic performers (of production owner - the ratio of gross or net income to capital cost, of employee - salary costs to working time, etc.), changes in the structure of contributions of different types of goods production in GDP and so on.

When modeling, you can change: price, salary and price elasticity of demand and supply for each product, the level of de- or revaluation, in- and deflation and the correlation between them, level and cost structure of each product (the share of wages, intermediate import, credit for working capital, etc.) rate loan, share of conditionally fixed costs, tax and deductions rate; exchange rate in the base period, the ratio between the cost of production, distribution and retail sales, population structure; the degree of influence of rate loan, excess emission, a shortage of working capital on the production; distribution of excess emission between the social sphere, "short" investments in the production of various goods (final or intermediate consumption or exports) and banks, the degree of import substitution of domestic products.

We proved the adequacy of the model by retrospective forecasting of the Ukrainian economy dynamics for 2008–2013: the resultative indicators calculated on the model have coincided with the real indicators with sufficient accuracy: deviations of the calculated legal GDP from the provided by the State Statistics Committee, are in [−0.7; 0.4%], of the gross profit - in [−1.8; 2.6%], of the issue - in [−9; 2.1%], of the salary - in [−2.9; 2%]. Therefore, there is reason to believe that the model calculations' accuracy is sufficient for further analysis and multivariate forecasts. The options below reflect the real-life dynamics of Ukraine's economy rather than the study of the Solow power function's mathematical properties, which is unlikely to reflect the real economy. Therefore, one has reason to trust our calculations.

To make the model suitable for other countries, you need to change its parameters and sometimes some its parts.

5. Analysis of the impact of credit on the economy of Ukraine

So, we are simulating the smooth development of the Ukrainian economy over 5 years at different levels of lending. In the sixth year, we stop lending and look how real GDP falls from the base year. If it falls hard, it will be a model of a financial crisis in a first approximation.

Due to the significant nonlinearity of the economy and our model, the same change of a particular factor in different economic situations (i.e., with varying values of other factors) gives different consequences. Therefore, we will get more accurate results if we study the element's influence in a situation close to real life. Therefore, we will accept the annual devaluation of the hryvnia by 10%, the annual growth of domestic prices for all goods and services 10%, external prices and physical volumes of legal and illegal exports and imports as in 2018. We will leave the distribution of loans between households and the non-financial sector, and their distribution for various purposes (to increase working capital, to invest in reducing material and/or labor costs per unit of traditional products; to invest in creating new varieties of goods that will be in high demand and sold at a higher price, for housing construction, etc.), as in 2018.

Scenario 1. Let the number of loans issued by the aggregate bank to the non-financial sector and households be four times less than the number of deposits placed by them in banks, be the same in all cycles, and is 5% of the 2018 outcome. In the coming years, the outcome will change, so loans share will be different.

Assume that there is no overproduction or underproduction, i.e., the full (including the shadow parts) production of consumer, intermediate, and investment goods and housing and their imports will exactly meet the demand for them.

Under such conditions, Ukraine's total real GDP (legal plus shadow) (TRGDP) will grow by 14.2% in 5 years; its amount will be UAH 16.03 trillion. The number of real wages in five years compared with the base year will increase by 16% (**Table 1**, first column). The latter indicator shows that while maintaining a constant real wage, the number of jobs will increase by 16% (if technological conditions permit).

To test if lending is excessive, that is, if it is not creating a financial bubble, we exclude borrowing in the sixth cycle. Compared to the base cycle, TRGDP not only did not fall but even increased by 3.5%. This means that the lending level of 5% of the issue (10.4% TRGDP) is safe, that is, one that does not create a dangerous financial bubble.

Scenario 2. Now let the number of loans is 10% of the 2018 issue (20.7% of the TRGDP). In the sixth cycle, when we excluded loans, TRGDP decreased by 0.8% compared to the base cycle (**Table 1**, second column). Even if stocks are sharply reduced, as was the case in 2008–2009, for example, by five times, the decrease will be only 0.81%. This is a tiny decrease, so the lending of 10% of output (20.7% of TRGDP) is safe.

Indicators	Scenarios				
	1	2	3	4	5
The share of loans in total output,%	5,0	10,0	15,0	20,0	25,0
Percentage of loans in TRGDP,%	10,4	20,7	31,1	41,5	51,8
The sum of TRGDP for five years, trillion UAH	16,03	17,37	18,72	20,07	22,98
Growth of TRGDP for five years,%	14,2	21,1	28,0	34,9	25,1
Inflation for five years,%	61	61	61	61	164
Increase in the number of investments for five years,%	6,9	9,9	12,8	15,8	19,8
Increase of the real salary for five years,%	16	23,1	31,8	39,9	11,9
The fall of TRGDP in the 6th cycle compared to the baseline,%	3,5	−0,8	−4,0	−6,3	−23,4

Table 1.
Development of Ukraine's economy at different levels of lending.

Scenario 3. Let the number of loans be 15% of the 2018 issue (31.1% of TRGDP). In the sixth cycle, when we excluded loans, TRGDP decreased by 4% (**Table 1**, third column). This is a small decrease, so the lending rate of 15% of output (31.1% of TRGDP) is also safe.

Scenario 4. If the amount of loans is 20% of the issue (41.5% of TRGDP), then in the sixth cycle TRGDP decreases by 6.3% (**Table 1**, fourth column). This is a considerable decrease but not catastrophic, so the lending of 20% of output (41.5% of TRGDP) can be considered more or less safe.

Scenario 5. If the amount of loans is 25% of the issue (51.8% of TRGDP), it exceeds the aggregate bank's number of deposits by 21.3%. Thus, the bank created an annual emission of 206595 million UAH. Due to the emission prices for all goods will rise. Inflation for five years will be 164%. We will accept the growth of the prices identical to all goods and services. In the first cycle, it will be 12.9%; in the following, it will be slightly lower due to the growth of output (and the volume of emissions remains the same for all cycles).

Due to rising prices, TRGDP will increase by only 25% in 5 years, while scenario four was increased by 35%. The increase in GDP for most goods and services will increase by 10–52%, but for exports of final and intermediate goods, it will decrease by 18.5% and 16.9%, respectively. The sum of real wages will increase by only 12%, while in scenario four they increased by 40% (**Table 1**, first column).

In the sixth cycle, TRGDP will decrease by 23.4% compared to the base cycle (**Table 1**, fifth column). Thus, the level of lending at 25% of output (51.8% of TRGDP) will undoubtedly create a devastating financial crisis. Thus, we can draw two conclusions: 1) the negative results from the increase in loans beat the positive ones, and 2) the negative effect from the price increase due to emission is the largest for export goods.

Scenario 6. Let in scenario 2 deposits become 20% less than loans. This creates an annual emission of UAH 78,377 million, which causes prices to increase by 5.43%. Because of this, inflation for five years will not be 61%, as in scenario 2, but 99%, TRGDP for five years will increase not by 21.1%, but only by 9.2%, the

Indicators	**Scenarios**				
	6	**7**	**8**	**9**	**10**
	Deposits fewer loans by 20%	**The technical productivity of investment is 1**			**The productivity 2**
The share of loans in total output,%	10,0	15,0	10,0	25,0	25,0
Percentage of loans in TRGDP,%	20,7	31,1	20,7	51,8	51,8
The sum of TRGDP for five years, trillion UAH	17,5	20,8	18,39	23,97	25,18
Growth of TRGDP for five years,%	9,2	6,5	31,4	29,5	32,0
Inflation for five years,%	99	224	61	164,3	165,5
Increase in the number of investments for five years	10,0	14,5	10,1	19,9	20,1
Increase of the real salary for five years,%	4,7	−9,7	35,0	16,5	22,3
The fall of TRGDP in the 6th cycle compared to the baseline,%	−12,2	−21,6	9,6	−19,6	−12,9

Table 2.
Development of Ukraine's economy in different economic situations.

number of real wages for five years will grow not by 23.1%, but only at 4.7%. In the sixth cycle, TRGDP will decrease by 12.2% instead of 0.8% compared to the base cycle (**Table 2**, first column).

Scenario 7. Let in scenario 6, the number of loans is 15% of the 2018 issue (31.1% of the TRGDP). Compared to scenario 3, the situation worsened similarly to scenario 6 (**Table 2**, second column). Thus, the price increases due to emission worsen the economic situation more than loans improve. The risk of a financial crisis occurs at lower levels of lending.

These seven scenarios took place at very low technical productivity of investment (0.2) and investment efficiency, typical for Ukraine. The low efficiency of investments in Ukraine is noted by all experts, for example, I. Yu. Egorov: « ... the country from year to year lost the positions in the markets of high-tech goods and services which in the modern world develop most dynamically ... In the total amount of costs for production and sale of industrial products, innovation did not exceed 1.0–1.6% (in 2005–2012). With such volumes of funding, it is almost impossible to expand innovative technological reproduction of industrial production and restructure the economy based on scientific and technical achievements ... Relevant indicators in OECD countries in 2012 were: 3.2% (Germany), 4.5% (South Korea), 5.8% (Canada), 6.7% (Sweden) (Egorov, 2015).

Let us repeat some of the scenarios at the technical productivity of investment 1.

Scenario 8. We repeat Scenario 2. As we can see from **Table 2** (column 3), the situation has improved: in the sixth cycle, after excluding loans, TRGDP increased by 9.6%, so at high the technical productivity of investment, the level of lending 10% of the issue (20.7% of TRGDP) is safe. Note that the efficiency of investment by GDP increased significantly - in the fifth cycle, it reached 15%, while at the efficiency of investment, 0.2 was negative.

Scenario 9. We repeat Scenario 3. The situation has improved again (**Table 2**, fourth column). However, even with the technical productivity of investment 1, the lending level of 15% of the issue (31.1% of TRGDP) cannot be considered safe. The safety requires that the technical productivity of investment be 2 (**Table 2**, fifth column).

Indicators	**Scenarios**	
	10	**11**
	Every year the loan grows by 1%	**In the first year the credit grows by 10.16%**
The share of loans in total output,%	10,0	10,16
Percentage of loans in TRVVP,%	20,7	21,1
The amount of TRVVP for five years, trillion. UAH	17,44	17,42
Growth of TRVVP for five years,%	21,8	21,3
Inflation for five years	61	61
Increase in the number of investments for five years,%	10,0	10,0
Increase of the real salary for five years,%	22,8	22,5
The fall of TRVVP in the 6th cycle compared to the baseline,%	−1,4	−1,2

Table 3.
Development of Ukraine's economy with a gradual and abrupt credit the methods.

Indicators	Scenarios			
	14	15	16	17
	Every year exports grow by 2%		Every year exports fall by 2%	
The share of loans in total output,%	15,0	25,0	20,0	25,0
Percentage of loans in TRVVP,%	31,1	51,8	41,5	51,8
The amount of TRVVP for five years, trillion. UAH	20,2	23,4	18,9	22,39
Growth of TRVVP for five years,%	49,0	31,9	21,9	21,4
Inflation for five years	61,1	163,0	61,1	163
Increase in the number of investments for five years,%	13,1	20,4	15,5	20,1
Increase of the real salary for five years,%	50,5	14,7	23,4	4,3
The fall of TRVVP in the 6th cycle compared to the baseline,%	23,1	−13,3	−16,7	−23,8

Table 4.
Development of Ukraine's economy at different levels of lending with changes in exports.

Scenario 11. We repeat Scenario 3 but the credit increases by 1% for each cycle. TRGDP for five cycles will increase slightly more than in Scenario 3 - by 9.4% but a decrease in the sixth cycle more - by 20.5 and 13% (**Table 3**, first column).

To quickly bring the economy out of the recession, a policy of monetary expansion is used, that is, they rapidly increase the money supply. Let us try to make such an expansion in credit.

Scenario 12. Let the amount of loans in the first cycle is immediately 10.16% of the issue, i.e., up to 10% is added the number of loan increments for five years. In subsequent cycles, the amount of loans is 10% of output. The situation has improved: although the TRGDP increased by only 2.7% in 5 years-cycles, in the sixth cycle, it fell much less - by 11.4 and 9% (**Table 3**, second column). So expansion helps.

In all previous scenarios, exports were unchanged.

Scenarios 13 and 14. Let us repeat scenarios 3 and 5 in conditions when exports grow by 2% every year. As can be seen from **Table 4** (columns 1–2), the situation has significantly improved: the growth of TRGDP and the amount of real wages for five cycles has increased more than in scenarios 3 and 5; in the sixth cycle, when we excluded loans, TRGDP at the lending level of 15% of output did not fall but increased by 23%, and at the lending level of 25% of output increased not by 23.4% but by 13.3%, so the increase in exports allows higher levels of lending without risk of crisis.

Scenarios 15 and 16. Let us repeat Scenarios 4 and 5 in conditions where exports decrease by 2% each year. The situation has significantly deteriorated: the increases in TRGDP and the number of real wages for five cycles have become less than in Scenarios 4 and 5; in the sixth cycle, TRGDP fell more (**Table 4**, columns 3–4), so the decline in exports brings the risk of crisis closer to lower levels of lending.

6. Conclusions

1. The use of the simulation approach has significantly increased our model's adequacy to the Ukrainian economy and significantly expanded its analytical capabilities.

2. The application of the model developed by the author to analyze the impact of excessive credit on the economy made it possible to significantly expand the variety of economic situations and research options, the intervals of changes in factors.

3. The introduction of the concept of technical productivity of investment provided an opportunity to characterize them from a macroeconomic point of view, namely - their ability to reduce the rate of material and/or labor costs, which increases the enterprise's competitiveness. The concept of the technical productivity of investment links it with the technical progress, and with GDP growth.

4. The proposal to measure the efficiency of investments in terms of all added-value provided an opportunity to reflect their effectiveness for the state and reflect the business owner's far-sighted interests in minimizing the turnover of skilled workers.

5. The proposal to measure the level of lending by the share of the loan amount not in GDP but in output increased the analysis's accuracy.

6. For the current state of Ukraine's economy, lending of 20% of output in 2018 (41.5% of TRGDP) is almost dangerous, and 25% of production (51.8% of TRGDP) certainly creates a devastating financial crisis.

7. If the loan amount exceeds the number of deposits received by the aggregated bank, creating emissions. The price increase due to emission is harmful, especially for export goods. The risk of a financial crisis occurs at lower levels of lending.

8. Increasing the technical productivity of investment improves the situation. The risk of a financial crisis occurs at higher levels of lending.

9. Increasing the number of loans in the first cycle gives better results than "stretching" them during all cycles.

10. Increased exports allow the use of higher levels of credit without the risk of crisis. With declining exports, the risk of a crisis occurs at lower levels of credit.

Conflict of interest

The authors declare no conflict of interest.

Author details

Yuriy V. Vasylenko
Economic Sciences, Kyiv, Ukraine

*Address all correspondence to: yuvas4009@gmail.com

References

[1] Cappiello Lorenzo, Kadareja Arjan, Sørensen Christoffer Kok and Protopapa, Marco. Do bank loans and credit standards have an effect on output? A panel approach for the euro area. ECB. working paper series, no 1150/January 2010.

[2] Manaresi Francesco and Pierri Nicola. Credit Supply and Productivity Growth. BIS Working Papers No 711. March 2018. https://www.bis.org/publ/work711.pdf

[3] Majeed Sadaf, Iftikhar Syed Faizan, Atiq Zeeshan. Modeling the impact of banking sector credit on growth performance: An empirical evidence of credit to household and enterprise in Pakistan, International Journal of Financial Engineering, 2019. 10.1142/S2424786319500129, (1950012)

[4] Bezemer Dirk, Samarina Anna, Zhang Lu. Does mortgage lending impact business credit? Evidence from a new disaggregated bank credit data set, Journal of Banking & Finance, 2020. 10.1016/j.jbankfin.2020.105760, (105760).

[5] Stiglitz J.E. Steep Dive: America and the New Economic Order after the Global Crisis. transl. from English M.: Eksmo, 2011.512 p. (in Russian)

[6] Laina Patrizio, Nyholm Juho and Sarlin Peter. Leading indicators of systemic banking crises: Finland in a panel of EU countries. ECB. Working Paper Series. Macroprudential Research Network. No 1758/February 2015 https://www.ecb.europa.eu/pub/pdf/ scpwps/ecbwp1758.en.pdf, Measuring the probability of a financial crisis

[7] Alessi Lucia and Detken Carsten. Identifying excessive credit growth and leverage.Journal of Financial Stability. Volume 35, April 2018, Pages 215-225.

[8] Balls Andrew. Do Lending Booms Lead to Financial Crises? NBER. September 16, 2020. https://www.nber.org/digest/sep01/w8249.html).

[9] Vasylenko Yu. Economic development model. Saarbrücken: Lambert Academic Publishing, 2016.

[10] Solow R.M. A Contribution to the Theory of Economic Growth. Quarterly Journal of Economics, 70, 1956. pp. 65-94.

[11] Harrod R. Economic Dynamics. London – New York, 1973.

[12] Neumann J. fon, Morgenstern, O. Game theory and economic behavior. - Moscow: Science. 1970. (In Russian)

[13] Samuelson P. Foundations of Economic Analysis. Cambridge, Harvard University Press. 1947.

[14] Blaug M. The Cambridge Revolution: Success or Failure? London: Published by The Institute of Economic Affairs. 1975.

[15] Krugman P. How Did Economists Get It So Wrong? The New York Times, September 2009, Annex.

[16] Easterly William. The Ghost of Financing Gap The qhost of a nq-dead arowth model still haunts aid How the Harrod-Domar to developing countries. Growth Model Still Haunts Development Economics. The World Bank Development Research Group, August 1997, http://documents1.worldbank.org/curated/en/494271468739201862/pdf/multi-page.pdf

[17] Koopmans T. C. On the Concept of Optimal Economic Growth, Cowles Foundation Discussion Paper, December 1963.

[18] Cass D. Optimum Growth in an Aggregative Model of Capital Accumulation. Review of Economic Studies 32(3), 1965: 233-240.

[19] Acemoglu Daron. Introduction to Modern Economic Growth. Princeton University Press. 2008. ISBN 9781400835775

[20] Acemoglu Daron & Robinson, James. Why Nations Fail: The Origins of Power, Prosperity, and Poverty. NY: Crown Publishers. 2012.

[21] Ramsey F. P. A Mathematical Theory of Saving, The Economic Journal 38, 1928: 543-559.

[22] Accounts of institutional sectors of the economy for 2018. State Statistics Service of Ukraine. 2020 [Data set] http://www.ukrstat.gov.ua/ (In Ukrainian)

[23] Pindyck R. Climate change policy: What do the models tell us? Journal of Economic Literature 51(3), 2013: 860-872.

[24] Nordhouse W. & Moffat E. A Survey of Global Impacts of Climate Change: Replication, Survey Methods, and a Statistical Analysis. Cowles foundation for research in economics Yale University Box 208281 New Haven, Connecticut 06520-8281. 2017 http://cowles.yale.edu/

[25] Aghion Philippe & Howitt Peter. A Model of Growth Through Creative Destruction. Econometrica 60(2), 1992: 323-352.

[26] Vasylenko Yu. The impact of shadow activities on the development of Ukraine's economy. Economics and forecasting 3, 2015: 89-103. (In Ukrainian)

[27] Basu N., Pryor R. and Quint T. ASPEN: A Microsimulation Model of the Economy. Computational Economics, 1998, №12, Issue 3, p. 223-241.

[28] Giles D.E.A. (1997) Causality between the measured and underground economies in New Zealand. Appl. Economics letters 4(1): 63–67.

[29] Lalitha N. (2000) Unorganized manufacturing sector in the national economy: an analysis of its growth dynamics and contribution to national income. – New Delhi: National Council of Applied Economic Research.

[30] Lasko M. (2000) Hidden economy – an unknown quantity? Comparative analysis of hidden economies in transition countries, 1989–1995. Economics of transition, Oxford 8(1): 117–145.

[31] Leontief Wassily. Input-Output Economics. 2nd ed.//New York: Oxford University Press. 1986.

[32] Frenkel J., & Mussa M. Asset Markets, Exchange Rates and the Balance of Payments. In R. Jones, & P. Kenen (Eds). Handbook of International Economics (Vol. I, Book 3) Amsterdam, N.Y, Oxford: North-Holland. 1985.

[33] Vasylenko Yu. Optimization of the distribution of net income in the collective farms. Proceedings of Academy of Sciences of SSSR, Economic Series, 2, 1983 (Russian).